# Brain Tumor Drug Development: Current Advances and Strategies

## *(Part 2)*

Edited by

**Prashant Tiwari**
*Department of Pharmacology*
*College of Pharmaceutical Sciences*
*Dayananda Sagar University*
*Bengaluru, Karnataka, India*

**Pankaj Kumar Singh**
*Department of Pharmaceutics*
*National Institute of Pharmaceutical*
*Education and Research (NIPER)*
*Hyderabad, Telangana, India*

&

**Sunil Kumar Kadiri**
*Department of Pharmacology*
*College of Pharmaceutical Sciences*
*Dayananda Sagar University*
*Bengaluru, Karnataka, India*

**Brain Tumor Drug Development: Current Advances and Strategies *(Part 2)***

Editors: Prashant Tiwari, Pankaj Kumar Singh and Sunil Kumar Kadiri

ISBN (Online): 979-8-89881-174-7

ISBN (Print): 979-8-89881-175-4

ISBN (Paperback): 979-8-89881-176-1



Published by Bentham Science Publishers Pte. Ltd. Singapore, in collaboration with Eureka Conferences, USA. 

First published in 2025.

**Bentham Science Publishers Pte. Ltd.**
No. 9 Raffles Place
Office No. 26-01
Singapore 048619
Singapore
Email: subscriptions@benthamscience.net

BENTHAM
SCIENCE

# CONTENTS

# FOREWORD

In recent years, brain tumor research has made remarkable strides, bringing to light new pathways, mechanisms, and potential therapeutic strategies. However, the complexity of the brain and the heterogeneity of brain tumors, particularly glioblastomas, continue to present formidable challenges in developing effective treatments. Against this backdrop, the book "Brain Tumor Drug Development: Current Advances and Strategies," edited by Dr. Prashant Tiwari, Dr. Pankaj Kumar Singh, and Dr. Sunil Kumar Kadiri, emerges as a critical resource for both experienced researchers and new entrants into the field.

The editors, each with extensive expertise in oncology, pharmacology, and drug development, have compiled a remarkable collection of insights that encompass the full spectrum of current advances in brain tumor therapeutics. From the latest molecular and genomic approaches to emerging targeted therapies, this book encapsulates a dynamic and evolving landscape. The emphasis on precision medicine, immunotherapy, and the repurposing of existing drugs highlights the cutting-edge strategies that are driving the next generation of brain tumor treatments.

In particular, the book addresses some of the most pressing questions facing the field: How can we overcome the blood-brain barrier to deliver effective drugs? What role do molecular biomarkers play in predicting patient outcomes? And how can we harness the immune system to combat tumors that have historically been resistant to conventional treatments?

The editors have ensured that this book not only reflects the state-of-the-art in brain tumor drug development but also points toward the future. By exploring the potential of novel therapeutic agents, innovative delivery systems, and combination treatments, this book provides readers with a comprehensive understanding of the strategies that are being employed to tackle one of the most aggressive forms of cancer.

I hope that this book will inspire and inform the next wave of discoveries, bridging the gap between fundamental research and clinical application. Whether you are a researcher, clinician, or industry professional, Brain Tumor Drug Development: Current Advances and Strategies is an invaluable resource that provides both a broad overview and in-depth discussions of the ongoing efforts to develop effective therapies for brain tumors.

I congratulate the editors and contributors for compiling such a timely and impactful book, and I am confident that it will make a lasting contribution to the field of brain tumor research.

**Shashi Alok**
Department of Pharmacognosy
Institute of Pharmacy
Bundelkhand University, Jhansi (U.P.)
India

# PREFACE

Brain tumors, particularly malignant forms such as glioblastoma multiforme, present formidable challenges in oncology due to their aggressive nature, complex biology, and resistance to conventional therapies. Despite progress in surgical resection, radiotherapy, and chemotherapy, survival rates remain dismal, underscoring the urgent need for innovative drug development strategies. This book addresses the evolving landscape of brain tumor therapeutics, focusing on molecularly targeted agents, immunotherapies, and advanced delivery systems designed to overcome the limitations posed by the blood-brain barrier and tumor heterogeneity.

Recent advances in genomics and proteomics have enabled the identification of actionable molecular targets, such as EGFR amplification, IDH mutations, and MGMT methylation status, which inform the development of personalized treatment regimens. Immunotherapeutic approaches, including checkpoint inhibitors, CAR-T cell therapy, and tumor vaccines, are gaining traction for their potential to harness the immune system against brain tumors. Furthermore, nanotechnology-based drug delivery platforms and focused ultrasound techniques are being explored to enhance therapeutic penetration and precision within the central nervous system.

This compilation brings together contributions from leading researchers, clinicians, and regulatory experts to provide a comprehensive overview of current strategies and future directions in brain tumor drug development. It emphasizes translational research, adaptive clinical trial designs, and the integration of real-world evidence to accelerate the bench-to-bedside transition. The role of pharmacogenomics, biomarker-driven therapy, and combination regimens is also discussed, offering insights into optimizing efficacy while minimizing toxicity.

As the field advances, interdisciplinary collaboration and regulatory harmonization will be essential to translate scientific discoveries into clinically meaningful outcomes. This book aims to serve as a valuable resource for academic investigators, pharmaceutical developers, and healthcare professionals committed to improving the prognosis and quality of life for patients with brain tumors. Through continued innovation and strategic integration of emerging technologies, the future of neuro-oncology holds promise for more effective and personalized therapeutic solutions.

**Prashant Tiwari**
Department of Pharmacology
College of Pharmaceutical Sciences, Dayananda Sagar University
Bengaluru, Karnataka-560111, India

**Pankaj Kumar Singh**
Department of Pharmaceutics
National Institute of Pharmaceutical Education and Research (NIPER)
Hyderabad, Telangana-500037, India

&

**Sunil Kumar Kadiri**
Department of Pharmacology, College of Pharmaceutical Sciences
Dayananda Sagar University, Bengaluru, Karnataka-560111, India

# List of Contributors


| | |
|---|---|
| **Ajmer Singh Grewal** | Guru Gobind Singh College of Pharmacy, Yamunanagar, Haryana, India |
| **Anish Dhuri** | Department of Pharmaceutics, National Institute of Pharmaceutical Education and Research (NIPER), Hyderabad-500037, India |
| **Anshita Gupta** | Department of Pharmacy, Guru Ghasidas Vishwavidyalaya (A Central University), Koni, Bilaspur-495009, Chhattisgarh, India |
| **Amudha Senthamaraikannan** | Department of Pharmaceutics, School of Pharmacy, Sathyabama Institute of Science and Technology, Chennai, India |
| **Anindita Behera** | Amity Institute of Pharmacy, Amity University, Kolkata, West Bengal, India |
| **Chaitanay Vinayak Narayan** | Department of Pharmacy, I.T.S. College of Pharmacy, Muradnagar, Ghaziabad, India |
| **Deependra Soni** | Faculty of Pharmacy, MATS University Campus, Aarang, Raipur, Chhattisgarh, India |
| **Dhaval J. Patel** | Department of Pharmaceutics, Saraswati Institute of Pharmaceutical Sciences, At & Po-Dhanap, Gandhinagar, Gujarat, India |
| **Gaurav Chaudhary** | Department of Pharmacy, I.T.S. College of Pharmacy, Muradnagar, Ghaziabad, India |
| **Gaurav Tiwari** | PSIT-Pranveer Singh Institute of Technology (Pharmacy), Kanpur-Agr--Delhi National, Kanpur, Uttar Pradesh-209305, India |
| **Gajanan Kalyankar** | Maliba Pharmacy College, Uka Tarsadia University, Tarsadi, Dist-Surat, Gujarat, India |
| **Hema Kumar C.** | Department of Biotechnology, Dayananda Sagar College of Engineering (Affiliated to Visveshwaraya Technological University, Belagavi), Kumar Swamy Layout, Bangalore-560111, Karnataka, India |
| **Khushi Rode** | Department of Pharmaceutics, National Institute of Pharmaceutical Education and Research (NIPER), Hyderabad-500037, India |
| **Manasvi Saini** | College of Pharmacy, Shivalik Campus, Dehradun, India |
| **M. Bharath** | Department of Pharmaceutics, National Institute of Pharmaceutical Education and Research (NIPER), Hyderabad-500037, India |
| **Manoj Kumar Sabnani** | Aakha Biologics, Frisco Square Blvd, Frisco, USA |
| **Neha Singh** | Department of Microbiology, Pt. Jawaharlal Nehru Medical College, Raipur, Chhattisgarh-492001, India |
| **Nishigandha Sa** | School of Pharmaceutical Sciences, Siksha' O' Anusandhan Deemed to be University, Bhubaneswar, Odisha, India |
| **Nikita Panchal** | Maliba Pharmacy College, Uka Tarsadia University, Tarsadi, Dist-Surat, Gujarat, India |
| **Prashant Tiwari** | College of Pharmaceutical Sciences, Dayananda Sagar University, Bengaluru-562112, India |

| | |
|---|---|
| **Pankaj Kumar Singh** | Department of Pharmaceutics, National Institute of Pharmaceutical Education and Research (NIPER), Hyderabad-500037, India |
| **Pankaj Sharma** | Department of Pharmaceutics, ShriRam College of Pharmacy, Banmore, Morena, (M.P.), India |
| **Renjil Joshi** | Raigarh College of Pharmacy, Raigarh, Chhattisgarh, India |
| **Ravindranath B.S.** | Manipal Institute of Technology, Manipal Academy of Higher Education (MAHE), Manipal, 576104, Karnataka, India |
| **Swati Verma** | Department of Pharmacy, Banasthali Vidyapith, Banasthali, Rajasthan, India<br>Department of Pharmacy, I.T.S. College of Pharmacy, Muradnagar, Ghaziabad, India |
| **Shashi Ravi Suman Rudrangi** | Product Development and Commercialisation, Kent Pharma UK Limited, Kent, United Kingdom |
| **Samatha Rudrangi** | Lifecare Pharmacy of Austin, Nogalitos, San Antonio, Texas, United States of America |
| **Sandesh Lodha** | Maliba Pharmacy College, Uka Tarsadia University, Tarsadi, Dist-Surat, Gujarat, India |
| **Sudhanshu Sekhar Rout** | School of Pharmaceutical Sciences, Siksha' O' Anusandhan Deemed to be University, Bhubaneswar, Odisha, India |
| **Tanmoy Kanp** | Department of Pharmaceutics, National Institute of Pharmaceutical Education and Research (NIPER), Hyderabad-500037, India |
| **Ujala Gupta** | Department of Pharmaceutics, National Institute of Pharmaceutical Education and Research (NIPER), Hyderabad-500037, India |
| **Unnati V. Patel** | Quality Control Specialist, Analytical Method and Development and Validation Team, Avanti Polar Lipids, Part of Croda International Plc, Alabaster, AL-35007, USA |
| **Vaibhavkumar B. Patel** | Department of Pharmacology and Pharmacy Practice, Saraswati Institute of Pharmaceutical Sciences, At & Po-Dhanap, Gandhinagar, Gujarat, India |
| **Vitthal Gajanan Kuchake** | Department of Pharmacy, OJAS College of Pharmacy, Jalna, Maharashtra, India |
| **Varsha Deva** | Department of Pharmacy, Glocal University Pharmacy College, Uttar Pradesh, India |
| **Vidhi V. Mandoth** | Department of Biotechnology, Dayananda Sagar College of Engineering (Affiliated to Visveshwaraya Technological University, Belagavi), Kumar Swamy Layout, Bangalore-560111, Karnataka, India |

# CHAPTER 1

# Phytopharmaceuticals for Brain Tumor: Opportunities and Challenges

**Swati Verma**[1,2], **Chaitanay Vinayak Narayan**[2], **Ajmer Singh Grewal**[3,*], **Manasvi Saini**[4] and **Gaurav Chaudhary**[2]

[1] *Department of Pharmacy, Banasthali Vidyapith, Banasthali, Rajasthan, India*

[2] *Department of Pharmacy, I.T.S. College of Pharmacy, Muradnagar, Ghaziabad, India*

[3] *Guru Gobind Singh College of Pharmacy, Yamunanagar, Haryana, India*

[4] *College of Pharmacy, Shivalik Campus, Dehradun, India*

**Abstract:** Phytochemicals from plants have provided humans with numerous health benefits. Still, the special properties of the BBB (Blood-Brain Barrier) impede their access to the targeted tissue, and therefore, prevent their therapeutic use in treating brain disorders. In response to the increasing number of fatal diseases, scientists and medical experts have created strict new therapies. The concerning rise in the incidence of fatal neurodegenerative illnesses has brought attention to the need for efficient treatment alternatives. However, the conventional medications that are used to treat them already have disadvantages that make them less than optimal. The most sought-after alternative to traditional medications is now traditional herbal remedies, which try to offset the disadvantages of synthetic treatments. Their accessibility, cost, and, most of all, their safety and effectiveness have had a major impact on their success against some of the most serious kinds of neurological illness as well as consumer dependence. Clinical and laboratory data indicate that they offer the potential to treat stroke, Alzheimer's, Parkinson's, and other illnesses. For the treatment of neurological illnesses, researchers are considering nanosystems based on herbs to enhance the transport and bioavailability of phytochemical compounds. Neurodegenerative diseases are the primary research targets for this investigation into polyherbalism, herbal-medicinal drugs, and herbal-based nanosystems. This chapter explores phytochemicals and their potential medicinal uses.

**Keywords:** Brain tumor, BBB, Bioactive, Nano-delivery, Nanosystem, Phytopharmaceuticals.

* **Corresponding author Ajmer Singh Grewal:** Guru Gobind Singh College of Pharmacy, Yamunanagar, Haryana, India; Tel: +91-94167004301 E-mails: ajmergrewal2007@gmail.com; ajmer.singh@gnkgei.ac.in

**Prashant Tiwari, Pankaj Kumar Singh & Sunil Kumar Kadiri (Eds.)**

## INTRODUCTION

Modern medicine would not exist without the use of natural phytochemicals, which have contributed to the development of well-known medicinal compounds. Various naturally occurring compounds derived from plants are collectively known as phytochemicals. The medicinal benefits of plants were first recognized by humans *via* their deliberate or accidental usage, leading to the advancement of medical science over herbology [1]. Modern pharmaceutical engineers have successfully utilized compounds derived from plants in the development of medications. The usual process begins with the isolation and purification of target phytochemicals. Then, their pharmacological capabilities and pharmacodynamics are evaluated in the lab, culminating in a druggability evaluation and the initiation of their development. To make them more effective, certain phytochemicals are derivatized or undergo structural optimization [2].

Using phytochemicals, a small-molecule library was also created for computer-based virtual drug screening and high-throughput drug screening. They offer new possibilities for pharmacological research and library enhancement due to their highly complex architectures [3]. Furthermore, significant study areas in current organic chemistry include obtaining precursors that can be derivatized in various ways, reducing synthesis costs, increasing yield and purity, and fully synthesizing naturally occurring phytochemicals of great economic value. Medications derived from plants can reach the human body through various pathways. Nasal feeding, oral delivery, and intravenous injection are the primary modalities of administration currently used [4, 5].

Because many phytochemicals exhibit pharmacological properties in food, they can be taken orally and have excellent intestinal absorption and tolerance to stomach acid. The low bioavailability of orally administered phytochemicals is attributed to the Blood-Brain Barrier (BBB) and limited targeted organ accumulation [6]. Since the direct therapeutic effects of many phytochemical candidates were inadequate, they either remained in the laboratory or were developed into healthcare products. Although it is still limited to brain targeting, intravenous administration increases the concentration of medicine in the blood and reduces drug loss in the gastrointestinal system [7].

Brain disorders are not contagious but may run in families and are characterized by abnormalities in normal brain structure and function that arise from birth defects or genetic malfunctions. Several illnesses and disorders can affect the brain, including malformations, tumors, infections, and traumas [8]. The increasing prevalence of brain illnesses due to the world's aging population calls for immediate action in the form of improved diagnostic tools, effective treatments, and effective prevention measures. The blood-brain barrier, a special capillary control system that keeps most substances from the outside world from entering the brain and spinal cord, has eluded researchers. Unlike other organs, the brain is not accessible *via* the circulatory system for over 98% of neurotherapeutic chemicals and more than 100% of larger therapeutic molecules [9].

Many innovative brain delivery platforms have emerged as a consequence of nanomedicine's rapid development; some have even received regulatory approval after demonstrating encouraging results in animal studies. In addition, research on nasal delivery as a means to enhance drug absorption in the brain is ongoing [10]. The use of lipid-soluble solvents in combination with nasal administration of certain medication formulations has the potential to reduce drug loss in blood circulation and improve medicine uptake by the brain *via* the olfactory bulb. Unfortunately, many drugs still don't work on humans because our olfactory bulbs are much smaller than those of mice, in comparison to the brain. Nanodelivery formulations may also improve BBB translocation efficiency after nasogastric injection [11].

Despite the widespread recognition of phytochemicals as potential therapeutic agents, there is a limited understanding of the molecular mechanisms by which these substances exert their effects. Unlike healthcare commodities, clinical medications require precise dosing, clear indications, and a comprehensive understanding of their metabolic and pathological interactions. Further study on its working mechanisms, particularly in conjunction with computational drug design and structural biology, may lead to the identification of additional potential *de novo* candidates derived from phytochemicals [12].

Nanocarriers (NCs) may enter the majority of cells by endocytosis and transcytosis, as determined by the research study. There has been a lot of focus lately on the use of NCs for administering drugs to the brain. To facilitate better BBB crossing, drugs were formerly combined with surfactants and co-solvents such as ethyl alcohol, polysorbate 80, dimethyl sulfoxide, and so on [13]. Nonetheless, these substances seriously jeopardize the BBB's integrity and defensive activities. The unique qualities of NCs, including low toxicity, selectivity, biocompatibility, small size, and solubility, combined with their ease of transport across the BBB, have made them a promising therapeutic option for brain diseases and disorders [14]. However, the therapeutic efficiency of conventional drugs used to treat brain disorders is limited due to their low absorption in brain tissue and several serious side effects on healthy tissue. The secondary metabolites of plants that have garnered recent attention include phytochemicals and phytonutrients. This covers substances such as phenolic acids, phytoprostanes/furanes, saponins, isothiocyanates, phytosterols, and indoles. For a long time, people have turned to phyto-bioactive components as a means of treating physical and mental disorders. To deliver phytochemicals to the intended site (the brain) more effectively, nanocarriers were developed. The carrier material determines whether hydrophilic or hydrophobic molecules are contained therein [15, 16].

To improve the efficacy of treatments for neurological diseases, several nanocarriers have been investigated recently. These include Magnetic nanocarriers, nanoliposomes, dendrimers, niosomes, polymeric nanoparticles, and so on. Several phytochemicals have shown encouraging therapeutic efficacy when delivered using nanocarriers based on polymers or lipids [17]. Important phyto-components delivered by nanocarriers to improve treatment effectiveness in brain diseases are discussed in this chapter. This chapter covered potential targets, phytoconstituents, methods for delivering medication to the brain, and nanocarrier systems for disease management and therapy. Given the current level of understanding of the therapeutic advantages of both permeable and non-permeable blood-brain barrier phytochemicals in brain treatment, this chapter focuses on introducing a nano delivery platform that may maximize these advantages [18].

## Overview of Brain Tumor

The uncontrolled expansion of brain tissue is known as a brain tumor. Due to the pressure it creates in the skull, it disrupts the brain's normal functioning. Benign brain tumors are those that do not spread cancer, while malignant brain tumors are those that do. For instance, malignant brain tumors may metastasize and grow rapidly, causing damage to normal brain structures. Brain tumors may be classified into four main types [19]:

Tumors of grade I do not metastasize and progress at a moderate pace. These may be surgically eliminated almost entirely and are associated with a higher chance of better order. Pilocytic astrocytomas are one kind of tumor. Grade II tumors not only grow over time but may also spread to other tissues and progress to higher grades. The patient may still be able to identify these cancers after treatment has been administered. Oligodendroglioma tumors are characterized by their ability to increase in size steadily. Class III tumors have progressed more rapidly than class II cancers and may have metastasized (spread to other tissues). Because surgery is insufficient for the treatment of certain cancers, chemotherapy or radiation therapy administered after surgery is necessary. Malignant tumors of grade IV are very hazardous and have the potential to metastasize. To hasten their maturation, they could even utilize blood vessels [20].

For patients to get sufficient treatment and have a chance of survival, brain tumors must be identified and classified quickly. Multiple weaknesses, including tumor shapes, sizes, appearance, locations, scanning parameters, and modalities, make brain tumor detection a challenging endeavor [21].

## BRAIN DISEASES AND THE BBB

The Blood-Brain Barrier is a continually evolving cell wall that lies between the CNS and the remainder of the body's circulatory system. Pericytes, Astrocytes, and extracellular matrix components are essential for the survival of brain microvessel endothelial cells (Fig. **1**) [8]. The BBB not only prevents hazardous chemicals, viruses, and cells from entering the brain *via* the circulation, but it also guarantees that neurons have access to the oxygen, nutrients, and minerals they need to carry out their everyday functions. Comparable to ECs in other tissues, Brain ECs (BECs) exhibit variations in their intricate molecular transport mechanisms. Paracellular and transcellular molecular exchanges of approximately 98% of commonly used chemicals are limited by the BECs' low transcytosis rate and the absence of apertures in their integrated, continuous intercellular tight junctions [8]. Only tiny molecules that are soluble in lipids or have fewer than eight hydrogen bonds can cross [9, 22]. As a result, very few specialized transporters can move fluids, ions, peptides, proteins, and nutrients across the blood-brain barrier. More precisely, aquaporin and the abluminal sodium pump are primarily responsible for transporting water and ions across the BBB. Ethyl and other small lipophilic compounds may easily diffuse over the BBB. Solute carrier-mediated transport, driven by a concentration gradient, can carry metabolites, nucleotides, and other substrates. In contrast, receptor-mediated transcytosis is the primary means of transporting peptides and proteins [11]. Several immune cells can penetrate the BBB and access the brain because BECs do not generate many signals that may be utilized to detect immune cells, including leukocyte adhesion molecules. It follows that the CNS's immune surveillance system is unique [12].

Taken as a whole, these factors limited our understanding of neurological disorders. Several recent studies have linked BBB to aging-related brain alterations. Damage to tissues, injuries, infections, or pathological molecular processes can cause the BBB to collapse, disrupting the transport networks. This disruption can lead to a range of brain illnesses [10]. A damaged blood-brain barrier also causes chemical flux dysregulations and greater immune cell invasion. Inflammation and other diseases may result from problems with immune surveillance and effector responses to brain infections. Among other pathological changes, the onset of neurodegenerative diseases has been linked to microbleeds, decreased p-glycoprotein function, decreased glucose transport, pericyte degeneration, endothelial degeneration, aberrant angiogenesis, capillary leaks, and other similar changes [12].

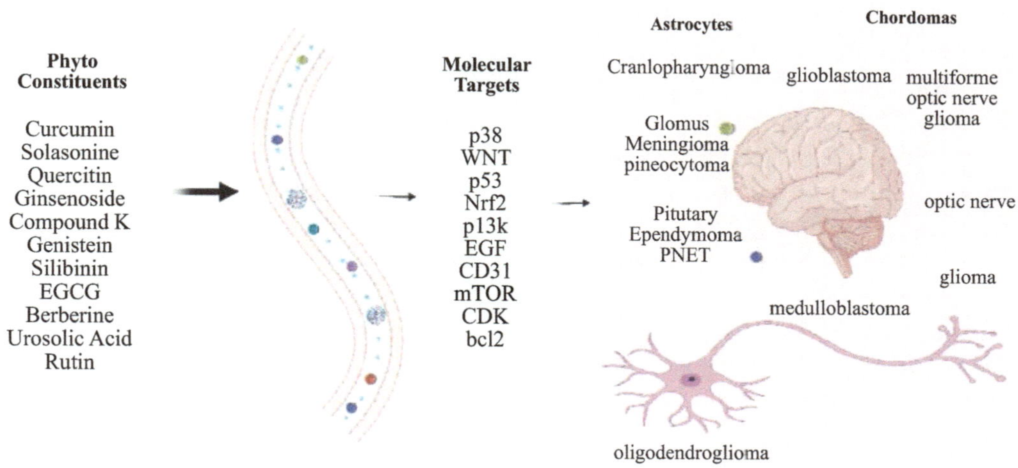


**Fig. (1).** Transporting phytochemicals-encased nanoparticles across the BBB to attack molecular targets in the fight against various brain cancers — with improved absorption and stability [23].

## Brain Diseases and P-gp

Studies have shown that the biggest challenge in treating brain illnesses and diseases is getting medications across the blood-cerebrospinal fluid and blood-brain barriers. It has been demonstrated that the downregulation of the protein P-gp, which is responsible for controlling the efflux of chemicals across the BBB, is associated with neurological diseases and the progression of cancer [24]. Medicines have a greater chance of crossing the BBB and having an impact when P-gp is inhibited. Scientists found that poly(butyl cyanoacrylate) NCs reduced phenytoin tolerance in rats, and P-gp mediated this effect. Furthermore, compared to free medicines, solid lipid NCs containing the neuroprotective medicine andrographolide had a higher probability of penetrating the BBB. Collectively, the findings point to the possibility that NCs control p-gp to enhance predicted drug penetration and pinpointability [25].

# ANTICANCER PROPERTIES OF PHYTOMEDICINE

## Herbal Remedies for Brain Tumor Therapy

Numerous plant bioactive substances have shown promise as anticancer agents: coumarins, terpenoids, alkaloids, fatty acids, phenols, flavonoids, cucurbitacins, iridoids, lignans, limonoids, diarylheptanoids, naphthoquinones, oligorhamnosides, physalins, sesquiterpenoids, phenanthrene derivatives, triterpenoids, stilbenoids, polyacetylenes, and many more (Table **1**). The usefulness of flavonoids as anticancer agents has been demonstrated by numerous studies, both *in vitro* and *in vivo*. Flavonoids from whole foods like oleuropein, genistein, apigenin, resveratrol, naringenin, and quercetin block aromatases and function *via* many different pathways [26]. These include kinase inhibitory, antiproliferative, anti-angiogenic, anti-metastatic, antioxidant, drug efflux inhibitory, anti-inflammatory, cytoskeletal protein and spindle fiber modulating, and modulating various cytoplasmic and nuclear proteins. Researchers have discovered that plant-derived products have anticancer properties. These properties make them effective adjuvants and chemopreventive agents that target various stages of carcinogenesis. For example, they can aid in carcinogen metabolism, DNA repair, free-radical scavenging, and detoxification. Additionally, they can suppress proliferation, and increase apoptosis, differentiation, inflammation, immunity, and gene expression, among other things [27].

**Table 1. Phytopharmaceuticals with their active constituents used for the treatment of brain tumor.**

| S. No. | Plant Source | Phytochemicals | Family | Type of Cancer | Refs. |
|---|---|---|---|---|---|
| 1 | *Gossypium hirsutum* | Gossypol | Malvaceae | Brain | [28, 29] |
| 2 | *Podophyllum hexandrum* | Podophyllin | Berberidaceae | Brain | [30] |
| 3 | *Schizophyllum commune* | Schizophyllan | Schizophyllaceae | head | [31] |
| 4 | *Podophyllum hexandrum, Podophyllum peltatum* | Podophyllotoxin | Berberidaceae | Brain | [30 - 32] |
| 5 | *Oldenlandia diffusa, Prunella vulgaris* | Ursolic acid | Lamiaceae, Rubiaceae | Brain | [33] |

## Nano-Phytopharmaceuticals in Neurological Disorders

An intricate web of astrocytes, microglia, neurons, and other ancillary structures floats in a medium of cerebrospinal fluid, making up the CNS. Nerve connections between the spinal cord and the brain are sensitive and may be readily disrupted by exogenous shocks or endogenous degenerative illnesses.

Parkinson's Disease (PD), Huntington's Disease (HD), Amyotrophic Lateral Sclerosis (ALS), tumors such as neuroblastoma and glioblastoma, Alzheimer's Disease (AD), Lewy Body Dementia (LBD), and various kinds of dementia are all examples of conditions that affect the CNS [34 - 36]. The development of plaques with entangled neuro fibers composed of amyloid and tau proteins is a hallmark of AD, a neurodegenerative disorder [37]. A deficiency of dopamine neurons in the substantia nigra is a hallmark of Parkinson's disease. It has been linked to an increase in ubiquitinated synuclein in dopamine-producing neurons. About one percent of the population over sixty-five has Parkinson's disease, the primary cause of Parkinsonism. Tremors, bradykinesia, and akinesia are common in advanced Parkinson's disease [38]. Huntington's disease is characterized by aberrant brain activity and neuronal death; therefore, the patient develops chorea, or abnormal movement [39]. The fundamental reason for Huntington's disease is an inherited trait that runs in families. Neuroinflammation, the formation of aberrant protein aggregates, and neuronal death are all components of the pathophysiology of central nervous system diseases [40].

There is a substantial global health burden due to the CNS and related disorders [41]. The BBB is a big hurdle when trying to treat illnesses that impact the central nervous system. The BBB presents a similar barrier for the majority of currently available medications when trying to reach the central nervous system and specifically target an infection [8]. On the other hand, the BBB gradually permits the passage of smaller molecules associated with nutrients and other essential CNS functions [42].

Long-term attempts to get pharmaceutical medications into brain tissue are hampered by the BBB, a natural barrier composed of pericytes, tight junctions, and capillary endothelial cells. The nerve tissue can only be penetrated by very tiny particles (<400 Da) due to the almost impermeable barrier [43]. Less than 5% of the common therapeutic drugs can pass this physiological barrier at different phases of pharmacological research. Thus, the possibilities for natural product compositions in nanocarriers that target the central nervous system are limitless [44]. Neurological disorders may find potential treatment in nutraceuticals and phytochemicals due to their neuroprotective, anti-inflammatory, and antioxidant characteristics (Fig. **2**) [45]. Neurological disorders may still be treatable with the help of nanoscale delivery methods that enhance solubility and retention rates and the capacity to traverse the BBB [46]. Curcumin and ginseng are two natural compounds that have great promise for use in neurotherapeutics when encased in nanoparticles (Table **2**) [47].

Curcumin, a polyphenol derived from the root of turmeric (*Curcuma longa*), has shown encouraging results in treating inflammatory illnesses and cancer, according to nutraceutical research. The medicinal uses of this yellow spice have deep roots in Ayurvedic practice and date back to bygone cultures. However, due to its low water solubility and rapid digestive clearance, traditional drug distribution is hindered, which presents an opportunity to develop nano-formulations [48].

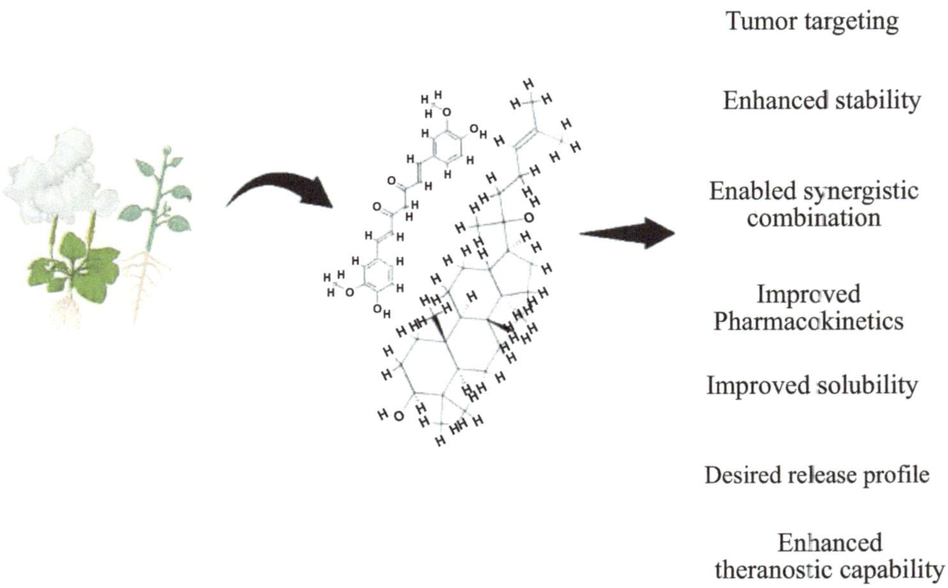


**Fig. (2).** Nanopharmaceuticals, phytoconstituents, and their relevance.

**Table 2. Curcumin and ginseng extract nanoformulations for neurological diseases.**

<table>
<tr><th>Phytopharmaceutical</th><th>Nanoformulation</th><th>Effect</th><th>References</th></tr>
<tr><td>Curcumin</td><td>Cesium oxide nanoparticles coated with dextran</td><td>Caspases enhance apoptosis in neuroblastoma cells.</td><td>[50]</td></tr>
<tr><td>Elixir of fresh ginseng leaves</td><td rowspan="2">Nanoparticles of Ag and Au</td><td>Activity against oxidative stress, cancer, and inflammation in the following cell lines: 3T3-L1 pre-adipocyte, HaCaT, A549 lung, B16BL6 skin melanoma, and RAW 264.7.</td><td rowspan="2">[66, 71]</td></tr>
<tr><td>Red ginseng water extract</td><td>Reduced intracellular oxidative stress and enhanced synaptic impulse transmission have a cytoprotective effect on PC-12 cells, which resemble neurons.</td></tr>
</table>

*(Table 2) cont.....*

| Phytopharmaceutical | Nanoformulation | Effect | References |
|---|---|---|---|
| **Curcumin** | The CUR/HP-β-CD inclusion complex consists of chitosan-coated poly (lactic-c--glycolic acid) nanoparticles and hydropropyl-β-cyclodextrin-encapsulated CUR. | Improved bioavailability in living organisms and anti-inflammatory effects in BV-2 cells. | [59] |
| **Curcumin** | Nanoparticles encased in Tween 80 | The anti-inflammatory effects were observed in BV-2 cells and improved bioavailability in live animals. Curcumin accumulates more in Wistar rats and is more soluble and distributed to important organs in the plasma. | [80] |
| **Combination of curcumin and piperine (CPZChN)** | Zein–chitosan nanoparticles | Reduced the viability of SH-SY5Y neuroblastoma cells by half. | [52] |
| **Curcumin** | Superparamagnetic iron oxide particles linked with curcumin; magnetic nanoparticles | MRI-based, less invasive AD diagnosis made possible by co-localization of particles with amyloid plaques. | [61] |
| **Combination of ginsenoside Rg3 and paclitaxel (Rg3-PT--LPs)** | Carrier for liposomes | Apoptotic suppression of C6 glioma cells; in C6-positive mice and rats, stimulation of the growth of CD8 + TT-cell populations. | [48] |
| **Curcumin (LipoCur)** | | Human microglia and astrocytes exhibit reduced proliferation and responsiveness, resulting in less glial scarring. | |
| **Curcumin** | Scaffolded silk fibroin micelles | Neuroblastoma cells with enhanced tumor-specific toxicity. | [50] |

*(Table 2) cont.....*

| Phytopharmaceutical | Nanoformulation | Effect | References |
|---|---|---|---|
| **An inhibitor of Aβ production and curcumin** | Innovative PLGA nanoparticles loaded with a cyclic CRTIGPSVC peptide that penetrates the blood-brain barrier. | Spatial memory and "new-object" recognition were both improved in transgenic Alzheimer's disease mice. | [81] |

One of the many new methods to highlight the importance of nanoparticulate systems in glioblastoma treatment is to increase their diffusion over the BBB. In a glioblastoma study, curcumin encapsulated in liposomal carriers (LipoCur) was shown to have lower toxicity to astrocytes and human microglia, exhibiting greater proliferation and responsiveness compared to the free material [49]. A similar reduction in glial scarring was seen with LipoCur in organotypic brain slices of mice exposed to lipopolysaccharide (LPS) for eight days. Studies on neuroblastoma mice have shown that the anticancer properties of curcumin were considerably enhanced after its distribution in nanocarriers. Curcumin nanosuspensions showed more solubilization and more intravenous plasma dispersion to vital organs than traditional curcumin. The medication is suspended in stable nanoscale droplets along with biopolymers and surfactants. Nanoparticles coated with Tween 80 were shown to transfer more curcumin to the brains of male Wistar rats compared to a solution of curcumin. Curcumin regulates neuroblastoma cell signaling, apoptosis, and mitochondrial dysfunction by controlling PTEN-Akt, NF-κB, and p53 [48].

An *in vitro* neuroblastoma model has been shown to be significantly induced to die by curcumin-containing cerium oxide nanoparticles (CNP-Cur) coated with dextran. In populations not amplified by MYCN, the formulation retained populations; nevertheless, it increased oxidative stress and induced caspase-dependent death in cancer cells [50]. The silkworm *Bombyx mori* produces silk fibroin, one potential component for nanoparticle shells. The biodegradable and biocompatible formulation of curcumin-loaded silk fibroin NPs seemed attractive due to its ability to improve and target tumor-specific cytotoxicity in neuroblastoma cells. The focus was not just on pure curcumin, but also on herbal combinations that showed synergistic benefits [51].

Curcumin had an encapsulation efficiency of 89%, and piperine had an efficiency of 87%. Because the constituent chemicals in complexes with many bioactive compounds, such as curcumin and piperine, are polar, the synthesis of these complexes is difficult. To overcome this, antisolvent precipitation, stacking of zein-hyaluronic acid, and chitosan coating were employed to encapsulate the two components in core-shell nanoparticles [52]. Scientists are paying a lot of attention to chronic degenerative diseases, including Alzheimer's disease and cancers of the CNS. The disease is famously complex to diagnose and cure, as symptoms don't appear until long after the degenerative changes have begun. It is characterized by tau protein neurofibrillary tangles and "senile" amyloid-β (Aβ) plaques in the brain. Research into Natural Products (NPs) may focus on neuro-inflammation since it is known to be a major mechanism in neuronal degeneration. Compared to free compounds and SLNs, which had an entrapment efficiency of 83% ± 0.50%, and Nanostructured Lipid Carriers (NLCs), which had an entrapment efficiency of 96.0 ± 0.75%, NLCs caused the greatest amount of curcumin to accumulate in the rats' brains [51 - 55].

Biocompatible polymer NPs and poly(D, L-lactic acid-co-glycolic acid) (PLGA) have been linked to a decrease in neuroinflammation in another 5xFAD mouse AD model. Huang *et al.* obtained encouraging results by co-administering curcumin with an Aβ production inhibitor and employing innovative PLGA NPs augmented with a BBB-penetrating peptide (cyclic CRTIGPSVC). After receiving treatment, transgenic AD mice demonstrated improved 'new-object recognition' and spatial memory scores [56 - 58]. Reducing β-amyloid aggregation in AD models has been demonstrated by combining phytol and selenium with PLGA NPs. Two formulations were tested for intranasal distribution by Zhang *et al.* [59]. Curcumin magnetic nanoparticles (Cur-MNP), which are biocompatible, are another type of nanoparticle [60]. These nanoparticles are composed of curcumin and superparamagnetic iron oxide (SPIO) particles, each with a diameter of less than 100 nm. The administration of Cur-MNPs to the brains of Tg2576 animals resulted in particle co-localization with amyloid plaques, making the use of MRI for AD diagnosis less invasive. Neurotheranostics NPs encapsulated in curcumin have piqued the interest of the pharmaceutical industry; nevertheless, more clinical trials are required to confirm this [61].

A further "super herb," called the Ginseng plant (*Panax ginseng*), with a long history of therapeutic usage dates back to the ancient civilizations of Korea and China. Common uses include enhancing mental activity, memory, and well-being. Neurotherapeutic drug development is based on its alleged effects on the CNS and ginsenoside, gintonin, and Compound K derived from ginseng. These compounds have been studied for their potential to treat and prevent neurodegenerative diseases, such as depression, PD, and AD [62, 63]. Mechanistic studies of Alzheimer's disease have shown inhibitors of amyloid beta aggregation, neurogenesis, antioxidants, and anti-apoptosis. The neuroprotective effects of a combination of Ginsenoside Rg3 and Thioflavin T, a diagnostic marker for Alzheimer's disease, encapsulated in biodegradable PLGA nanoparticles, were investigated [64, 65].

The efficacy of BBB penetration and the molecular routes of bioactivity were evaluated using an *in vitro* model. When searching for a nano-delivery approach to treat neurological illnesses, they concluded that PLGA-Rg3 NPs may be effective. Since ginsenoside possesses antioxidant properties, regulates glutathione levels, inhibits iron formation in the substantia nigra, and mediates ROS-mediated NF-κB signaling, it has been investigated as a potential treatment for Parkinson's Disease (PD). The biotransformation of the ginsenosides Rb1, Rb2, and Rc produces the metabolite Compound K (CK), which other groups have studied for its bioactivity [66]. Rd, Re, and Rg131 are three other ginsenosides that have been extensively researched. Being the most bioavailable part of ginsenoside metabolism, its bioavailability has been the main focus of studies on its neuroprotective and cognitive-enhancing properties. Clinical studies on CK have been limited, relying on fermented red ginseng, which is high in CK. However, CK is generally well-tolerated and safe. The efflux, poor water solubility, and membrane permeability of CK were shown to be overcome by delivering it *via* nanocarriers, which allowed research to move into clinical trials [67, 68].

Kim *et al.* conducted DPPH, ORAC, and ABTS tests to determine the antioxidant activity of a red ginseng water extract formulation (WERGGN) that included gold nanoparticles. Majorly, a reduction in intracellular oxidative stress was responsible for WERGGN's cytoprotective effects in neuron-like PC-12 cells [69]. The pharmacokinetics of ginseng extract when taken with synthetic drugs, such as selegiline, are dose-dependent; lower doses induce CYP1A2 and result in poorer bioavailability, while higher concentrations inhibit CYP [70].

While the root of Panax ginseng is typically used to extract the bioactive components, the leaves could be just as effective. The gold and silver nanoparticles were studied and confirmed by Singh *et al.* using an extract from fresh ginseng leaves [71]. Researchers have demonstrated that polymer-ginsenoside nanoconjugates exhibit anti-carcinogenic effects by enhancing drug bioavailability in cancerous areas, facilitating efficient drug release, and targeting oncogenes [72]. Zhu *et al.* created Rg3-LPs, a multipurpose liposomal system based on ginsenoside Rg3, in a glioblastoma spheroid model. These LPs, when combined with paclitaxel (PTX), inhibited C6 glioma cells more effectively than C-LPs, which were based on cholesterol [73].

Rg3-PTX-LPs increased $CD8^+$ T-cell populations in C6-bearing mice and rats, thereby enhancing immune surveillance in the fight against cancer. The researchers hypothesized that ginsenoside Rg3 provides a delivery pathway for neurological malignancies and works in conjunction with synthetic chemotherapy drugs [72, 73]. Ultimately, more research into Panax ginseng derivatives is warranted due to their tremendous potential in the treatment of neurological disorders and a wide variety of brain tumors [74].

It should be mentioned that nanomaterials could have some restrictions and that there is proof of negative effects on the CNS [75 - 79].

## OPPORTUNITIES OF PHYTOPHARMACEUTICALS FOR BRAIN TUMOR

### Therapeutic Significance of Phytotherapeutics

To alleviate the symptoms of illness, many individuals opt to use herbs instead of traditional medications. Consequently, modern pharmaceutical companies have focused on discovering novel, plant-based medical molecules with therapeutic potential. The majority of natural novel molecular entities approved by the FDA are derived from plants. Morphine, a benzylisoquinoline alkaloid found in Papaver Somniferum and authorized for use in 1827, is the first NME developed from a plant. The primary mechanism by which it influences the nervous system is by increasing the inhibition of neurotransmitter release from presynaptic neurons. Paclitaxel, an alkaloid isolated from the Pacific yew plant (*Taxus brevifolia*), was one of the most potent chemicals found after morphine. Paclitaxel binds to β-tubulin, stabilizing the microtubule polymer. Proper mitotic spindle formation and chromosomal segregation are essential processes in cell division, although these processes can be inhibited. Several types of cancer are being treated with paclitaxel, including pancreatic, ovarian, breast, lung, and herpesvirus-8-induced *Kaposi sarcoma.* Vincristine and Vinblastine were similarly isolated from the *Catharanthus roseus* plant. In contrast to paclitaxel, these anticancer drugs target β-tubulin at a different site, allowing them to impede microtubule polymerization and halt the cell cycle in the M-phase [61].

Low solubility and instability are two examples of the physicochemical constraints that hinder the efficient delivery of phytotherapeutics. Furthermore, clinical trials may encounter difficulties due to pharmacokinetic factors, such as inadequate absorption and restricted bioavailability. Poor effectiveness and batch-to-batch variability are two other significant obstacles to integrating phyto-therapeutics into the pharmaceutical industry. The few examples above demonstrate the importance and challenge of phytochemicals in preventing and treating many diseases. Due to this issue, providing it efficiently and safely has become a top priority. When compared to more conventional ways, the efficacy of novel delivery systems for phytomedicinal substances, such as nanocapsules, phytosomes, liposomes, and nanoparticles, is superior. These administration strategies lengthened the half-life by increasing distribution time, bioavailability, chemical and physical stability, and pharmacological activity [66].

Poor solubility and instability are two examples of the physicochemical disadvantages that make the effective delivery of phytotherapeutics more difficult. Furthermore, clinical studies might encounter difficulties due to pharmacokinetic factors, such as inadequate absorption and limited bioavailability. Another major obstacle is the translation of phytotherapeutics into the pharmaceutical industry, where issues such as batch-to-batch inconsistency and poor efficacy persist [6]. We can see the importance of herbal therapy and the difficulties it faces in treating serious ailments from the few instances given above. The focus has shifted to ensuring its safe and effective delivery because of this issue. These days, it has been shown that newer delivery systems—such as liposomes, phytosomes, nanoparticles, and nanocapsules—are more effective at delivering phytomedicinal components than the older ones [7].

## Phytochemical Drugs with Promising Brain Therapeutic Potential to Cross the BBB

Although the findings of numerous studies suggest that specific phytochemicals are beneficial for treating brain diseases, there have been few successful clinical translations of these findings. The existence of the BBB significantly limits the therapeutic potential of certain phytochemicals by preventing their passage into the brain and, consequently, their use in the treatment of brain diseases. Very high-dose administration nevertheless poses a biocompatibility risk to ensure proper brain accumulation of phytochemical medications. The hydrophobic polyphenol curcumin, which is derived from the dried rhizomes of *Curcuma longa* L., has shown anti-inflammatory, cytokine-reduction, and NF-kB signaling pathway inhibitory effects in APP/PS1 mice [82]. Curcumin may improve AD outcomes by reducing Ab42 oligomer development and disaggregation of generated fibrils, according to other findings. Curcumin has significant neuroprotective properties, but it has limited use in treating neurodegenerative diseases and brain tumors because of its low brain bioavailability [83]. It has been demonstrated that salidroside, a compound present in Rhodiola rosea, can cross the blood-brain barrier, modulate thioredoxin and thioredoxin-interacting proteins, as well as NF-κB pathway proteins, thereby enhancing cognitive recovery in rats with Alzheimer's disease [84].

Salidroside reduced Ab1-42 buildup and microglial activation, providing therapeutic advantages in a model of accelerated aging in mice when administered at a dosage of 49 mg/kg/day for three months [85]. Salidroside induces an increase in telomerase reverse transcriptase expression through the PI3K/Akt pathway, as further investigations on the treated animals have shown [86 - 90]. In a senescence-accelerated mouse model initiated at 3 months of age, erinacines from Hericium erinaceus [91], administered orally for 11 weeks at a dose of 108 mg/kg/day, significantly improved learning and memory capacity. Similarly, in a model of Alzheimer's disease, a 30-day dosage of 300 mg/kg/day reduced the Ab plaque burden [92]. Because of its high solubility in water and its ability to pass across the BBB, paeoniflorin suppresses neuroinflammation *via* the NF-$_k$B and GSK-3b signaling pathways, as well as the NLRP3 inflammasome [93]. One of the polyphenols found in wine, resveratrol, also possesses anti-inflammatory, neuroprotective, antioxidant, and anti-aging effects. Baicalin, wogonin, and Baicalein are flavonoid compounds that may cross the blood-brain barrier within 30 minutes [94]. These chemicals are primarily derived from the roots of Scutellaria baicalensis Georgi (Lamiaceae). Baicalein and baicalin are similar in that they inhibit cancer growth in living organisms and kill tumor cells in humans [95]. Blocking 12-lipoxygenase, inducing apoptosis, suppressing metastasis, stopping the cell cycle, and activating autophagy are the mechanisms that work together to decrease tumor growth. In rats modeled after Parkinson's disease caused by rotenone, baicalein (400 mg/kg/day for 42 days) increased the quantity of $TH^+$ neurons [96].

In addition, the main flavanol in green tea, epigallocatechin-3-gallate (EGCG), restored the learning abilities of brain-accelerated senescence model mice (SAMP10) when given 20 mg/kg/day for two weeks, demonstrating its positive effects on cognitive dysfunctions. Nevertheless, the high dosage is still limiting its clinical application, even though EGCG can penetrate the blood-brain barrier thanks to organic anion-transporting polypeptides (SLC21A3) on the BBB [97, 98]. Cannabidiol (CBD) and tetrahydrocannabinol (THC) are the two most famous and medicinally useful compounds found in the cannabis plant. CBD and THC, two cannabinoid components, inhibit the production of T-helper type 1 (Th1) cytokines and stimulate the development of a Th2 immune response in both laboratory and living organism studies. Their ability to bind to various receptors in the body allows them to serve multiple purposes. These include receptors in the endocannabinoid system, for example, cannabinoid receptors 1 and 2, as well as GPR55, PPAR, glycine, 5-HT, TRP, and opioid receptors [99, 100].

Analytical chemistry made strides thanks to the practical parts of phytochemical extraction and identification [101]. The current pharmaceutical industry's screening library is lacking in structural complexity, which means that most compounds that are easy to synthesize are already in the screening bank. However, these compounds are no longer sufficient for high-throughput screening, so the library needs to be updated with more complex structures. Potentially useful in this metamorphosis are the phytochemical derivatives [102].

One advantage of phytochemical derivatives is their high biological reactivity, which is due to their endogeneity, and another is their low biosynthetic costs.

Research into medicinal phytochemicals through *de novo* chemical synthesis has recently gained importance in synthetic chemistry as a means to reduce costs and increase purity [103]. This means that phytochemicals and their derivatives will likely continue to shed light on future medical studies.

## Achieving Phytochemical Brain Therapies with Nano-delivery Systems

Solubility, stability, target selectivity, effective accumulation, and transit across the BBB are just a few of the issues that nanoparticles might address as a potent delivery platform for phytochemical medicinal applications. Materials or structures existing at the nanoscale are known as nanoparticles. The ability to cross cell membranes is a key property of nanoparticles. Liposomes, dendrimers, micelles, synthetic biomaterials, and inorganic nanocarriers are several types of therapeutic nanoparticles [104]. Although several nanoparticle formulations exist, each with its own set of advantages and disadvantages, the most flexible method allows for the *in vivo* optimization of phytochemicals or their clinical use in the management of brain illnesses through the application of simple physical and chemical evaluations [105]. In addition, the customized transport and release capabilities of nanoparticles may significantly reduce the dosage of medicine, whether taken orally or intravenously. The release of nanomedicines at their injection sites is crucial to their efficacy. Conditional stimulus-responsive techniques are used to achieve effective controlled release [106]. Scientists have developed several methods to utilize nanoparticles in the treatment of phytochemicals for the brain. The first tactic is to load nano-delivery devices with phytoactive compounds that may regulate inflammation and oxidative stress, two factors that are believed to be crucial in brain diseases. Research suggests that nanocarriers laden with phytochemicals may have a greater impact on brain and CNS diseases like neurodegenerative disorders, rheumatoid arthritis, and glioblastoma [107]. The goal of this approach is to increase the permeability of the BBB to the active phytochemicals that were previously characterized as BBB-impermeable [108]. In rats with Alzheimer's disease, resveratrol increases antioxidant activity and crosses the blood-brain barrier. Spherical nanoparticles loaded with resveratrol were used to treat intracerebral hemorrhage damage [109 - 111]. A different study found that administering 10 mg/kg intravenously every two days for six days to C6 orthotopic rats and U87MG-xenograft mice loaded with resveratrol-polyethylene glycol-polylactic acid nanoparticles considerably decreased tumor volume and improved life expectancy [112].

Some of the most important transporters involved in drug bioavailability and disposal are ABC (ATP Binding Cassette) and SLC (Solute carrier) transporters, which are expressed on the luminal side of BBB *via* active efflux from the endothelium into the blood. Nanocarriers for transcytosis may be adorned with phytochemicals, especially those of smaller molecular sizes, as they can permeate the BBB. The Glut-1 transporter is highly expressed on the luminal side of the blood-brain barrier, and a well-known example is glucose. This carbohydrate can be used to facilitate nanoparticle transcytosis across the BBB [113].

Other translocating processes that use the BBB's unique influence on phytochemical absorption are currently being studied by researchers. For example, menthol's ability to significantly enhance cell membrane fluidity and, by extension, lower membrane potential, as well as disrupt tight junction integrity, may facilitate the translocation of nanoparticles across the BBB [114, 115]. It is also possible to functionalize phytochemicals into nanoparticles, which would allow them to be more easily absorbed across the BBB and then used therapeutically once released into the brain microenvironment. A research study utilized functionalized EGCG to mask the original compound, employing a one-step polyphenolic condensation process to create EGCG nanoparticles [116].

## Glioblastoma Therapy and BBB Permeability

When it comes to adult CNS tumors, glioblastoma is by far the utmost prevalent and deadly kind. With such a bleak outlook, the median survival time was less than 15 months [117]. Temozolomide (TMZ) is the only oral medicine permitted by the US Food and Drug Administration (FDA) for the treatment of gastrointestinal bowel cancer [117]. Chemotherapeutic chemicals used in this therapy are also quite limited. TMZ is a chemical that alkylates DNA, which accelerates the process of DNA double-strand fragmentation and, ultimately, cell death [118]. TMZ, the only treatment now available for GBM, was administered to GBM xenograft mouse models using a variety of techniques [119]. For the best outcomes in mice GBM xenograft models, either inject 50 mg/kg intraperitoneally three times over two weeks or administer 600 mg/kg orally once [120]. It takes a considerable amount of preclinical (in animal models) and clinical research to determine an effective dosage of chemotherapeutic drugs like TMZ [121]. Even more important than the recommended dosage for GBM therapy is the amount of active drug at the site (*i.e.*, brain tumor tissue) [122, 123]. We still need many more studies to learn about the *in-situ* active concentration of both currently licensed drugs and potential future treatments in the clinic [124].

At least half of patients treated with TMZ do not respond, which is a major setback for the standard-of-care chemotherapeutic therapy for GBM due to drug resistance. Therefore, research into new chemotherapeutic drugs with improved efficacy and fewer side effects is ongoing [125]. Among all-natural items, phytochemicals are the most desired. Several types of cancer have shown promise in clinical trials for treatment using phytochemicals. For instance, pancreatic, colon, and liver cancers that have not responded to surgery have been treated with *Allium sativum*. However, the BBB still blocks most extracellular chemicals from reaching the brain, as mentioned before [126]. The permeability of BBB decreases by a factor of 100 when the molecular weight of a drug increases from 150 to 450 Da [126, 127], and BBB transport is lowered by a factor of ten for every pair of hydrogen bonds on the solute. The BBB is thought to be impermeable to phytochemicals due to their often more complex chemical structures, which limits its potential therapeutic application against brain disorders (with a few notable exceptions, such as phytochemical therapy for GBM) [128].

To achieve this, one study utilized quercetin, which, when administered orally at a dose of 20 mg/kg/day for 14 days, inhibited Hsp27 and COX-2 in both laboratory and animal studies, thereby sensitizing GBM to t-AUCB. However, a different group discovered that *Angelica sinensis* had an anti-GBM impact; however, this experiment used a higher dose of 500 mg/kg/day. At the same time, most research on GBM is performed in a controlled laboratory setting [129].

Epigenetic changes, particularly those affecting DNA-bound histone proteins, have recently garnered interest as potential novel therapeutic targets in cancer therapy. Clusters of changes in histone acetylation state concentration are the most striking. Enzymes belonging to the classes of histone deacetylases (HDACs) and histone acetylases (HATs) compete with one another to control histone acetylation [130]. Histone deacetylases (HDACs) deacetylate histone proteins, resulting in a more compact chromatin structure and reduced transcription of DNA. Histone antigens (HATs), a similar family of proteins, promote chromatin structural relaxation. Several cancers have an abnormal histone acetylation pattern, which alters HDAC recruitment and synthesis. An inhibitory combination of polyphenols and HDACs may induce senescence in glioblastoma and other malignancies [131, 132]. These polyphenols influence the activity of histone deacetylases, which are one of numerous proteins that control the progression of cancer cells. Sodium butyrate, an inhibitor of histone deacetylase, amplifies the anti-aging effects of quercetin and resveraterol in glioma and human cells [133].

## WHY IS PHYTOMEDICINE SUPERIOR TO SYNTHETIC MEDICATION?

The recent surge in the success of phytomedicines in the pharmaceutical industry is due to several reasons. In comparison to chemically manufactured medications, the therapeutic efficacy of plant extracts is higher due to the synergistic action of their medicinal components. When compared to synthetic chemotherapeutics, phytomedicines are bioequivalent. There has been a trend towards phytotherapeutics, as the concept of multidrug and multitarget action has shown greater promise than monodrug therapy. There has been a shift away from conventional mono-drug procedures and towards combination therapy for many medical illnesses, including AIDS, malaria, cancer, hypertension, diabetes, *etc.* This technique is more effective than treating the ailment with a single pharmaceutical component, according to numerous studies and observations, as it addresses multiple disorders and their underlying causes simultaneously [134]. Herbal therapy is seen as a safer, more accessible, and less expensive alternative to medications made from chemicals, with no negative side effects when used as directed. According to research, synthetic drugs are recognized to reduce symptoms in the majority of patients with the condition. In contrast, herbal remedies work by bolstering the immune system's capacity to repair itself [135].

Herbal medicines are gentle since they repair broken or insufficient approaches and procedures, which helps eliminate the abnormality. As a trade-off for the benefits of pharmaceutical prescriptions to human health, side effects are often considered. Medications are designed to induce a specific response to a particular physiological abnormality [136]. The synergistic action mechanism of herbal drugs, on the other hand, permits them to provide therapeutic benefits with minimal side effects. The chemical-free composition of herbal medicines, their extensive appeal, and the lack of side effects have astounded people all over the globe. People are intrigued because it has the potential to provide personalized therapy while also addressing the issue of side effects associated with chemical medications. In the past, people have turned to it to help with non-life-threatening health issues. In cases when conventional treatments have been ineffective, such as in advanced cancer or other mysterious diseases, its use has recently increased. Herbal remedies also have a reputation for being much less harmful, more natural, and less poisonous than conventional pharmaceuticals [134].

## CHALLENGES OF PHYTOPHARMACEUTICALS FOR BRAIN TUMOR

### Limited Phytochemicals can cross BBB

Phytochemicals' current therapeutic potential for brain diseases is largely dependent on their capacity to permeate the BBB, and the amount needed to achieve this result is still a matter of discussion. Whether a therapy has high therapeutic efficacy depends on several elements, not the least of which is the prescribed dosage. However, we attempted to fill in the gaps in the existing data by gathering more examples of phytochemicals that, when administered at doses below or above 10 mg/kg/day, exhibit more likely BBB permeability, respectively. All current techniques undergo significant loss through blood circulation or gastrointestinal absorption; thus, this dosage estimation is merely a recommendation and does not account for the provided approaches.

Phytochemicals, particularly polyphenols [137] and flavonoids [138], have been the subject of extensive research due to their potential neuroprotective and broad-spectrum disease therapy benefits. These diseases include cancer, metabolic syndrome, cardiovascular disease, and metabolic syndrome. Regrettably, the exact process by which these particular phytochemicals may traverse the BBB is still a mystery.

Bioactive phytochemicals have the potential to treat brain illnesses, but our understanding of their mechanisms of action is limited. In a mouse model of traumatic brain damage, the flavonoid lutein, which is rich in *Capsicum annuum* [139, 140] leaves, was shown to strongly inhibit the accumulation of amyloid-b (Ab), activation of glycogen synthase-3 (GSK-3), phospho-tau, and pro-inflammatory cytokines [141]. The neuroprotective effects of luteolin in traumatic brain injury have been associated with the activation of the Nrf2-ARE pathway, and luteolin also promotes the translocation of nuclear erythroid 3-related factor 2 (Nrf2) to the nucleus in both *in vivo* and *in vitro* situations [141]. In addition, a natural compound called icariin (ICA) came from the herb *Herba Epimedii* [142]. It was found to have neuroprotective effects on dopaminergic neurons in a mouse model of Parkinson's disease. The effects may have been mediated through the PI3K/Akt and MEK/ERK pathways [143 - 147]. Additionally, research has shown that withanamides derived from the fruit extract of *Withania somnifera* L. may decrease lipid peroxidation. To demonstrate BBB permeability, withanamides were given to mice intraperitoneally (i.p.) once at a dosage of 5 mg/kg [148]. To inhibit cancer development and cell death, withanamides preferentially activate p53 in tumor cells. A rise in g-H2AX and the percentage of cells expressing the phosphorylated form [149] demonstrate that cellular toxicity is addressed by reducing both mitochondrial respiration [150] and DNA damage [147, 151]. *Magnolia obovata* contains the biphenolic compound obovatol, which exhibits neuroprotective effects against neuroinflammation [152 - 155].

## Passive *v/s* Active Targeting

Relying solely on the EPR effect, passively targeted nanocarriers often fall short in reaching the tumor site. More thorough research into how NCs interact with physiological barriers is required, with the findings informing the creation of complicated new solutions [156]. The use of tumor-infiltrating lymphocytes and the utilization of cells' natural tropism toward certain tissues are two other techniques for boosting NC accumulation in tumors. Some physiological constraints have been lifted as a result of recent advances in NC engineering. Methods to study the immune system's interaction with NCs are now under development, which will allow future clinical trials to predict NC-mediated toxicity [157].

There is a huge complexity difference between active and passive targeting tactics. The complex engineering and design of these systems compound the challenges already present due to physiological barriers and tumor heterogeneity, making it even more difficult to develop new pharmaceuticals, scale them up for production according to Good Manufacturing Practice (GMP), and reduce the total cost of therapy. One example of an active targeted NC is SGT-94, developed by SynerGene Therapeutics and now in Phase I studies for genitourinary malignancies. It contains cationic liposomes that confine the tumor suppressor gene RB94 and are decorated with a portion of the anti-transferrin receptor (TfR) single-chain antibody [158, 159].

Despite all the problems, actively targeted NCs could change the game when it comes to treating hematological cancers and metastatic lesions where EPR doesn't matter by reaching all the different parts of the body. To ensure that preclinical tumor models accurately reflect actual illness, including solid and hematological malignancies, and the therapeutic effectiveness of actively targeted NCs, careful evaluation and funding of these models are essential [160].

The development of supplementary diagnostic imaging techniques for assessing the targeting efficacy of NCs is equally critical for passive and active targeting strategies. Patients must be carefully selected in advance, and therapies must be personalized to enhance tumor accumulation, treatment effectiveness, and therapeutic outcomes. Reducing the number of unnecessary treatments and unfavorable responses is just one way that weeding out unsuitable individuals may save money for health insurance companies and government agencies [161].

## Regulatory and Industry Barriers

Along with the aforementioned problems, the marketing and approval of NCs by regulatory bodies are now the biggest roadblocks to bridging the bench-bed divide. The GLP circumstances are rare, and concerns exist about the validity and repeatability of academic research, making it challenging to collaborate with the pharmaceutical industry. The GLP guidelines were developed in response to concerns raised by numerous toxicity studies conducted in the 1970s regarding data integrity and study quality. Both the FDA and the EMA have issued rules defining baseline requirements for pre-clinical laboratory testing to speed up the commercialization of drugs that their respective bodies have authorized. In fully working pre-clinical models, GLP is a key factor in deciding the technology's translation potential. Since GLP adoption may significantly increase the total expenses of a preclinical investigation, a thorough evaluation of the study's aims is necessary to determine if GLP is necessary. Only GLPs with proven viability and translation readiness should get support from funding organizations [162].

The structural and chemical complexity of NCs makes it unlikely that conventional regulatory procedures can ascertain their safety and effectiveness. Furthermore, the need for placebo-controlled treatment regimens might be questioned. The FDA has already launched the Nanotechnology Regulatory Science Research Plan to address significant information gaps concerning the data, methodologies, and instruments needed to conduct regulatory evaluations of NCs and other nanomaterials. Clinical trial design and patient selection are significant considerations for assessing NC-based therapy. When tested against advanced cervical and head and neck cancer and non-small cell lung cancer, the PSMA-targeted docetaxel nanoparticle BIND-014 failed to meet the primary endpoints in both studies. In light of this setback, we believe that the current approach to using NCs in cancer clinical trials has to be drastically changed. Patients should be preselected based on their EPR level, tumor heterogeneity, existence of target receptors, capacity to actively target NC binding, and need for companion diagnostics to enhance treatment results [163].

Aiming to determine if the buildup of ferumoxytol iron nanoparticles in tumors, as assessed by quantitative MRI, could be used as a predictor of nanoliposomal irinotecan response, Merrimack Pharmaceuticals recently conducted groundbreaking clinical research [164]. Tumors with significant amounts of FMX were more effectively targeted by nal-IRI, according to the results. These techniques, combined with patient selection, aim to bridge the bench-bed gap by promoting the development of well-characterized NCS with comprehensive safety and effectiveness profiles, as well as an enhanced clinical trajectory. This means that NCs have a higher chance of really making it to the clinic and keeping their word.

## Tumor Microenvironment (TME)

A thorough understanding of the features and activities of the tumor microenvironment (TME) is essential for comprehending the biology and therapy of brain tumors [165]. Research organizations studying neurodegeneration and other non-malignant brain diseases have spent decades developing a comprehensive understanding of the biology of the extracellular matrix (ECM), brain-resident cells, and blood vessels [166, 167].

The vascular and immunological components of the tumor microenvironment (TME) are anticipated to play a pivotal role in enhancing the therapy of brain tumors. Some examples include the growing amount of evidence that immunotherapies may successfully heal brain malignancies and the discovery of a brain lymphatic or immunological system [168, 169]. Two types of myeloid cells, microglia and tumor-associated macrophages, comprise the majority of a brain tumor microenvironment (TME). To successfully treat brain tumors using immunotherapies, several significant challenges must be overcome. A more thorough understanding of tumor biology is essential for minimizing treatment-related side effects and selecting individuals who will reap the greatest therapeutic benefits [170 - 172].

## Develop Pre-clinical Models

When it comes to predicting how medications will work in humans, preclinical models have a history of inaccuracy. The therapy of brain tumors has not progressed because of this. Due to the small number of patients eligible to participate in clinical trials, preclinical drug development pipelines that reliably predict which medications will be effective are of the utmost importance in the treatment of rare diseases like brain tumors [173, 174]. Present brain tumor therapy pathways are inadequate for several reasons. To begin with, rather than using orthotopic models of brain tumors, they rely on *in vitro* systems or subcutaneous tumor xenografts that are not well-characterized. Second, these pathways do not permit the comparison of the efficacy of novel treatments to that of the standard neurosurgery, radiation, and chemotherapy regimen in terms of overall survival rates. Last but not least, these pipelines fail to find better ways to treat brain cancers. Thus, to treat all patients with brain tumors, a new strategy is needed, one that relies on our growing knowledge of neurology and the biology of brain tumors to create and evaluate more precise preclinical models [175].

A greater understanding of the mechanisms that drive the formation and progression of brain tumors would enable the creation of more accurate genetically modified mouse models and xenografts produced from patients. The brain tumor community is encouraged to establish a central registry that includes all current and future preclinical models of brain cancers that have undergone extensive genetic, imaging, and histological assessments [176]. The models in this collection should be made publicly available for free, allowing anybody to access verified models that accurately represent human cancers. To facilitate better comparisons of study outcomes and basic requirements for model characterization, the community should establish baseline standards for preclinical investigations. In contrast to the complicated multimodality treatment that brain tumor patients get, preclinical studies on mice often utilize monotherapies, which are seldom as successful as the current standard of care for the studied malignancy [177]. The next logical step is to establish experimental "mouse hospitals" to test novel drugs alongside established treatments, such as neurosurgery, fractionated radiation, and conventional chemotherapy. Researchers having access to these types of preclinical hospitals should do thorough biomarker, pharmacokinetic, and pharmacodynamic investigations for new drugs to increase the probability and speed of clinical translation success. Testing imaging biomarkers, tumors that penetrate the blood-brain barrier, and liquid biopsy should be the primary goals of these investigations [178, 179].

## CONCLUSION

The unique characteristics of the BBB and their interference with phytochemical usage in brain diseases are discussed in this chapter. Despite phytochemicals' limited ability to cross the BBB, they show great promise as a therapy for brain disorders. A practical nano delivery system was suggested to optimize the utilization of phytochemicals, whether they are permeable or not, across the BBB. This opens up a new field of potential therapeutic uses for phytochemicals in brain disorders. Nanotechnology has had a significant impact on public health. Nanoparticles, when combined with novel drug delivery technologies, have the potential to alleviate many of the problems associated with conventional, allopathic medicine and the limitations of plant-based medicine. Future research needs to address and overcome several limitations, despite the significant advancements in nanotechnology. Negative effects may result, for instance, from the inorganic components of nanostructures being unevenly distributed throughout the brain. In addition, these nanostructures hinder mitochondrial activity, autophagy, apoptosis, and neuronal inflammation, suggesting that their presence in the brain may cause neurotoxicity. Nanomaterials, as demonstrated by *in vivo* and preclinical studies, enhance the effectiveness of plant-based therapies by increasing their distribution efficiency to target cells or tissues. Coordination based on comparable *in vitro* and *in vivo* procedures would be necessary, as the *in vitro* and *in vivo* effects of many herbal medications are not reproducible, which prevents them from progressing to the clinical trial phase. Due to their efficacy and safety, the phytoconstituents in question require extensive preclinical pharmacodynamic and pharmacokinetic studies.

## FUTURE PROSPECTIVE

Despite intensive treatment, the median survival time for patients with brain tumors is still rather low. Therefore, studies to find effective substitutes for traditional treatment are continuing. It should be noted that none of these phytochemical substances have a license to treat glioblastoma; however, some studies show they might delay the formation of glioblastoma cells. Understanding the function of the blood-brain barrier and the feasibility of delivering therapeutic doses of these natural compounds to the primary tumor location is, hence, of the utmost importance. Bioavailability is another aspect that impacts how well these natural substances work as cancer treatments. Thus, there needs to be a heavy focus on early and preclinical research on effective drug delivery methods to increase the efficacy of recognized natural compounds. Only recently have researchers begun to understand the role of cancer stem cells in chemoresistance and disease recurrence. Therefore, further focused research is required to identify naturally occurring chemicals that target CSCs and have therapeutic effectiveness. Extensive testing is still necessary before new medicines targeting cancer stem cells become available.

Natural compound-based glioblastoma treatments show the best promise when used in combination. It is well recognized that combination medications are more successful than monotherapies, and the aggressiveness and intra- and inter-tumor variability of glioblastoma have contributed to the adoption of multimodal treatments as a therapy. Unfortunately, the cellular toxicity of chemotherapeutic medications limits their use; this is one area where natural chemicals might be beneficial. Incorporating naturally occurring phytochemicals into combinatorial therapeutic techniques for glioblastoma may lead to more effective and less harmful approaches.

## AUTHORS' CONTRIBUTIONS

Swati Verma: Writing – original draft, Investigation, Resources, Formal analysis, Chaitanay Vinayak Narayan, Manasvi Saini, Gaurav Chaudhary: Data curation, Writing – review and amp; editing, Dr. Ajmer Singh Grewal: Conceptualization, Supervision, Resources, Visualization.

## REFERENCES

[1] Cheng YT, Yang CC, Shyur LF. Phytomedicine—Modulating oxidative stress and the tumor microenvironment for cancer therapy. Pharmacol Res 2016; 114: 128-43. Available from: https://linkinghub.elsevier.com/retrieve/pii/S1043661816306570 [http://dx.doi.org/10.1016/j.phrs.2016.10.022] [PMID: 27794498]

[2] Homayun B, Lin X, Choi HJ. Challenges and recent progress in oral drug delivery systems for biopharmaceuticals. Pharmaceutics 2019; 11(3): 129. Available from: https://www.mdpi.com/1999-4923/11/3/129 [http://dx.doi.org/10.3390/pharmaceutics11030129] [PMID: 30893852]

[3] Agrawal U, Sharma R, Gupta M, Vyas SP. Is nanotechnology a boon for oral drug delivery? Drug Discov Today 2014; 19(10): 1530-46. Available from: https://linkinghub.elsevier.com/retrieve/pii/S1359644614001482 [http://dx.doi.org/10.1016/j.drudis.2014.04.011] [PMID: 24786464]

[4] Zheng M, Tao W, Zou Y, Farokhzad OC, Shi B. Nanotechnology-based strategies for siRNA brain delivery for disease therapy. Trends Biotechnol 2018; 36(5): 562-75. Available from: https://linkinghub.elsevier.com/retrieve/pii/S0167779918300271 [http://dx.doi.org/10.1016/j.tibtech.2018.01.006] [PMID: 29422412]

[5] Liu J, Chen C, Wei T, *et al.* Dendrimeric nanosystem consistently circumvents heterogeneous drug response and resistance in pancreatic cancer. Exploration 2021; 1(1): 21-34. Available from: https://onlinelibrary.wiley.com/doi/10.1002/EXP.20210003 [http://dx.doi.org/10.1002/EXP.20210003] [PMID: 37366462]

[6] Guo S, Li K, Hu B, *et al.* Membrane-destabilizing ionizable lipid empowered imaging-guided siRNA delivery and cancer treatment. Exploration 2021; 1(1): 35-49. Available from: https://onlinelibrary.wiley.com/doi/10.1002/EXP.20210008 [http://dx.doi.org/10.1002/EXP.20210008] [PMID: 37366466]

[7] Khan AR, Liu M, Khan MW, Zhai G. Progress in brain targeting drug delivery system by nasal route. J Control Release 2017; 268: 364-89. Available from: https://linkinghub.elsevier.com/retrieve/pii/S0168365917308258 [http://dx.doi.org/10.1016/j.jconrel.2017.09.001] [PMID: 28887135]

[8] Abbott NJ, Rönnbäck L, Hansson E. Astrocyte–endothelial interactions at the blood–brain barrier. Nat Rev Neurosci 2006; 7(1): 41-53. Available from: https://www.nature.com/articles/nrn1824 [http://dx.doi.org/10.1038/nrn1824] [PMID: 16371949]

[9] Pardridge WM. Targeted delivery of protein and gene medicines through the blood–brain barrier. Clin Pharmacol Ther 2015; 97(4): 347-61. Available from: https://ascpt.onlinelibrary.wiley.com [http://dx.doi.org/10.1002/cpt.18] [PMID: 25669455]

[10] Sweeney MD, Sagare AP, Zlokovic BV. Blood–brain barrier breakdown in Alzheimer disease and other neurodegenerative disorders. Nat Rev Neurol 2018; 14(3): 133-50. Available from: https://www.nature.com/articles/nrneurol.2017.188 [http://dx.doi.org/10.1038/nrneurol.2017.188] [PMID: 29377008]

[11] Bellettato CM, Scarpa M. Possible strategies to cross the blood–brain barrier. Ital J Pediatr 2018; 44(S2) (Suppl. 2): 131. Available from: https://ijponline.biomedcentral.com/articles/10.1186/s13052-018-0563-0 [http://dx.doi.org/10.1186/s13052-018-0563-0] [PMID: 30442184]

[12] Ransohoff RM, Engelhardt B. The anatomical and cellular basis of immune surveillance in the central nervous system. Nat Rev Immunol 2012; 12(9): 623-35. Available from: https://www.nature.com/articles/nri3265 [http://dx.doi.org/10.1038/nri3265] [PMID: 22903150]

[13] Hmingthansanga V, Singh N, Banerjee S, Manickam S, Velayutham R, Natesan S. Improved topical drug delivery: Role of permeation enhancers and advanced approaches. Pharmaceutics 2022; 14(12): 2818. Available from: https://www.mdpi.com/1999-4923/14/12/2818 [http://dx.doi.org/10.3390/pharmaceutics14122818] [PMID: 36559311]

[14] Silva P, Bonifácio B, Ramos M, Negri K, Maria Bauab T, Chorilli M. Nanotechnology-based drug delivery systems and herbal medicines: A review. Int J Nanomedicine 2013; 1(Dec): 1. Available from: http://www.dovepress.com/nanotechnology-based-drug-delivery-systems-and-herbal-medicines-a-revi -peer-reviewed-article-IJN [http://dx.doi.org/10.2147/IJN.S52634]

[15] Javed MN, Dahiya ES, Ibrahim AM, Alam MS, Khan FA, Pottoo FH. Recent advancement in clinical application of nanotechnological approached targeted delivery of herbal drugs.Nanophytomedicine. Singapore: Springer Singapore 2020; pp. 151-72. Internet Available from: http://link.springer.com/10.1007/978-981-15-4909-0_9 [http://dx.doi.org/10.1007/978-981-15-4909-0_9]

[16] Fernandes F, Dias-Teixeira M, Delerue-Matos C, Grosso C. Critical review of lipid-based nanoparticles as carriers of neuroprotective drugs and extracts. Nanomaterials (Basel) 2021; 11(3): 563. Available from: https://www.mdpi.com/2079-4991/11/3/563 [http://dx.doi.org/10.3390/nano11030563] [PMID: 33668341]

[17] Han H, Li S, Xu M, *et al.* Polymer- and lipid-based nanocarriers for ocular drug delivery: Current status and future perspectives. Adv Drug Deliv Rev 2023; 196: 114770. Available from: https://linkinghub.elsevier.com/retrieve/pii/S0169409X23000856 [http://dx.doi.org/10.1016/j.addr.2023.114770] [PMID: 36894134]

[18] Liu Y, Chen Z, Li A, Liu R, Yang H, Xia X. The phytochemical potential for brain disease therapy and the possible nanodelivery solutions for brain access. Front Oncol 2022; 12: 936054. Available from: https://www.frontiersin.org/articles/10.3389/fonc.2022.936054/full [http://dx.doi.org/10.3389/fonc.2022.936054] [PMID: 35814371]

[19] Gnanaselvan S, Yadav SA, Sowmya Priya Manoharan BP. Uncovering the anticancer potential of phytomedicine and polyherbal's synergism against cancer – a review. Biointerface Res Appl Chem 2022; 13(4): 356. Available from: https://biointerfaceresearch.com/wp-content/uploads/2022/09/BRIAC134.356.pdf [http://dx.doi.org/10.33263/BRIAC134.356]

[20] McFaline-Figueroa JR, Lee EQ. Brain tumors. Am J Med 2018; 131(8): 874-82. Available from: https://linkinghub.elsevier.com/retrieve/pii/S0002934318300317 [http://dx.doi.org/10.1016/j.amjmed.2017.12.039] [PMID: 29371158]

[21] Aamir M, Rahman Z, Ahmed Abro W, Aslam Bhatti U, Ahmed Dayo Z, Ishfaq M. Brain tumor classification utilizing deep features derived from high-quality regions in MRI images. Biomed Signal Process Control 2023; 85: 104988. Available from: https://linkinghub.elsevier.com/retrieve/pii/S1746809423004214 [http://dx.doi.org/10.1016/j.bspc.2023.104988]

[22] Banks WA. From blood–brain barrier to blood–brain interface: new opportunities for CNS drug delivery. Nat Rev Drug Discov 2016; 15(4): 275-92. Available from: https://www.nature.com/articles/nrd.2015.21 [http://dx.doi.org/10.1038/nrd.2015.21] [PMID: 26794270]

[23] Mehan S, Arora N, Bhalla S, *et al.* Involvement of phytochemical-encapsulated nanoparticles' interaction with cellular signalling in the amelioration of benign and malignant brain tumours. Molecules 2022; 27(11): 3561. Available from: https://www.mdpi.com/1420-3049/27/11/3561 [http://dx.doi.org/10.3390/molecules27113561] [PMID: 35684498]

[24] L. de Klerk O, J. Bosker F, Luurtsema G, M. Nolte I, Dierckx R, A. den Boer J. The role of p-glycoprotein in psychiatric disorders: A reliable guard of the brain? Cent Nerv Syst Agents Med Chem. 2011 Sep 1;11(3):197–209. Available from: http://www.eurekaselect.com/openurl/content.php?genre=article&issn=1871-5249&volume=11&issue=3&spage=197

[25] Rapposelli S, Digiacomo M, Balsamo A. P-gp transporter and its role in neurodegenerative diseases. Curr Top Med Chem 2009; 9(2): 209-17. Available from: http://www.eurekaselect.com/openurl/content.php?genre=article&issn=1568-0266&volume=9&issue=2&spage=209 [http://dx.doi.org/10.2174/156802609787521544] [PMID: 19200006]

[26] Yadav B, Bajaj A, Saxena M, Saxena AK. *In Vitro* anticancer activity of the root, stem and leaves of Withania Somnifera against various human cancer cell lines. Indian J Pharm Sci 2010; 72(5): 659-63. Available from: http://www.ijpsonline.com/text.asp?2010/72/5/659/78543 [http://dx.doi.org/10.4103/0250-474X.78543] [PMID: 21695006]

[27] Choudhari AS, Mandave PC, Deshpande M, Ranjekar P, Prakash O. Phytochemicals in cancer treatment: From preclinical studies to clinical practice. Front Pharmacol 2020; 10: 1614. Available from: https://www.frontiersin.org/article/10.3389/fphar.2019.01614/full [http://dx.doi.org/10.3389/fphar.2019.01614] [PMID: 32116665]

[28] Lan L, Appelman C, Smith AR, *et al.* Natural product (−)-gossypol inhibits colon cancer cell growth by targeting RNA-binding protein Musashi-1. Mol Oncol 2015; 9(7): 1406-20. Available from: https://febs.onlinelibrary.wiley.com/doi/10.1016/j.molonc.2015.03.014 [http://dx.doi.org/10.1016/j.molonc.2015.03.014] [PMID: 25933687]

[29] Zhan Y, Jia G, Wu D, Xu Y, Xu L. Design and synthesis of a gossypol derivative with improved antitumor activities. Arch Pharm (Weinheim) 2009; 342(4): 223-9. Available from: https://onlinelibrary.wiley.com/doi/10.1002/ardp.200800185 [http://dx.doi.org/10.1002/ardp.200800185] [PMID: 19340835]

[30] Liu YQ, Tian J, Qian K, *et al.* Recent progress on C-4-modified podophyllotoxin analogs as potent antitumor agents. Med Res Rev 2015; 35(1): 1-62. Available from: https://onlinelibrary.wiley.com/doi/10.1002/med.21319 [http://dx.doi.org/10.1002/med.21319] [PMID: 24827545]

[31] Park HJ. Current uses of mushrooms in cancer treatment and their anticancer mechanisms. Int J Mol Sci 2022; 23(18): 10502. Available from: https://www.mdpi.com/1422-0067/23/18/10502 [http://dx.doi.org/10.3390/ijms231810502] [PMID: 36142412]

[32] Choi JY, Hong WG, Cho JH, Kim EM, Kim J, Jung C-H. Podophyllotoxin acetate triggers anticancer effects against non-small cell lung cancer cells by promoting cell death *via* cell cycle arrest, ER stress and autophagy. Int J Oncol [Internet]. 2015 Oct; 47(4): 1257–65. Available from: https://www.spandidos-publications.com [http://dx.doi.org/10.3892/ijo.2015.3123]

[33] Woźniak Ł, Skąpska S, Marszałek K. Ursolic acid—a pentacyclic triterpenoid with a wide spectrum of pharmacological activities. Molecules 2015; 20(11): 20614-41. Available from: http://www.mdpi.com/1420-3049/20/11/19721 [http://dx.doi.org/10.3390/molecules201119721] [PMID: 26610440]

[34] Abbas M. Potential role of nanoparticles in treating the accumulation of amyloid-beta peptide in alzheimer's patients. Polymers (Basel) 2021; 13(7): 1051. Available from: https://www.mdpi.com/2073-4360/13/7/1051 [http://dx.doi.org/10.3390/polym13071051] [PMID: 33801619]

[35] Zhai K, Siddiqui M, Abdellatif B, Liskova A, Kubatka P, Büsselberg D. Natural compounds in glioblastoma therapy: Preclinical insights, mechanistic pathways, and outlook. Cancers (Basel) 2021; 13(10): 2317. Available from: https://www.mdpi.com/2072-6694/13/10/2317 [http://dx.doi.org/10.3390/cancers13102317] [PMID: 34065960]

[36] Khadka B, Lee JY, Park DH, Kim KT, Bae JS. The role of natural compounds and their nanocarriers in the treatment of CNS inflammation. Biomolecules 2020; 10(10): 1401. Available from: https://www.mdpi.com/2218-273X/10/10/1401 [http://dx.doi.org/10.3390/biom10101401] [PMID: 33019651]

[37] Barchet TM, Amiji MM. Challenges and opportunities in CNS delivery of therapeutics for neurodegenerative diseases. Expert Opin Drug Deliv 2009; 6(3): 211-25. Available from: http://www.tandfonline.com/doi/full/10.1517/17425240902758188 [http://dx.doi.org/10.1517/17425240902758188] [PMID: 19290842]

[38] Walker FO. Huntington's disease. Lancet 2007; 369(9557): 218-28. Available from: https://linkinghub.elsevier.com/retrieve/pii/S0140673607601111 [http://dx.doi.org/10.1016/S0140-6736(07)60111-1] [PMID: 17240289]

[39] Imarisio S, Carmichael J, Korolchuk V, *et al.* Huntington's disease: from pathology and genetics to potential therapies. Biochem J 2008; 412(2): 191-209. Available from: https://portlandpress.com/biochemj/article/412/2/191/44880/Huntington-s-disease-from-pathology-and-genetics [http://dx.doi.org/10.1042/BJ20071619] [PMID: 18466116]

[40] Dugger BN, Dickson DW. Pathology of neurodegenerative diseases. Cold Spring Harb Perspect Biol 2017; 9(7): a028035. Available from: http://cshperspectives.cshlp.org/lookup [http://dx.doi.org/10.1101/cshperspect.a028035] [PMID: 28062563]

[41] Gooch CL, Pracht E, Borenstein AR. The burden of neurological disease in the United States: A summary report and call to action. Ann Neurol 2017; 81(4): 479-84. Available from: https://onlinelibrary.wiley.com/doi/10.1002/ana.24897 [http://dx.doi.org/10.1002/ana.24897] [PMID: 28198092]

[42] Bergmann S, Lawler SE, Qu Y, *et al.* Blood–brain-barrier organoids for investigating the permeability of CNS therapeutics. Nat Protoc 2018; 13(12): 2827-43. Available from: https://www.nature.com/articles/s41596-018-0066-x

[http://dx.doi.org/10.1038/s41596-018-0066-x] [PMID: 30382243]

[43] Guerra M, Blázquez JL, Rodríguez EM. Blood–brain barrier and foetal-onset hydrocephalus, with a view on potential novel treatments beyond managing CSF flow. Fluids Barriers CNS 2017; 14(1): 19. Available from: http://fluidsbarrierscns.biomedcentral.com/articles/10.1186/s12987-017-0067-0 [http://dx.doi.org/10.1186/s12987-017-0067-0] [PMID: 28701191]

[44] Dong X. Current strategies for brain drug delivery. Theranostics 2018; 8(6): 1481-93. Available from: http://www.thno.org/v08p1481.htm [http://dx.doi.org/10.7150/thno.21254] [PMID: 29556336]

[45] Makkar R, Behl T, Bungau S, *et al.* Nutraceuticals in neurological disorders. Int J Mol Sci 2020; 21(12): 4424. Available from: https://www.mdpi.com/1422-0067/21/12/4424 [http://dx.doi.org/10.3390/ijms21124424] [PMID: 32580329]

[46] Uddin MS, Hossain MF, Mamun AA, *et al.* Exploring the multimodal role of phytochemicals in the modulation of cellular signaling pathways to combat age-related neurodegeneration. Sci Total Environ 2020; 725: 138313. Available from: https://linkinghub.elsevier.com/retrieve/pii/S004896972031826X [http://dx.doi.org/10.1016/j.scitotenv.2020.138313] [PMID: 32464743]

[47] Zhu FD, Hu YJ, Yu L, *et al.* Nanoparticles: A hope for the treatment of inflammation in CNS. Front Pharmacol 2021; 12: 683935. Available from: https://www.frontiersin.org/articles [http://dx.doi.org/10.3389/fphar.2021.683935] [PMID: 34122112]

[48] Schmitt C, Lechanteur A, Cossais F, *et al.* Liposomal encapsulated curcumin effectively attenuates neuroinflammatory and reactive astrogliosis reactions in glia cells and organotypic brain slices. Int J Nanomedicine 2020; 15: 3649-67. Available from: https://www.dovepress.com/liposomal-encapsulated-curcumin-effectively-attenuates-neuroinflammato-peer-reviewed-article-IJN [http://dx.doi.org/10.2147/IJN.S245300] [PMID: 32547020]

[49] Alphandéry E. Nano-therapies for glioblastoma treatment. Cancers (Basel) 2020; 12(1): 242. Available from: https://www.mdpi.com/2072-6694/12/1/242 [http://dx.doi.org/10.3390/cancers12010242] [PMID: 31963825]

[50] Kalashnikova I, Mazar J, Neal CJ, *et al.* Nanoparticle delivery of curcumin induces cellular hypoxia and ROS-mediated apoptosis *via* modulation of Bcl-2/Bax in human neuroblastoma. Nanoscale 2017; 9(29): 10375-87. Available from: https://xlink.rsc.org/?DOI=C7NR02770B [http://dx.doi.org/10.1039/C7NR02770B] [PMID: 28702620]

[51] Montalbán M, Coburn J, Lozano-Pérez A, Cenis J, Víllora G, Kaplan D. Production of curcumin-loaded silk fibroin nanoparticles for cancer therapy. Nanomaterials (Basel) 2018; 8(2): 126. Available from: http://www.mdpi.com/2079-4991/8/2/126 [http://dx.doi.org/10.3390/nano8020126] [PMID: 29495296]

[52] Baspinar Y, Üstündas M, Bayraktar O, Sezgin C. Curcumin and piperine loaded zein-chitosan nanoparticles: Development and *in-vitro* characterisation. Saudi Pharm J 2018; 26(3): 323-34. Available from: https://linkinghub.elsevier.com/retrieve/pii/S1319016418300069 [http://dx.doi.org/10.1016/j.jsps.2018.01.010] [PMID: 29556123]

[53] Ege D. Action mechanisms of curcumin in alzheimer's disease and its brain targeted delivery. Materials (Basel) 2021; 14(12): 3332. Available from: https://www.mdpi.com/1996-1944/14/12/3332 [http://dx.doi.org/10.3390/ma14123332] [PMID: 34208692]

[54] Sadegh Malvajerd S, Azadi A, Izadi Z, *et al.* Brain delivery of curcumin using solid lipid nanoparticles and nanostructured lipid carriers: Preparation, optimization, and pharmacokinetic evaluation. ACS Chem Neurosci 2019; 10(1): 728-39. Available from: https://pubs.acs.org/doi/10.1021/acschemneuro.8b00510 [http://dx.doi.org/10.1021/acschemneuro.8b00510] [PMID: 30335941]

[55] Maiti P, Paladugu L, Dunbar GL. RETRACTED ARTICLE: Solid lipid curcumin particles provide greater anti-amyloid, anti-inflammatory and neuroprotective effects than curcumin in the 5xFAD mouse model of Alzheimer's disease. BMC Neurosci 2018; 19(1): 7. Available from:

https://bmcneurosci.biomedcentral.com/articles/10.1186/s12868-018-0406-3 [http://dx.doi.org/10.1186/s12868-018-0406-3] [PMID: 29471781]

[56] Jeon SG, Cha MY, Kim J, *et al.* Vitamin D-binding protein-loaded PLGA nanoparticles suppress Alzheimer's disease-related pathology in 5XFAD mice. Nanomedicine 2019; 17: 297-307. Available from: https://linkinghub.elsevier.com/retrieve/pii/S1549963419300358 [http://dx.doi.org/10.1016/j.nano.2019.02.004] [PMID: 30794963]

[57] Huang N, Lu S, Liu XG, Zhu J, Wang YJ, Liu RT. PLGA nanoparticles modified with a BBB-penetrating peptide co-delivering Aβ generation inhibitor and curcumin attenuate memory deficits and neuropathology in Alzheimer's disease mice. Oncotarget 2017; 8(46): 81001-13. Available from: https://www.oncotarget.com/lookup/doi/10.18632/oncotarget.20944 [http://dx.doi.org/10.18632/oncotarget.20944] [PMID: 29113362]

[58] Sathya S, Shanmuganathan B, Balasubramaniam B, Balamurugan K, Devi KP. Phytol loaded PLGA nanoparticles regulate the expression of Alzheimer's related genes and neuronal apoptosis against amyloid-β induced toxicity in Neuro-2a cells and transgenic Caenorhabditis elegans. Food Chem Toxicol 2020; 136: 110962. Available from: https://linkinghub.elsevier.com/retrieve/pii/S0278691519307525 [http://dx.doi.org/10.1016/j.fct.2019.110962] [PMID: 31734340]

[59] Zhang L, Yang S, Wong LR, Xie H, Ho PCL. *In Vitro* and *In Vivo* Comparison of Curcumin-Encapsulated Chitosan-Coated Poly(lactic- *co* -glycolic acid) Nanoparticles and Curcumin/Hydroxypropyl-β-Cyclodextrin Inclusion Complexes Administered Intranasally as Therapeutic Strategies for Alzheimer's Disease. Mol Pharm 2020; 17(11): 4256-69. Available from: https://pubs.acs.org/doi/10.1021/acs.molpharmaceut.0c00675 [http://dx.doi.org/10.1021/acs.molpharmaceut.0c00675] [PMID: 33084343]

[60] Huo X, Zhang Y, Jin X, Li Y, Zhang L. A novel synthesis of selenium nanoparticles encapsulated PLGA nanospheres with curcumin molecules for the inhibition of amyloid β aggregation in Alzheimer's disease. J Photochem Photobiol B 2019; 190: 98-102. Available from: https://linkinghub.elsevier.com/retrieve/pii/S1011134418311655 [http://dx.doi.org/10.1016/j.jphotobiol.2018.11.008] [PMID: 30504054]

[61] Cheng KK, Chan PS, Fan S, *et al.* Curcumin-conjugated magnetic nanoparticles for detecting amyloid plaques in Alzheimer's disease mice using magnetic resonance imaging (MRI). Biomaterials 2015; 44: 155-72. Available from: https://linkinghub.elsevier.com/retrieve/pii/S0142961214012496 [http://dx.doi.org/10.1016/j.biomaterials.2014.12.005] [PMID: 25617135]

[62] Huang X, Li N, Pu Y, Zhang T, Wang B. Neuroprotective Effects of Ginseng Phytochemicals: Recent Perspectives. Molecules 2019; 24(16): 2939. Available from: https://www.mdpi.com/1420-3049/24/16/2939 [http://dx.doi.org/10.3390/molecules24162939] [PMID: 31416121]

[63] Mohd Sairazi NS, Sirajudeen KNS. Natural products and their bioactive compounds: Neuroprotective potentials against neurodegenerative diseases. Evidence-Based Complement Altern Med [Internet]. 2020 Feb 14;2020:1–30. Available from: https://www.hindawi.com/journals/ecam/2020/6565396/

[64] Rajabian A, Rameshrad M, Hosseinzadeh H. Therapeutic potential of *Panax ginseng* and its constituents, ginsenosides and gintonin, in neurological and neurodegenerative disorders: a patent review. Expert Opin Ther Pat 2019; 29(1): 55-72. Available from: https://www.tandfonline.com/doi/full/10.1080/13543776.2019.1556258 [http://dx.doi.org/10.1080/13543776.2019.1556258] [PMID: 30513224]

[65] Razgonova M, Veselov V, Zakharenko A, Golokhvast K, Nosyrev A, Cravotto G. Panax ginseng components and the pathogenesis of Alzheimer's disease (Review). Mol Med Rep [Internet]. 2019 Feb 19; Available from: http://www.spandidos-publications.com [http://dx.doi.org/10.3892/mmr.2019.9972]

[66] Oh J, Kim JS. Compound K derived from ginseng: neuroprotection and cognitive improvement. Food Funct 2016; 7(11): 4506-15. Available from: https://xlink.rsc.org/?DOI=C6FO01077F

[http://dx.doi.org/10.1039/C6FO01077F] [PMID: 27801453]

[67] Sharma A, Lee HJ. Ginsenoside compound K: Insights into recent studies on pharmacokinetics and health-promoting activities. Biomolecules 2020; 10(7): 1028. Available from: https://www.mdpi.com/2218-273X/10/7/1028 [http://dx.doi.org/10.3390/biom10071028] [PMID: 32664389]

[68] Li J, Huang Q, Chen J, *et al.* Neuroprotective potentials of panax ginseng against alzheimer's disease: A review of preclinical and clinical evidences. Front Pharmacol 2021; 12: 688490. Available from: https://www.frontiersin.org/articles/10.3389/fphar.2021.688490/full [http://dx.doi.org/10.3389/fphar.2021.688490] [PMID: 34149431]

[69] Kim JW, Cho CH, Hwang YG, Park WJ, Kang H, Kim DO. Protective effects of red ginseng treated with gold nanoparticles against $H_2O_2$-induced oxidative stress in neuronal PC-12 cells. Korean J Food Sci Technol 2017; 49(2): 222-7. Available from: http://koreascience.or.kr/journal/view.jsp?kj=SPGHB5&py=2017&vnc=v49n2&sp=222 [http://dx.doi.org/10.9721/KJFST.2017.49.2.222]

[70] Yang L, Li CL, Tsai TH. Preclinical herb–drug pharmacokinetic interaction of *panax ginseng* extract and selegiline in freely moving rats. ACS Omega 2020; 5(9): 4682-8. Available from: https://pubs.acs.org/doi/10.1021/acsomega.0c00123 [http://dx.doi.org/10.1021/acsomega.0c00123] [PMID: 32175515]

[71] Singh P, Singh H, Ahn S, *et al.* Pharmacological importance, characterization and applications of gold and silver nanoparticles synthesized by *Panax ginseng* fresh leaves. Artif Cells Nanomed Biotechnol 2017; 45(7): 1415-24. Available from: https://www.tandfonline.com/doi/full/10.1080/21691401.2016.1243547 [http://dx.doi.org/10.1080/21691401.2016.1243547] [PMID: 27855495]

[72] Mathiyalagan R, Yang DC. Ginseng nanoparticles: a budding tool for cancer treatment. Nanomedicine (Lond) 2017; 12(10): 1091-4. Available from: https://www.futuremedicine.com/doi/10.2217/nnm-2017-0070 [http://dx.doi.org/10.2217/nnm-2017-0070] [PMID: 28447867]

[73] Zhu Y, Liang J, Gao C, *et al.* Multifunctional ginsenoside Rg3-based liposomes for glioma targeting therapy. J Control Release 2021; 330: 641-57. Available from: https://linkinghub.elsevier.com/retrieve/pii/S0168365920307604 [http://dx.doi.org/10.1016/j.jconrel.2020.12.036] [PMID: 33359582]

[74] Dai L, Zhu W, Si C, Lei J. "Nano-Ginseng" for enhanced cytotoxicity against cancer cells. Int J Mol Sci 2018; 19(2): 627. Available from: http://www.mdpi.com/1422-0067/19/2/627 [http://dx.doi.org/10.3390/ijms19020627] [PMID: 29473838]

[75] Pattan G, Kaul G. Health hazards associated with nanomaterials. Toxicol Ind Health 2014; 30(6): 499-519. Available from: http://journals.sagepub.com/doi/10.1177/0748233712459900 [http://dx.doi.org/10.1177/0748233712459900] [PMID: 23012342]

[76] Wang J, Chen C, Liu Y, Jiao F, Li W, Lao F. Potential neurological lesion after nasal instillation of TiO2 nanoparticles in the anatase and rutile crystal phases. Toxicol Lett [Internet]. 2008 Dec 15;183(1–3):72–80. Available from: https://linkinghub.elsevier.com/retrieve/pii/S0378427408012848

[77] Pisanic TR II, Blackwell JD, Shubayev VI, Fiñones RR, Jin S. Nanotoxicity of iron oxide nanoparticle internalization in growing neurons. Biomaterials 2007; 28(16): 2572-81. Available from: https://linkinghub.elsevier.com/retrieve/pii/S0142961207001317 [http://dx.doi.org/10.1016/j.biomaterials.2007.01.043] [PMID: 17320946]

[78] Zhao J, Xu L, Zhang T, Ren G, Yang Z. Influences of nanoparticle zinc oxide on acutely isolated rat hippocampal CA3 pyramidal neurons. Neurotoxicology 2009; 30(2): 220-30. Available from: https://linkinghub.elsevier.com/retrieve/pii/S0161813X08002416 [http://dx.doi.org/10.1016/j.neuro.2008.12.005] [PMID: 19146874]

[79] Cha KE, Myung H. Cytotoxic effects of nanoparticles assessed *in vitro* and *in vivo*. J Microbiol

Biotechnol 2007; 17(9): 1573-8. Available from: http://www.ncbi.nlm.nih.gov/pubmed/18062241 [PMID: 18062241]

[80] Dibaei M, Rouini MR, Sheikholeslami B, Gholami M, Dinarvand R. The effect of surface treatment on the brain delivery of curcumin nanosuspension: *in vitro* and *in vivo* studies. Int J Nanomedicine 2019; 14: 5477-90. Available from: https://www.dovepress.com/the-effect-of-surface-treatment--n-the-brain-delivery-of-curcumin-nano-peer-reviewed-article-IJN [http://dx.doi.org/10.2147/IJN.S199624] [PMID: 31409999]

[81] Pelaz B, Alexiou C, Alvarez-Puebla RA, *et al.* Diverse applications of nanomedicine. ACS Nano 2017; 11(3): 2313-81. Available from: https://pubs.acs.org/doi/10.1021/acsnano.6b06040 [http://dx.doi.org/10.1021/acsnano.6b06040] [PMID: 28290206]

[82] Liu ZJ, Li ZH, Liu L, *et al.* Curcumin attenuates beta-amyloid-induced neuroinflammation *via* activation of peroxisome proliferator-activated receptor-gamma function in a rat model of alzheimer's disease. Front Pharmacol 2016; 7: 261. Available from: http://journal.frontiersin.org/Article/10.3389/fphar.2016.00261/abstract [http://dx.doi.org/10.3389/fphar.2016.00261] [PMID: 27594837]

[83] Yang F, Lim GP, Begum AN, *et al.* Curcumin inhibits formation of amyloid β oligomers and fibrils, binds plaques, and reduces amyloid *in vivo*. J Biol Chem 2005; 280(7): 5892-901. Available from: https://linkinghub.elsevier.com/retrieve/pii/S0021925819630966 [http://dx.doi.org/10.1074/jbc.M404751200] [PMID: 15590663]

[84] Askarizadeh A, Barreto GE, Henney NC, Majeed M, Sahebkar A. Neuroprotection by curcumin: A review on brain delivery strategies. Int J Pharm 2020; 585: 119476. Available from: https://linkinghub.elsevier.com/retrieve/pii/S0378517320304609 [http://dx.doi.org/10.1016/j.ijpharm.2020.119476] [PMID: 32473377]

[85] Guo N, Hu Z, Fan X, *et al.* Simultaneous determination of salidroside and its aglycone metabolite p-tyrosol in rat plasma by liquid chromatography-tandem mass spectrometry. Molecules 2012; 17(4): 4733-54. Available from: http://www.mdpi.com/1420-3049/17/4/4733 [http://dx.doi.org/10.3390/molecules17044733] [PMID: 22525439]

[86] Barhwal K, Das SK, Kumar A, Hota SK, Srivastava RB. *Retracted:* Insulin receptor A and Sirtuin 1 synergistically improve learning and spatial memory following chronic salidroside treatment during hypoxia. J Neurochem 2015; 135(2): 332-46. Available from: https://onlinelibrary.wiley.com [http://dx.doi.org/10.1111/jnc.13225] [PMID: 26173704]

[87] Gao J, He H, Jiang W, *et al.* Salidroside ameliorates cognitive impairment in a d-galactose-induced rat model of Alzheimer's disease. Behav Brain Res 2015; 293: 27-33. Available from: https://linkinghub.elsevier.com/retrieve/pii/S0166432815300620 [http://dx.doi.org/10.1016/j.bbr.2015.06.045] [PMID: 26192909]

[88] Xie Z, Lu H, Yang S, *et al.* Salidroside attenuates cognitive dysfunction in senescence-accelerated mouse prone 8 (SAMP8) mice and modulates inflammation of the gut-brain axis. Front Pharmacol 2020; 11: 568423. Available from: https://www.frontiersin.org/articles/10.3389/fphar.2020.568423/full [http://dx.doi.org/10.3389/fphar.2020.568423] [PMID: 33362539]

[89] Barhwal K, Das SK, Kumar A, Hota SK, Srivastava RB. Retracted: Insulin receptor A and Sirtuin 1 synergistically improve learning and spatial memory following chronic salidroside treatment during hypoxia. J Neurochem [Internet]. 2015 Oct 27;135(2):332–46. Available from: https://onlinelibrary.wiley.com [http://dx.doi.org/10.1002/ptr.7235]

[90] Lv C, Wang L, Liu X, *et al.* Multi-faced neuroprotective effects of geniposide depending on the RAGE-mediated signaling in an Alzheimer mouse model. Neuropharmacology 2015; 89: 175-84. Available from: https://linkinghub.elsevier.com/retrieve/pii/S0028390814003268 [http://dx.doi.org/10.1016/j.neuropharm.2014.09.019] [PMID: 25261783]

[91] Hu JH, Li IC, Lin TW, *et al.* Absolute bioavailability, tissue distribution, and excretion of erinacine S in *Hericium erinaceus* mycelia. Molecules 2019; 24(8): 1624. Available from: https://www.mdpi.com/1420-3049/24/8/1624 [http://dx.doi.org/10.3390/molecules24081624] [PMID: 31022946]

[92] Lee L-Y, Chou W, Chen W-P, Wang M-F, Chen Y-J, Chen C-C. Erinacine a-enriched hericium erinaceus mycelium delays progression of age-related cognitive decline in senescence accelerated mouse prone 8 (SAMP8) mice. Nutrients [Internet]. 2021 Oct 19;13(10):3659. Available from: https://www.mdpi.com/2072-6643/13/10/3659

[93] Tzeng TT, Chen CC, Chen CC, *et al.* The cyanthin diterpenoid and sesterterpene constituents of hericium erinaceus mycelium ameliorate alzheimer's disease-related pathologies in app/ps1 transgenic mice. Int J Mol Sci 2018; 19(2): 598. Available from: http://www.mdpi.com/1422-0067/19/2/598 [http://dx.doi.org/10.3390/ijms19020598] [PMID: 29463001]

[94] Hu PY, Liu D, Zheng Q, Wu Q, Tang Y, Yang M. Elucidation of transport mechanism of paeoniflorin and the influence of ligustilide, senkyunolide i and senkyunolide a on paeoniflorin transport through mdck-mdr1 cells as blood–brain barrier *in vitro* model. Molecules 2016; 21(3): 300. Available from: http://www.mdpi.com/1420-3049/21/3/300 [http://dx.doi.org/10.3390/molecules21030300] [PMID: 26950101]

[95] Zhang HR, Peng JH, Cheng XB, Shi BZ, Zhang MY, Xu RX. Paeoniflorin atttenuates amyloidogenesis and the inflammatory responses in a transgenic mouse model of alzheimer's disease. Neurochem Res 2015; 40(8): 1583-92. Available from: http://link.springer.com/10.1007/s11064-015-1632-z [http://dx.doi.org/10.1007/s11064-015-1632-z] [PMID: 26068144]

[96] Gong W, Zhao Z, Liu B, Lu L, Dong J. Exploring the chemopreventive properties and perspectives of baicalin and its aglycone baicalein in solid tumors. Eur J Med Chem 2017; 126: 844-52. Available from: https://linkinghub.elsevier.com/retrieve/pii/S0223523416310066 [http://dx.doi.org/10.1016/j.ejmech.2016.11.058] [PMID: 27960146]

[97] Gregory J, Vengalasetti YV, Bredesen DE, Rao RV. Neuroprotective herbs for the management of alzheimer's disease. Biomolecules 2021; 11(4): 543. Available from: https://www.mdpi.com/2218-273X/11/4/543 [http://dx.doi.org/10.3390/biom11040543] [PMID: 33917843]

[98] Sowndhararajan K, Deepa P, Kim M, Park SJ, Kim S. Baicalein as a potent neuroprotective agent: A review. Biomed Pharmacother 2017; 95: 1021-32. Available from: https://linkinghub.elsevier.com/retrieve/pii/S0753332217340143 [http://dx.doi.org/10.1016/j.biopha.2017.08.135] [PMID: 28922719]

[99] Zhang X, Du L, Zhang W, Yang Y, Zhou Q, Du G. Therapeutic effects of baicalein on rotenone-induced Parkinson's disease through protecting mitochondrial function and biogenesis. Sci Rep 2017; 7(1): 9968. Available from: https://www.nature.com/articles/s41598-017-07442-y [http://dx.doi.org/10.1038/s41598-017-07442-y] [PMID: 28855526]

[100] Pervin M, Unno K, Nakagawa A, *et al.* Blood brain barrier permeability of (−)-epigallocatechin gallate, its proliferation-enhancing activity of human neuroblastoma SH-SY5Y cells, and its preventive effect on age-related cognitive dysfunction in mice. Biochem Biophys Rep 2017; 9: 180-6. Available from: https://linkinghub.elsevier.com/retrieve/pii/S2405580817300055 [http://dx.doi.org/10.1016/j.bbrep.2016.12.012] [PMID: 28956003]

[101] Roth M, Timmermann BN, Hagenbuch B. Interactions of green tea catechins with organic anion-transporting polypeptides. Drug Metab Dispos 2011; 39(5): 920-6. Available from: http://dmd.aspetjournals.org/lookup/doi/10.1124/dmd.110.036640 [http://dx.doi.org/10.1124/dmd.110.036640] [PMID: 21278283]

[102] Eisenstein TK, Meissler JJ. Effects of cannabinoids on t-cell function and resistance to infection. J Neuroimmune Pharmacol 2015; 10(2): 204-16. Available from:

https://link.springer.com/10.1007/s11481-015-9603-3
[http://dx.doi.org/10.1007/s11481-015-9603-3] [PMID: 25876735]

[103] Scalbert A, Andres-Lacueva C, Arita M, *et al.* Databases on food phytochemicals and their health-promoting effects. J Agric Food Chem 2011; 59(9): 4331-48. Available from: https://pubs.acs.org/doi/10.1021/jf200591d
[http://dx.doi.org/10.1021/jf200591d] [PMID: 21438636]

[104] Xia X, Shi B, Wang L, *et al.* From mouse to mouse-ear cress: Nanomaterials as vehicles in plant biotechnology. Exploration 2021; 1(1): 9-20. Available from: https://onlinelibrary.wiley.com/doi/10.1002/EXP.20210002
[http://dx.doi.org/10.1002/EXP.20210002] [PMID: 37366467]

[105] Li Z, Tan S, Li S, Shen Q, Wang K. Cancer drug delivery in the nano era: An overview and perspectives. Oncol Rep 2017; 38(2): 611-24. Available from: https://www.spandidos-publications.com/10.3892/or.2017.5718
[http://dx.doi.org/10.3892/or.2017.5718] [PMID: 28627697]

[106] Tu L, Liao Z, Luo Z, Wu YL, Herrmann A, Huo S. Ultrasound-controlled drug release and drug activation for cancer therapy. Exploration 2021; 1(3): 20210023. Available from: https://onlinelibrary.wiley.com/doi/10.1002/EXP.20210023
[http://dx.doi.org/10.1002/EXP.20210023] [PMID: 37323693]

[107] Tsai YM, Chien CF, Lin LC, Tsai TH. Curcumin and its nano-formulation: The kinetics of tissue distribution and blood–brain barrier penetration. Int J Pharm 2011; 416(1): 331-8. Available from: https://linkinghub.elsevier.com/retrieve/pii/S0378517311005837
[http://dx.doi.org/10.1016/j.ijpharm.2011.06.030] [PMID: 21729743]

[108] Chen Y, Pan L, Jiang M, Li D, Jin L. Nanostructured lipid carriers enhance the bioavailability and brain cancer inhibitory efficacy of curcumin both *in vitro* and *in vivo*. Drug Deliv 2016; 23(4): 1383-92. Available from: https://www.tandfonline.com/doi/full/10.3109/10717544.2015.1049719
[http://dx.doi.org/10.3109/10717544.2015.1049719] [PMID: 26066035]

[109] Seo EJ, Fischer N, Efferth T. Phytochemicals as inhibitors of NF-κB for treatment of Alzheimer's disease. Pharmacol Res 2018; 129: 262-73. Available from: https://linkinghub.elsevier.com/retrieve/pii/S1043661817311349
[http://dx.doi.org/10.1016/j.phrs.2017.11.030] [PMID: 29179999]

[110] Mo Y, Duan L, Yang Y, *et al.* Nanoparticles improved resveratrol brain delivery and its therapeutic efficacy against intracerebral hemorrhage. Nanoscale 2021; 13(6): 3827-40. Available from: https://xlink.rsc.org/?DOI=D0NR06249A
[http://dx.doi.org/10.1039/D0NR06249A] [PMID: 33565555]

[111] Zheng CQ, Fan HX, Li XX, Li JJ, Sheng S, Zhang F. Resveratrol alleviates levodopa-induced dyskinesia in rats. Front Immunol 2021; 12: 683577. Available from: https://www.frontiersin.org/articles/10.3389/fimmu.2021.683577/full
[http://dx.doi.org/10.3389/fimmu.2021.683577] [PMID: 34248967]

[112] Jhaveri A, Deshpande P, Pattni B, Torchilin V. Transferrin-targeted, resveratrol-loaded liposomes for the treatment of glioblastoma. J Control Release 2018; 277: 89-101. Available from: https://linkinghub.elsevier.com/retrieve/pii/S0168365918301287
[http://dx.doi.org/10.1016/j.jconrel.2018.03.006] [PMID: 29522834]

[113] Xiong S, Li Z, Liu Y, *et al.* Brain-targeted delivery shuttled by black phosphorus nanostructure to treat Parkinson's disease. Biomaterials 2020; 260: 120339. Available from: https://linkinghub.elsevier.com/retrieve/pii/S0142961220305858
[http://dx.doi.org/10.1016/j.biomaterials.2020.120339] [PMID: 32861017]

[114] Yang B, Du S, Lu Y, *et al.* Influence of paeoniflorin and menthol on puerarin transport across MDCK and MDCK-MDR1 cells as blood–brain barrier *in vitro* model. J Pharm Pharmacol 2018; 70(3): 349-60. Available from: https://academic.oup.com/jpp/article/70/3/349-360/6121741

[http://dx.doi.org/10.1111/jphp.12853] [PMID: 29238980]

[115] Aparicio-Blanco J, Romero IA, Male DK, Slowing K, García-García L, Torres-Suárez AI. Cannabidiol enhances the passage of lipid nanocapsules across the blood–brain barrier both *in vitro* and *in vivo*. Mol Pharm 2019; 16(5): 1999-2010. Available from: https://pubs.acs.org/doi/10.1021/acs.molpharmaceut.8b01344 [http://dx.doi.org/10.1021/acs.molpharmaceut.8b01344] [PMID: 30865462]

[116] Tian M, Chen G, Xu J, *et al.* Epigallocatechin gallate-based nanoparticles with reactive oxygen species scavenging property for effective chronic periodontitis treatment. Chem Eng J 2022; 433: 132197. Available from: https://linkinghub.elsevier.com/retrieve/pii/S1385894721037761 [http://dx.doi.org/10.1016/j.cej.2021.132197]

[117] Linz U. Commentary on effects of radiotherapy with concomitant and adjuvant temozolomide *versus* radiotherapy alone on survival in glioblastoma in a randomised phase III study: 5-Year analysis of the EORTC-NCIC trial (*Lancet Oncol.* 2009; 10: 459-466). Cancer 2010; 116(8): 1844-6. Available from: https://acsjournals.onlinelibrary.wiley.com/doi/10.1002/cncr.24950 [http://dx.doi.org/10.1002/cncr.24950] [PMID: 20151424]

[118] Yin J, Kim SS, Choi E, *et al.* ARS2/MAGL signaling in glioblastoma stem cells promotes self-renewal and M2-like polarization of tumor-associated macrophages. Nat Commun 2020; 11(1): 2978. Available from: https://www.nature.com/articles/s41467-020-16789-2 [http://dx.doi.org/10.1038/s41467-020-16789-2] [PMID: 32532977]

[119] Stupp R, Mason WP, van den Bent MJ, *et al.* Radiotherapy plus concomitant and adjuvant temozolomide for glioblastoma. N Engl J Med 2005; 352(10): 987-96. Available from: http://www.nejm.org/doi/abs/10.1056/NEJMoa043330 [http://dx.doi.org/10.1056/NEJMoa043330] [PMID: 15758009]

[120] Davis M. Glioblastoma: Overview of disease and treatment. Clin J Oncol Nurs 2016; 20(5) (Suppl.): S2-8. Available from: http://cjon.ons.org/cjon/20/5/supplement/glioblastoma-overview-disease-and-treatment [http://dx.doi.org/10.1188/16.CJON.S1.2-8] [PMID: 27668386]

[121] Viel T, Monfared P, Schelhaas S, *et al.* Optimizing glioblastoma temozolomide chemotherapy employing lentiviral-based anti-MGMT shRNA technology. Mol Ther 2013; 21(3): 570-9. Available from: https://linkinghub.elsevier.com/retrieve/pii/S1525001616306268 [http://dx.doi.org/10.1038/mt.2012.278] [PMID: 23319055]

[122] Portnow J, Badie B, Chen M, Liu A, Blanchard S, Synold TW. The neuropharmacokinetics of temozolomide in patients with resectable brain tumors: potential implications for the current approach to chemoradiation. Clin Cancer Res 2009; 15(22): 7092-8. Available from: https://aacrjournals.org/clincancerres/article/15/22/7092/74928/The-Neuropharmacokinetics-of-Temozolomide-in [http://dx.doi.org/10.1158/1078-0432.CCR-09-1349] [PMID: 19861433]

[123] Ozdemir-Kaynak E, Qutub AA, Yesil-Celiktas O. Advances in glioblastoma multiforme treatment: new models for nanoparticle therapy. Front Physiol 2018; 9: 170. Available from: http://journal.frontiersin.org/article/10.3389/fphys.2018.00170/full [http://dx.doi.org/10.3389/fphys.2018.00170] [PMID: 29615917]

[124] Tan SK, Jermakowicz A, Mookhtiar AK, Nemeroff CB, Schürer SC, Ayad NG. Drug repositioning in glioblastoma: A pathway perspective. Front Pharmacol 2018; 9: 218. Available from: http://journal.frontiersin.org/article/10.3389/fphar.2018.00218/full [http://dx.doi.org/10.3389/fphar.2018.00218] [PMID: 29615902]

[125] Hombach-Klonisch S, Mehrpour M, Shojaei S, *et al.* Glioblastoma and chemoresistance to alkylating agents: Involvement of apoptosis, autophagy, and unfolded protein response. Pharmacol Ther 2018; 184: 13-41. Available from: https://linkinghub.elsevier.com/retrieve/pii/S016372581730267X [http://dx.doi.org/10.1016/j.pharmthera.2017.10.017] [PMID: 29080702]

[126] Pardridge WM. Drug transport across the blood-brain barrier. J Cereb Blood Flow Metab 2012; 32(11): 1959-72. Available from: http://journals.sagepub.com/doi/10.1038/jcbfm.2012.126 [http://dx.doi.org/10.1038/jcbfm.2012.126] [PMID: 22929442]

[127] Li J, Tang C, Li L, Li R, Fan Y. Quercetin sensitizes glioblastoma to t-AUCB by dual inhibition of Hsp27 and COX-2 *in vitro* and *in vivo*. J Exp Clin Cancer Res 2016; 35(1): 61. Available from: http://jeccr.biomedcentral.com/articles/10.1186/s13046-016-0331-1 [http://dx.doi.org/10.1186/s13046-016-0331-1] [PMID: 27039073]

[128] Fong D, Yeh A, Naftalovich R, Choi TH, Chan MM. Curcumin inhibits the side population (SP) phenotype of the rat C6 glioma cell line: Towards targeting of cancer stem cells with phytochemicals. Cancer Lett 2010; 293(1): 65-72. Available from: https://linkinghub.elsevier.com/retrieve/pii/S0304383509007344 [http://dx.doi.org/10.1016/j.canlet.2009.12.018] [PMID: 20089354]

[129] Zhang Y, Yang H, Wei D, *et al.* Mitochondria-targeted nanoparticles in treatment of neurodegenerative diseases. Exploration 2021; 1(3): 20210115. Available from: https://onlinelibrary.wiley.com/doi/10.1002/EXP.20210115 [http://dx.doi.org/10.1002/EXP.20210115] [PMID: 37323688]

[130] Marks PA, Xu WS. Histone deacetylase inhibitors: Potential in cancer therapy. J Cell Biochem 2009; 107(4): 600-8. [http://dx.doi.org/10.1002/jcb.22185] [PMID: 19459166]

[131] Duvic M, Talpur R, Ni X, *et al.* Phase 2 trial of oral vorinostat (suberoylanilide hydroxamic acid, SAHA) for refractory cutaneous T-cell lymphoma (CTCL). Blood 2007; 109(1): 31-9. [http://dx.doi.org/10.1182/blood-2006-06-025999] [PMID: 16960145]

[132] Campas-Moya C. Romidepsin for the treatment of cutaneous T-cell lymphoma. Drugs Today (Barc) 2009; 45(11): 787-95. [http://dx.doi.org/10.1358/dot.2009.45.11.1437052] [PMID: 20126671]

[133] Vargas JE, Filippi-Chiela EC, Suhre T, Kipper FC, Bonatto D, Lenz G. Inhibition of HDAC increases the senescence induced by natural polyphenols in glioma cells. Biochem Cell Biol 2014; 92(4): 297-304. [http://dx.doi.org/10.1139/bcb-2014-0022] [PMID: 25070040]

[134] Srivastava A, Srivastava P, Pandey A, Khanna VK, Pant AB. Phytomedicine.New Look to Phytomedicine. Elsevier 2019; pp. 625-55. Internet Available from: https://linkinghub.elsevier.com/retrieve/pii/B9780128146194000252

[135] O. O. Pharmacokinetics and drug interactions of herbal medicines: A missing critical step in the phytomedicine/drug development process. In: Readings in Advanced Pharmacokinetics - Theory, Methods and Applications [Internet]. InTech; 2012.Available from: http://www.intechopen.com/books/readings-in-advanced-pharmacokinetics-theory-methods-and-applications/pharmacokinetics-and-drug-interactions-of-herbal-medicines-a-missing-criti-al-step-in-phytomedicine-

[136] Kum WF, Durairajan SSK, Bian ZX, Man SC, Lam YC, Xie LX. Treatment of idiopathic parkinson's disease with traditional chinese herbal medicine: a randomized placebo-controlled pilot clinical study. Evidence-Based Complement Altern Med [Internet]. 2011;2011:1–8. Available from: http://www.hindawi.com/journals/ecam/2011/724353/ [http://dx.doi.org/10.1093/ecam/nep116]

[137] Durazzo A, Lucarini M, Souto EB, Cicala C, Caiazzo E, Izzo AA. Polyphenols: A concise overview on the chemistry, occurrence, and human health. Phyther Res [Internet]. 2019 Sep 29;33(9):2221–43. Available from: https://onlinelibrary.wiley.com/doi/10.1002/ptr.6419 [http://dx.doi.org/10.1002/ptr.6419]

[138] Perez-Vizcaino F, Fraga CG. Research trends in flavonoids and health. Arch Biochem Biophys 2018; 646: 107-12. Available from: https://linkinghub.elsevier.com/retrieve/pii/S0003986118301474

[http://dx.doi.org/10.1016/j.abb.2018.03.022] [PMID: 29580946]

[139] Nabavi SF, Braidy N, Gortzi O, *et al.* Luteolin as an anti-inflammatory and neuroprotective agent: A brief review. Brain Res Bull 2015; 119(Pt A): 1-11. Available from: https://linkinghub.elsevier.com/retrieve/pii/S0361923015300307 [http://dx.doi.org/10.1016/j.brainresbull.2015.09.002] [PMID: 26361743]

[140] Onishi S, Tebayashi S, Hikichi Y, Sawada H, Ishii Y, Kim CS. Inhibitory effects of luteolin and its derivatives on osteoclast differentiation and differences in luteolin production by *Capsicum annuum* varieties. Biosci Biotechnol Biochem 2021; 85(11): 2224-31. Available from: https://academic.oup.com/bbb/article/85/11/2224/6358067 [http://dx.doi.org/10.1093/bbb/zbab149] [PMID: 34435616]

[141] Xu J, Wang H, Ding K, *et al.* Luteolin provides neuroprotection in models of traumatic brain injury *via* the Nrf2–ARE pathway. Free Radic Biol Med 2014; 71: 186-95. Available from: https://linkinghub.elsevier.com/retrieve/pii/S0891584914001269 [http://dx.doi.org/10.1016/j.freeradbiomed.2014.03.009] [PMID: 24642087]

[142] Zhang B, Wang G, He J, *et al.* Icariin attenuates neuroinflammation and exerts dopamine neuroprotection *via* an Nrf2-dependent manner. J Neuroinflammation 2019; 16(1): 92. Available from: https://jneuroinflammation.biomedcentral.com/articles/10.1186/s12974-019-1472-x [http://dx.doi.org/10.1186/s12974-019-1472-x] [PMID: 31010422]

[143] Chen WF, Wu L, Du ZR, *et al.* Neuroprotective properties of icariin in MPTP-induced mouse model of Parkinson's disease: Involvement of PI3K/Akt and MEK/ERK signaling pathways. Phytomedicine 2017; 25: 93-9. Available from: https://linkinghub.elsevier.com/retrieve/pii/S0944711316302653 [http://dx.doi.org/10.1016/j.phymed.2016.12.017] [PMID: 28190476]

[144] Zheng J, Hu S, Wang J, *et al.* Icariin improves brain function decline in aging rats by enhancing neuronal autophagy through the AMPK/mTOR/ULK1 pathway. Pharm Biol 2021; 59(1): 181-9. Available from: https://www.tandfonline.com/doi/full/10.1080/13880209.2021.1878238 [http://dx.doi.org/10.1080/13880209.2021.1878238] [PMID: 33556283]

[145] Chen FJ, Liu B, Wu Q, *et al.* Icariin delays brain aging in senescence-accelerated mouse prone 8 (SAMP8) model *via* inhibiting autophagy. J Pharmacol Exp Ther 2019; 369(1): 121-8. Available from: http://jpet.aspetjournals.org/lookup/doi/10.1124/jpet.118.253310 [http://dx.doi.org/10.1124/jpet.118.253310] [PMID: 30837279]

[146] Lee YJ, Choi DY, Choi IS, *et al.* Inhibitory effect of 4-O-methylhonokiol on lipopolysaccharide-induced neuroinflammation, amyloidogenesis and memory impairment *via* inhibition of nuclear factor-kappaB *in vitro* and *in vivo* models. J Neuroinflammation 2012; 9(1): 35. Available from: https://jneuroinflammation.biomedcentral.com/articles/10.1186/1742-2094-9-35 [http://dx.doi.org/10.1186/1742-2094-9-35] [PMID: 22339795]

[147] Choi IS, Lee YJ, Choi DY, *et al.* 4-O-methylhonokiol attenuated memory impairment through modulation of oxidative damage of enzymes involving amyloid-β generation and accumulation in a mouse model of Alzheimer's disease. J Alzheimers Dis 2011; 27(1): 127-41. Available from: https://www.medra.org/servlet/aliasResolver?alias=iospress&doi=10.3233/JAD-2011-110545 [http://dx.doi.org/10.3233/JAD-2011-110545] [PMID: 21799245]

[148] Durazzo A, Lucarini M, Souto EB, Cicala C, Caiazzo E, Izzo AA. Polyphenols: A concise overview on the chemistry, occurrence, and human health. Phyther Res [Internet]. 2019 Sep 29;33(9):2221–43. Available from: https://onlinelibrary.wiley.com [http://dx.doi.org/10.1002/ptr.5118]

[149] Widodo N, Kaur K, Shrestha BG, *et al.* Selective killing of cancer cells by leaf extract of Ashwagandha: identification of a tumor-inhibitory factor and the first molecular insights to its effect. Clin Cancer Res 2007; 13(7): 2298-306. Available from: https://aacrjournals.org/clincancerres/article/13/7/2298/182440/Selective-Killing-of-Cancer-Cells-by-Leaf-Extract [http://dx.doi.org/10.1158/1078-0432.CCR-06-0948] [PMID: 17404115]

[150] Dutta R, Khalil R, Green R, Mohapatra SS, Mohapatra S. *Withania somnifera* (ashwagandha) and withaferin A: Potential in integrative oncology. Int J Mol Sci 2019; 20(21): 5310. Available from: https://www.mdpi.com/1422-0067/20/21/5310
[http://dx.doi.org/10.3390/ijms20215310] [PMID: 31731424]

[151] Hahm ER, Lee J, Huang Y, Singh SV. Withaferin a suppresses estrogen receptor-α expression in human breast cancer cells. Mol Carcinog 2011; 50(8): 614-24. Available from: https://onlinelibrary.wiley.com/doi/10.1002/mc.20760
[http://dx.doi.org/10.1002/mc.20760] [PMID: 21432907]

[152] Socała K, Szopa A, Serefko A, Poleszak E, Wlaź P. Neuroprotective effects of coffee bioactive compounds: A review. Int J Mol Sci 2020; 22(1): 107. Available from: https://www.mdpi.com/1422-0067/22/1/107
[http://dx.doi.org/10.3390/ijms22010107] [PMID: 33374338]

[153] Wong AD, Ye M, Levy AF, Rothstein JD, Bergles DE, Searson PC. The blood-brain barrier: An engineering perspective. Front Neuroeng 2013; 6: 7. Available from: http://journal.frontiersin.org/article/10.3389/fneng.2013.00007/abstract
[http://dx.doi.org/10.3389/fneng.2013.00007] [PMID: 24009582]

[154] Grabska-Kobylecka I, Kaczmarek-Bak J, Figlus M, *et al.* The presence of caffeic acid in cerebrospinal fluid: Evidence that dietary polyphenols can cross the blood-brain barrier in humans. Nutrients 2020; 12(5): 1531. Available from: https://www.mdpi.com/2072-6643/12/5/1531
[http://dx.doi.org/10.3390/nu12051531] [PMID: 32466115]

[155] Vardi N, Parlakpinar H, Ates B. Beneficial effects of chlorogenic acid on methotrexate-induced cerebellar Purkinje cell damage in rats. J Chem Neuroanat 2012; 43(1): 43-7. Available from: https://linkinghub.elsevier.com/retrieve/pii/S0891061811001475
[http://dx.doi.org/10.1016/j.jchemneu.2011.09.003] [PMID: 21946024]

[156] Semple SC, Akinc A, Chen J, *et al.* Rational design of cationic lipids for siRNA delivery. Nat Biotechnol 2010; 28(2): 172-6.
[http://dx.doi.org/10.1038/nbt.1602] [PMID: 20081866]

[157] Tam Y, Chen S, Cullis P. Advances in lipid nanoparticles for siRNA delivery. Pharmaceutics 2013; 5(3): 498-507.
[http://dx.doi.org/10.3390/pharmaceutics5030498] [PMID: 24300520]

[158] Ramishetti S, Landesman-Milo D, Peer D. Advances in RNAi therapeutic delivery to leukocytes using lipid nanoparticles. J Drug Target 2016; 24(9): 780-6.
[http://dx.doi.org/10.3109/1061186X.2016.1172587] [PMID: 27030014]

[159] Mi Y, Wolfram J, Mu C, Liu X, Blanco E. Enzyme-responsive multistage vector for drug delivery to tumor tissue. Pharmacol. Res. 2016;113(Pt A):92–99.
[http://dx.doi.org/10.1016/j.phrs.2016.08.024]

[160] Tasciotti E, Liu X, Bhavane R, Plant K, Leonard AD. Mesoporous silicon particles as a multistage delivery system for imaging and therapeutic applications. Nat. Nano. 2008; 3(3):151–157 (2008).
[http://dx.doi.org/10.1038/nnano.2008.34]

[161] Poon Z, Chang D, Zhao X, Hammond PT. Layer-by-layer nanoparticles with a pH-sheddable layer for *in vivo* targeting of tumor hypoxia. ACS Nano 2011; 5(6): 4284-92.
[http://dx.doi.org/10.1021/nn200876f] [PMID: 21513353]

[162] Baldeshwiler AM. History of FDA good laboratory practices. Qual. Assur. 2003; J(7):157–161.
[http://dx.doi.org/10.1002/qaj.228]

[163] Sanhai WR, Spiegel J, Ferrari M. A critical path approach to advance nanoengineered medical products. Drug Discov Today Technol 2007; 4(2): 35-41.
[http://dx.doi.org/10.1016/j.ddtec.2007.10.003] [PMID: 24980838]

[164] Ramanathan RK, Korn RL, Raghunand N, *et al.* Correlation between ferumoxytol uptake in tumor

lesions by MRI and response to nanoliposomal irinotecan in patients with advanced solid tumors: a pilot study. Clin Cancer Res 2017; 23(14): 3638-48. [http://dx.doi.org/10.1158/1078-0432.CCR-16-1990] [PMID: 28159813]

[165] Quail DF, Joyce JA. The microenvironmental landscape of brain tumors. Cancer Cell 2017; 31(3): 326-41. [http://dx.doi.org/10.1016/j.ccell.2017.02.009] [PMID: 28292436]

[166] Quail DF, Joyce JA. Microenvironmental regulation of tumor progression and metastasis. Nat Med 2013; 19(11): 1423-37. [http://dx.doi.org/10.1038/nm.3394] [PMID: 24202395]

[167] Griveau A, Seano G, Shelton SJ, *et al.* A glial signature and Wnt7 signaling regulate glioma-vascular interactions and tumor microenvironment. Cancer Cell 2018; 33(5): 874-889.e7. [http://dx.doi.org/10.1016/j.ccell.2018.03.020] [PMID: 29681511]

[168] Phoenix TN, Patmore DM, Boop S, *et al.* Medulloblastoma genotype dictates blood brain barrier phenotype. Cancer Cell 2016; 29(4): 508-22. [http://dx.doi.org/10.1016/j.ccell.2016.03.002] [PMID: 27050100]

[169] Louveau A, Harris TH, Kipnis J. Revisiting the mechanisms of CNS immune privilege. Trends Immunol 2015; 36(10): 569-77. [http://dx.doi.org/10.1016/j.it.2015.08.006] [PMID: 26431936]

[170] Tivnan A, Heilinger T, Lavelle EC, Prehn JHM. Advances in immunotherapy for the treatment of glioblastoma. J Neurooncol 2017; 131(1): 1-9. [http://dx.doi.org/10.1007/s11060-016-2299-2] [PMID: 27743144]

[171] Zacharakis N, Chinnasamy H, Black M, *et al.* Immune recognition of somatic mutations leading to complete durable regression in metastatic breast cancer. Nat Med 2018; 24(6): 724-30. [http://dx.doi.org/10.1038/s41591-018-0040-8] [PMID: 29867227]

[172] Campbell BB, Light N, Fabrizio D, *et al.* Comprehensive analysis of hypermutation in human cancer. Cell 2017; 171(5): 1042-1056.e10. [http://dx.doi.org/10.1016/j.cell.2017.09.048] [PMID: 29056344]

[173] Northcott PA, Jones DTW, Kool M, *et al.* Medulloblastomics: the end of the beginning. Nat Rev Cancer 2012; 12(12): 818-34. [http://dx.doi.org/10.1038/nrc3410] [PMID: 23175120]

[174] Pajtler KW, Witt H, Sill M, *et al.* Molecular classification of ependymal tumors across all CNS compartments, histopathological grades, and age groups. Cancer Cell 2015; 27(5): 728-43. [http://dx.doi.org/10.1016/j.ccell.2015.04.002] [PMID: 25965575]

[175] Hubert CG, Rivera M, Spangler LC, *et al.* A three-dimensional organoid culture system derived from human glioblastomas recapitulates the hypoxic gradients and cancer stem cell heterogeneity of tumors found *in vivo*. Cancer Res 2016; 76(8): 2465-77. [http://dx.doi.org/10.1158/0008-5472.CAN-15-2402] [PMID: 26896279]

[176] Toledo CM, Ding Y, Hoellerbauer P, *et al.* Genome-wide CRISPR-Cas9 screens reveal loss of redundancy between PKMYT1 and WEE1 in glioblastoma stem-like cells. Cell Rep 2015; 13(11): 2425-39. [http://dx.doi.org/10.1016/j.celrep.2015.11.021] [PMID: 26673326]

[177] Atkinson JM, Shelat AA, Carcaboso AM, *et al.* An integrated *in vitro* and *in vivo* high-throughput screen identifies treatment leads for ependymoma. Cancer Cell 2011; 20(3): 384-99. [http://dx.doi.org/10.1016/j.ccr.2011.08.013] [PMID: 21907928]

[178] Housden BE, Li Z, Kelley C, *et al.* Improved detection of synthetic lethal interactions in *Drosophila* cells using variable dose analysis (VDA). Proc Natl Acad Sci USA 2017; 114(50): E10755-62.

[http://dx.doi.org/10.1073/pnas.1713362114] [PMID: 29183982]

[179] Pitter KL, Tamagno I, Alikhanyan K, *et al.* Corticosteroids compromise survival in glioblastoma. Brain 2016; 139(5): 1458-71.
[http://dx.doi.org/10.1093/brain/aww046] [PMID: 27020328]

# CHAPTER 2

# Current Innovation in *In Vitro*, *Ex Vivo* and *In Vivo* Screening of Brain Cancer

**Tanmoy Kanp**[1], **Anish Dhuri**[1], **M. Bharath**[1], **Ujala Gupta**[1], **Khushi Rode**[1], **Prashant Tiwari**[3], **Neha Singh**[2] and **Pankaj Kumar Singh**[1,*]

[1] *Department of Pharmaceutics, National Institute of Pharmaceutical Education and Research (NIPER), Hyderabad-500037, India*

[2] *Department of Microbiology, Pt. Jawaharlal Nehru Medical College, Raipur, Chhattisgarh-492001, India*

[3] *College of Pharmaceutical Sciences, Dayananda Sagar University, Bengaluru-562112, India*

**Abstract:** In recent years, brain tumors have become one of the leading causes of cancer-related mortality in children. Among all brain cancers, glioblastoma (GBM) is considered one of the most aggressive and lethal malignancies in oncology. The difficulty in replicating the complex structure of the brain and its surrounding microenvironment is a major obstacle to understanding potential therapeutic methods for this condition. To overcome the challenges of treating brain tumors, researchers have focused on studying different models of brain tumors. These models offer valuable insights that can aid scientists in developing more effective treatments for brain tumors. *In vitro*, tumor models are essential for understanding the molecular and cellular mechanisms underlying the malignant progression of tumor blood vessels. *Ex vivo* tumor models involve the study of malignant tissues removed from living organisms in a controlled laboratory environment. These models utilize tissue slices, organoids, and brain tumor cell cultures. Furthermore, *in vivo* models can be useful for preclinical testing of potential new treatment approaches and for investigating the molecular mechanisms of brain tumors and their associated microenvironments. However, this chapter discusses brain cancer, including its types, symptoms, and diagnostic characteristics. It explains the differences in cell-signaling mechanisms between glial and non-glial brain cancers. The chapter also covers the latest advancements in the prognosis of GBM. Additionally, it includes a discussion of different types of pre-clinical models used to overcome the limitations of current therapeutic approaches, such as *in vitro* 2D and 3D cellular models. Furthermore, the chapter addresses *in vivo* animal models of brain cancer created through chemical and genetic methods, as well as *ex vivo* models like patient-derived models, xenografts, and organoids.

**Keywords:** Brain tumor, *In vitro* models, *In vivo* models, Signaling pathways.

* **Corresponding author Pankaj Kumar Singh:** Department of Pharmaceutics, National Institute of Pharmaceutical Education and Research (NIPER), Hyderabad-500037, India; E-mail: pankajksingh3@gmail.com

**Prashant Tiwari, Pankaj Kumar Singh & Sunil Kumar Kadiri (Eds.)**

# INTRODUCTION

Brain tumors represent a broad and diverse category of cancers that affect the central nervous system [1]. Among these, glioblastoma (GBM) is considered one of the most aggressive and lethal primary brain malignancies in oncology. The patients show a terrible prognosis and exhibit a poor quality of life as the disease progresses [2]. GBM has been stated as grade IV gliomas by the World Health Organization (WHO), displaying a 5-year survival of 7.2%. It often affects the cerebral hemispheres, with a peak incidence between 45 and 70 years old. It has a high genetic heterogeneity and an infiltrative nature. The protective presence of the blood-brain barrier has hugely contributed to the worsened treatment efficacy [3]. The current standard of care for brain tumors—primarily involving surgical resection followed by prolonged chemotherapy—offers limited improvement in patient survival. Despite advancements in neurosurgical techniques, radiation therapy, and chemotherapeutic regimens, the prognosis for patients with high-grade gliomas, particularly glioblastoma multiforme (GBM), remains dismal. GBM's highly proliferative and infiltrative nature poses a significant challenge, as complete surgical eradication is virtually impossible. This biological aggressiveness renders conventional therapeutic approaches largely ineffective [4]. Modern cancer diagnosis increasingly relies on genomic, molecular, and histopathological analyses, which provide critical insights into tumor classification and guide personalized treatment strategies [5]. To address the limitations of existing therapies, researchers have developed various 2D and 3D *in vitro* models. Among these, three-dimensional models offer superior regulation of cell morphology, and enable advanced techniques such as live imaging and antibody staining, making them particularly valuable for preclinical research [4]. Additionally, *ex vivo* and *in vivo* models are widely employed to investigate tumor angiogenesis, growth kinetics, and therapeutic responses, offering translational relevance in understanding tumor biology and evaluating novel interventions [6]. This book chapter provides a comprehensive overview of brain cancer subtypes, associated clinical symptoms, and diagnostic features. It aims to enhance readers' understanding of the cell-signaling pathways involved in both glial and non-glial brain tumors, thereby laying the foundation for exploring emerging therapeutic strategies and precision oncology approaches.

Recent advancements in the prognosis of glioblastoma (GBM) will also be discussed. In addition, various two-dimensional (2D) and three-dimensional (3D) *in vitro* cellular preclinical models that address the limitations of current treatment modalities are reviewed. *ex vivo* models, such as patient-derived xenografts and organoids, along with chemically and genetically induced *in vivo* animal models of brain cancer, are also examined.

## BRAIN CANCER- CLASSIFICATIONS, SYMPTOMS, AND DIAGNOSIS

### Classification

Brain tumors are broadly classified into two categories: malignant and benign. Malignant tumors consist of cancerous cells that can infiltrate adjacent tissues and, in some cases, metastasize to distant organs. In contrast, benign tumors typically exhibit slow growth, are less invasive, and have a lower likelihood of recurrence following treatment. Tumors are further categorized based on their origin: Primary brain tumors originate within the brain or Central Nervous System (CNS). Secondary brain tumors, or brain metastases, result from the spread of cancer from other parts of the body to the brain [7 - 9]. The World Health Organization (WHO) classifies brain tumors using a grading system that reflects their biological behavior and aggressiveness: Grade I tumors are generally benign, slow-growing, and most commonly observed in pediatric populations. These tumors are often curable with surgical resection. Grade II tumors, more prevalent in adults, include oligodendrogliomas, oligoastrocytomas, and astrocytomas. They are low-grade gliomas with a potential for progression. Grade III tumors, such as anaplastic oligodendrogliomas, anaplastic oligoastrocytomas, and anaplastic astrocytomas, exhibit greater cellular atypia and invasiveness than Grade II tumors. Grade IV tumors, notably glioblastoma multiforme (GBM), represent the most aggressive form of glioma, characterized by rapid proliferation, extensive infiltration, and poor prognosis. High-grade gliomas encompass Grade III and IV tumors, which are malignant and challenging to surgically remove due to their diffuse invasion into surrounding brain tissue. In contrast, low-grade gliomas include Grade I and II tumors, which are typically benign and may be amenable to complete surgical excision [10].

### Symptoms

In addition to being a typical sign of disease, headaches may also be an indication of a brain tumor. Roaring or quivering muscles in the limbs are symptoms of seizures, which may affect as many as 80% of patients with brain tumors. It is unusual to feel unwell, particularly when you are moving abruptly. An increase in intracranial pressure may affect the brain, leading to drowsiness or unconsciousness. Vision loss, floating objects, or tunnel vision might result from worsening eye conditions. Confusion, short-term memory loss, and abnormal thought processes could result from changes in behavior & personality, and leakage of milk from the breasts are some of the clinical manifestations involving brain tumors [8, 11, 12].

## Diagnosis

For patients suspected of having a brain tumor, a comprehensive diagnostic workup is essential. This typically includes neuroimaging, a detailed medical history, and a neurological examination. Depending on the tumor's location and associated symptoms, targeted physical assessments—such as visual field testing and audiometric evaluations may provide additional diagnostic insights [13]. The initial imaging modality often involves computed tomography (CT) or magnetic resonance imaging (MRI) to evaluate the presence, size, and anatomical location of the tumor, as well as any associated cerebral edema. While CT is superior for detecting acute intracranial hemorrhage, MRI is the preferred technique due to its high spatial resolution and ability to guide surgical planning. MRI protocols typically include FLAIR sequences, contrast-enhanced T1-weighted imaging, and T1- and T2-weighted sequences.

Advanced MRI techniques offer further diagnostic precision, including diffusion-weighted imaging (DWI), which assesses cellular density and tumor aggressiveness. Perfusion imaging evaluates vascularity and treatment response. Magnetic resonance spectroscopy (MRS) offers molecular-level insights into tissue composition, providing information without contrast or radiation, and complements structural MRI findings [14]. Positron emission tomography (PET) is another valuable tool, particularly for assessing tumor metabolism and metabolic activity. Fluorodeoxyglucose (FDG)-PET scans can differentiate between treatment-related changes and tumor progression, as malignant cells typically exhibit increased glucose uptake compared to normal brain tissue [15]. Despite the utility of imaging, definitive diagnosis of tumor type and grade requires histopathological analysis of tissue samples obtained *via* stereotactic biopsy or open craniotomy. These samples are essential for accurate classification and treatment planning. Certain brain tumors such as ependymoma, germ cell tumors, pineoblastoma, medulloblastoma, and primary CNS lymphoma may disseminate to the spinal cord, necessitating cerebrospinal fluid (CSF) analysis for malignant cells. Additionally, tumors like germinomas may secrete tumor markers detectable in blood or CSF, aiding in diagnosis and monitoring [16].

# CELL SIGNALING PATHWAYS IN BRAIN TUMORS

## Signaling Pathways in Glial Tumors

### *PI3K/AKT/mTOR Pathway*

The PI3K/Akt/mTOR (PAM) signaling pathway plays a pivotal role in regulating cell cycle progression, and its dysregulation actively contributes to tumor initiation and progression. In isocitrate dehydrogenase (IDH)-wildtype glioblastoma multiforme (GBM), mutations affecting components of the PAM pathway have been frequently identified. Upon activation of Akt, cyclin D1 binds to cyclin-dependent kinases (CDKs), initiating cell division through downstream activation of mTOR. Elevated cyclin D1 expression during the G1-to-S phase transition is associated with enhanced proliferative capacity and carcinogenesis. Under normal conditions, P27^kip1 functions as a CDK inhibitor, arresting the cell cycle and suppressing growth. However, Akt-mediated phosphorylation of P27^kip1 neutralizes its inhibitory effect, thereby bypassing this critical checkpoint and permitting uncontrolled cell proliferation and differentiation. The PAM pathway also supports rapid cancer cell growth by regulating the biosynthesis of macromolecules, including proteins, nucleotides, and lipids, essential for tumor expansion [17, 18].

Beyond cell cycle control, the PI3K/mTOR axis plays a crucial role in cell motility, growth, survival, and metabolic regulation. Phosphoinositide 3-kinase (PI3K) belongs to a family of lipid kinases that generate phosphatidylinositol-3,4,5-trisphosphate (PIP3) upon activation by receptor tyrosine kinases (RTKs) [19]. The tumor suppressor gene PTEN encodes a phosphatase that degrades PIP3, thereby inhibiting the PI3K/Akt pathway. Loss of PTEN function results in elevated Akt activity, which in turn activates mTOR, promoting cell survival and proliferation [20]. RTKs activate Class IA PI3Ks either directly or through insulin receptor substrate (IRS) proteins, which act as adaptor molecules to facilitate signal transduction. This interaction leads to the production of PIP3 and the subsequent activation of PIP3-dependent serine/threonine kinases, including PDK1 and PDK2. These kinases activate Akt, linking extracellular growth factor signals to intracellular pathways governing proliferation, metabolism, and survival [21]. In summary, the RTK/PI3K/Akt/mTOR signaling cascade is constitutively active in glioma cells, driving tumor growth, metabolic reprogramming, and resistance to apoptosis, making it a critical target for therapeutic intervention.

### *MAPK/Ras Pathway*

The MAPK/ERK signaling pathway, also known as the RAS/MAPK pathway, plays a critical role in the initiation and progression of various malignancies, including brain tumors. This pathway comprises a cascade of proteins that transmit signals from cell surface receptors to the nuclear DNA, ultimately influencing gene expression. The signaling cascade is initiated when growth factors bind to receptor tyrosine kinases (RTKs) on the cell membrane, leading to the activation of RAS, a small GTPase. Activated RAS subsequently stimulates RAF kinase, which phosphorylates and activates MEK. MEK then activates ERK, which translocates into the nucleus to regulate genes involved in cell proliferation, differentiation, and survival [22]. Mutations in key components of the RAS/MAPK pathway such as RAS, BRAF, and MEK can result in constitutive pathway activation, contributing to uncontrolled cell division and resistance to apoptosis. These alterations are frequently observed in gliomas and other brain tumors, promoting tumor growth and invasiveness [23].

Given its central role in oncogenesis, the RAS/MAPK pathway has become a prominent therapeutic target in neuro-oncology. Clinical trials are actively investigating MEK and RAF inhibitors for their potential to suppress aberrant signaling and inhibit tumor progression. However, therapeutic resistance remains a significant challenge. Resistance mechanisms may include: Secondary mutations in pathway components. Activation of compensatory signaling pathways, such as PI3K/Akt or JAK/STAT. Tumor heterogeneity, which limits uniform drug response. Overcoming resistance will require combination therapies, biomarker-driven patient stratification, and adaptive treatment strategies to enhance efficacy and durability of response.

### *P53 Pathway*

The p53 gene, also known as tumor protein p53 (TP53), encodes a nuclear protein that plays a pivotal role in regulating cell cycle progression, DNA repair, and apoptosis. In response to genotoxic stress, p53 can induce cell cycle arrest at the G0/G1 phase and trigger programmed cell death, thereby acting as a critical tumor suppressor [17]. Mutations in the TP53 gene are strongly associated with the development and progression of various cancers, including those associated with Li-Fraumeni syndrome. In gliomas, particularly glioblastoma multiforme (GBM), dysfunction of the TP53 gene is a common feature. This can occur through direct mutations, or *via* MDM2/MDM4-mediated inactivation, leading to uncontrolled cell proliferation. Abnormalities in p53 signaling are observed in approximately 87% of sporadic gliomas, with a higher prevalence in secondary GBMs and low-grade astrocytic gliomas that often co-occur with IDH mutations. In contrast, primary GBMs exhibit widespread genomic instability, with frequent TP53 mutations clustering at codons 248 and 272, which may reflect the chaotic nature of the tumor microenvironment. In GBM, p53 inactivation arises through multiple mechanisms: TP53 gene mutations. Loss of ARF (~5%). Amplification of MDM2 (~11%) and MDM4 (~4%). Homozygous deletion of CDKN2A/ARF, the most commonly disrupted component of the p53 pathway, occurring in ~60% of cases—particularly in younger patients [18, 19].

The ARF protein exerts its tumor-suppressive function by promoting MDM2 degradation, thereby stabilizing and preserving p53 activity [20]. Given the high frequency and functional significance of TP53 mutations in GBM, therapeutic strategies targeting p53 are of considerable interest. Restoring or reactivating wild-type p53 (wt-p53) function has shown promise across various cancer types, including GBM. Several approaches have been explored, ranging from small-molecule inhibitors to gene therapy aimed at suppressing mutant p53 or restoring wild-type p53 activity [21]. One notable gene therapy candidate is SGT-53, currently under clinical evaluation (NCT02340156). SGT-53 consists of a cationic liposome encapsulating a plasmid encoding wt-p53, designed to cross the blood-brain barrier (BBB) and deliver functional p53 directly to tumor cells. This adenovirus-mediated p53 gene therapy represents a novel and promising approach to overcome TP53 dysfunction in GBM.

### $R_b$ *Pathway*

The retinoblastoma protein (Rb) is a crucial tumor suppressor that regulates cell cycle progression. Mutations or functional disruptions in the Rb pathway have been implicated in the development of various cancers. The activity of Rb is tightly controlled by phosphorylation, which determines its ability to suppress cell proliferation [24]. In its hypophosphorylated state, Rb actively inhibits cell cycle progression by binding to and inactivating E2F transcription factors, thereby preventing the expression of genes required for S-phase entry. This interaction occurs during the early G1 phase of the cell cycle. Conversely, hyperphosphorylated Rb becomes functionally inactive, allowing E2F-mediated transcription and promoting cellular proliferation. The phosphorylation status of Rb is regulated by cyclin-dependent kinases (CDKs), particularly CDK4 and CDK6, which phosphorylate Rb and disrupt its interaction with E2F proteins [25].

CDK inhibitors, such as CDKN2A (p16^INK4a) and CDKN2B (p15^INK4b), play a protective role by preventing CDK-mediated Rb phosphorylation, thereby maintaining Rb's tumor-suppressive function and stabilizing the Rb-E2F complex [26, 27]. Dysregulation of this pathway—through loss of CDK inhibitors or overactivation of CDKs can lead to unchecked cell cycle progression and tumorigenesis. Although no therapies directly targeting the Rb pathway have been approved to date, ongoing research is exploring targeted interventions for patients with Rb pathway abnormalities. One promising candidate is PD 0332991 (palbociclib), a small molecule inhibitor of CDK4/6, currently under clinical investigation (NCT01227434). By inhibiting CDK4/6 activity, palbociclib aims to restore Rb function and suppress tumor cell proliferation, offering a potential therapeutic strategy for cancers with intact but functionally suppressed Rb signaling.

## Signaling Pathways in Non-glial Cells

The growth and progression of non-glial tumors, like those found in other parts of the body, are similarly governed by intricate signaling networks. The cytoplasmic build-up of β-catenin characterizes the wnt/β-catenin pathway. Then, β-catenin moves into the nucleus, where it functions as a transcriptional coactivator for genes involved in the development, advancement, and proliferation of cells. This pathway, which affects functions including body axis patterning and cell fate specification, is crucial for healthy embryonic development. Because the Wnt/β-catenin pathway regulates cell proliferation and differentiation, dysregulation of this route has been associated with cancer [28]. The pathway is triggered when Wnt proteins attach to Frizzled family receptors on the cell surface. Because of this contact, β-catenin is prevented from breaking down, which allows it to build up and enter the nucleus [29]. Many other non-glial pathways are involved, including the Hedgehog Pathway and the Notch Signaling Pathway. Abnormal activation of these pathways can contribute to tumor progression, resistance to chemotherapeutic agents, and metastasis [30, 31].

## NEOTERIC ADVANCEMENTS IN EXPERIMENTAL MODELS FOR SCREENING OF BRAIN CANCER

### *In Vitro* Models

#### *2D Cellular Pre-Clinical Model*

Two-dimensional (2D) cell models have long served as foundational tools in brain tumor research, offering a simple and cost-effective platform for studying tumor initiation, progression, and drug response. In these models, tumor cells are cultured as a flat monolayer, allowing for easy manipulation, microscopic observation, and high-throughput screening. Their accessibility and affordability have made them a popular choice for preliminary drug testing and basic cancer biology investigations [32, 33]. However, it is important to acknowledge the limitations of 2D models. The *in vivo* tumor microenvironment is highly complex, characterized by three-dimensional architecture, cell–cell and cell–matrix interactions, nutrient gradients, and mechanical cues features that are not accurately replicated in 2D cultures. Tumor cells grown in 2D lack spatial continuity, which may lead to misrepresentation of tumor behavior, molecular heterogeneity, and drug sensitivity profiles. As a result, findings derived from 2D models may not fully translate to clinical settings, undermining their predictive validity [34].

In contrast, three-dimensional (3D) *in vitro* models are increasingly being adopted to simulate the physiological conditions of brain tumors better. These models utilize GBM cell lines cultured within extracellular matrix-mimicking environments, such as 3D scaffolds, for structural support. Microfluidic devices to replicate fluid dynamics and nutrient exchange. Co-culture systems incorporating multiple cell types (*e.g.*, astrocytes, endothelial cells) to mimic tumor–stroma interactions. Bioreactors and perfusion systems to maintain continuous nutrient supply and waste removal. Advanced imaging techniques and functional assays enable real-time observation of cell behavior, migration, and response to therapies. These models are particularly valuable for: Biomarker discovery, Drug efficacy and toxicity testing, and Studying tumor heterogeneity and resistance mechanisms. Compared to conventional 2D cultures, 3D models offer a more physiologically relevant platform for investigating brain tumor biology and evaluating therapeutic strategies, bridging the gap between *in vitro* experimentation and *in vivo* outcomes [35, 36].

### *3-D Cellular Pre-Clinical Model*

Three-dimensional (3D) imaging and culture models play a pivotal role in both the diagnosis and therapeutic evaluation of brain tumors. To accurately replicate the mechanism of action of candidate drugs in human patients, it is essential to develop 3D models that closely simulate the tumor microenvironment. Compared to traditional two-dimensional (2D) cultures, 3D systems offer superior platforms for drug assessment by generating tissue-scale solute concentration gradients, local pericellular gradients, and complex extracellular matrices, all of which influence solute diffusion and cellular response [37]. In the context of glioblastoma multiforme (GBM), 3D models, particularly GBM organoids, provide enhanced physiological relevance. These organoids are derived from resected patient tumors, preserving the genetic and phenotypic heterogeneity of the original malignancy. The initial development of 3D GBM models began with sphere cultures, maintained in serum-free media supplemented with growth factors such as epidermal growth factor (EGF) and fibroblast growth factor (FGF). These conditions were designed to support and enrich GBM stem cell populations.

Functionally, GBM spheres exhibit aggressive behavior in assays such as transplantation, and they retain tumor heterogeneity more effectively than 2D cultures, which are prone to clonal selection and genetic drift over time. In one study, researchers developed a long-term 3D culture system using tumor organoids derived from various GBM sources—including patient-derived primary cultures, xenografts, genetically engineered glioma models, and clinical specimens. Cultured in Matrigel and supplemented with EGF and FGF, these organoids maintained their 3D architecture and proliferated for several months, reaching diameters of 3–5 mm. Notably, the organoids exhibited regional heterogeneity, with a hypoxic core containing quiescent cells, including cancer stem cells, and a proliferative rim enriched with SOX2+ stem-like cells [38]. In another study by Gomez-Roman *et al.*, patient-derived GBM stem cells were cultured on a polystyrene scaffold using advanced bioengineering techniques to create a novel *in vitro* 3D GBM model. The cells were maintained in a serum-free medium specifically formulated for stem cells and supplemented with vascular endothelial growth factor (VEGF), EGF, and FGF, effectively replicating the perivascular stem cell niche. The resulting model closely mimicked GBM biology, with outputs comparable to those observed in orthotopic xenograft models [39]. These findings underscore the value of 3D GBM models in preclinical drug testing, biomarker discovery, and tumor biology research, offering a more physiologically accurate and clinically relevant alternative to conventional 2D systems.

## *Ex Vivo* Models

In the era of personalized medicine, the most crucial step is to stratify patients effectively. Model systems produced from living tumor samples are necessary for effective *ex vivo* experiments to predict individual treatment responses. These model systems should closely reflect the features and environment of *in vivo* tumors. Advanced model systems allow for a range of *ex vivo* functional assays specifically designed to forecast a patient's response to targeted therapies.

### *Patient-Derived Xenografts (PDXs)*

Conventional cell-line-derived xenograft models have drawbacks, including an inadequate microenvironment and a lack of tumor heterogeneity; however, they have helped us understand how cancer develops and how drugs work. PDX models, on the other hand, provide a more individualized approach to medication therapy, as they are created by implanting tumor tissues from patients into immunodeficient mice, as shown in Fig. (**1A**). Patient-derived xenograft (PDX) models have been widely used for the preclinical evaluation of individualized treatment plans, biomarker identification, drug screening, and the investigation of resistance mechanisms. The histological traits, gene expression patterns, copy number variation, and metastatic behavior of the *in vivo* tumor are all preserved in the PDX models [40]. U-87 MG and patient-derived glioblastoma cells were administered into the brain parenchyma of Balb/c nude mice in order to show that intracranial tumor development is feasible. The patient-derived glioblastoma cells did not develop cerebral tumors until six weeks following tumor cell injection, in contrast to the U-87 MG cells. On the other hand, upon intravitreal injection, the patient-derived cells successfully developed intraocular tumors, developing from plaques at two weeks to masses at four weeks. The patient's initial and *in vivo* tumors showed identical immunopositivity for human mitochondria, GFAP, vimentin, and nestin. Additionally, cells extracted from the *in vivo* tumors demonstrated immunopositivity for the same markers and morphology as their parent cells. Overall, the intravitreal injection of tumor cells resulted in the establishment of a new PDX model of glioblastoma. This model will be crucial for researching and creating cutting-edge treatment options for glioblastoma [41].

### Patient-Derived Organoids (PDOs)

Patient-derived organoids (PDOs) are three-dimensional *in vitro* cultures that replicate the biological characteristics of the original tumor from which they are derived, as illustrated in Fig. (**1B**). Organoids are composed of cell types and structural features that closely resemble those found in native tissues, enabling the mimicry of physiological functions and tissue-specific architecture. In glioma research, brain organoids have emerged as powerful tools for modeling the tumor microenvironment, preserving tumor heterogeneity, and predicting patient-specific responses to anticancer therapies. Using human brain organoids as 3D tissue substrates, Liu *et al.* demonstrated that knockdown of the oncogenic gene lncGRS-1 in glioblastoma multiforme (GBM) selectively inhibited tumor growth and enhanced radiosensitivity of glioma cells [42]. This highlights the potential of organoid platforms for functional genomics and therapeutic optimization.

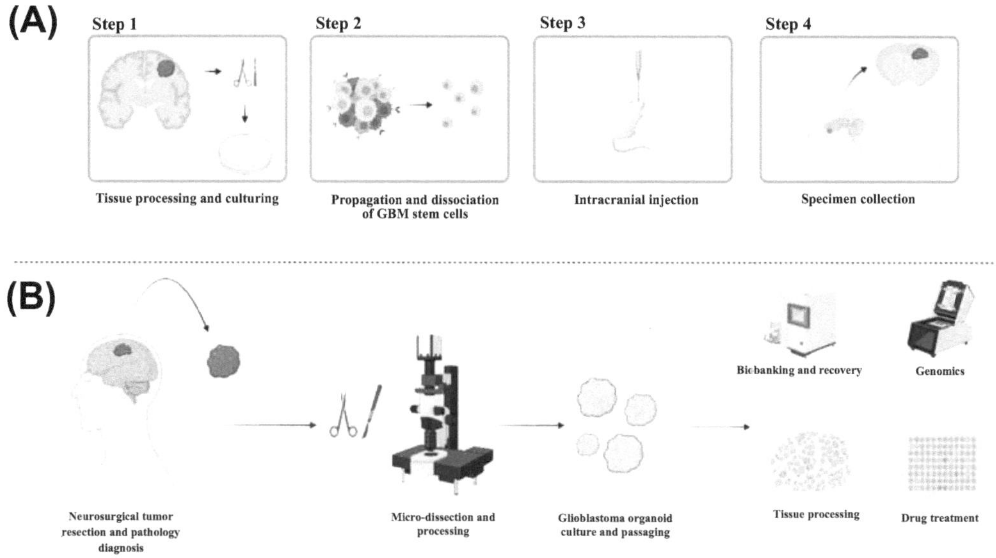


**Fig. (1). A**) Development of patient-derived Xenograft. **B**) Development of patient-derived organoids.

In another study, Oudin *et al.* harvested GBM tissue from patients and mechanically dissociated it into microscopic fragments (~0.5 mm × 0.5 mm) using a sterile scalpel at 20–22 °C within a biosafety cabinet. These fragments were cultured to generate tumor organoids, which were subsequently implanted intracortically into immunodeficient mice, establishing a PDO model of GBM for *in vivo* tumor propagation. PDOs derived from both primary and metastatic brain tumors have been shown to recapitulate clinical treatment responses, making them valuable for identifying effective therapeutic regimens and adjuvant strategies for patients with recurrent disease [43].

Further research using 3D GBM organoid models has explored mesenchymal gene signatures, treatment resistance, and self-renewal capacity in quiescent tumor cell populations. Through bioinformatics analysis and functional assays, hypoxia and TGF-β signaling were identified as key niche factors that support GBM quiescence, mesenchymal transition, and tumor-microenvironment interactions [44]. Collectively, these studies underscore the utility of PDOs in precision oncology, offering a robust platform for drug screening, mechanistic studies, and personalized treatment planning in brain tumor research.

## *In-Vivo* Models

### *Genetically Engineered Mouse Model*

The field of cancer research has significantly advanced through the use of genetically engineered mouse models (GEMMs). Unlike traditional cancer cell inoculation models, GEMMs develop tumors *de novo* within an immune-competent and physiologically relevant environment. Tumors arising in advanced GEMMs exhibit genetic heterogeneity, closely mimic the histological and molecular features of human cancers, and can progress spontaneously to metastatic disease. GEMMs encompass both transgenic and knockout mice, which are generated by introducing oncogenes or disrupting tumor suppressor genes into embryonic stem cells or zygotes. Beyond oncogenes and tumor suppressors, key molecules involved in tumor signaling pathways can also be manipulated to create disease-relevant models [32]. However, global deletion of tumor suppressor genes may result in systemic pathologies, complicating the study of their tissue-specific roles in brain tumorigenesis. For example, mice lacking the Cdkn2a locus (Ink4a/Arf) remain viable and fertile but develop lymphomas and fibrosarcomas, limiting their utility for brain-specific cancer studies [45].

To address this, inducible or conditional knockout systems have been developed, enabling spatial and temporal control over gene editing. The Cre/loxP recombination system is widely used for this purpose. In this system, Cre recombinase excises genes flanked by loxP sites, and its expression can be driven by tissue-specific promoters, enabling targeted gene deletion in specific cell populations. While germline GEMMs require extensive breeding programs that are costly and time-consuming, non-germline GEMMs offer a more efficient alternative. These models utilize virus-mediated gene transfer to deliver Cre recombinase directly to somatic cells, bypassing the need for complex breeding while retaining control over gene expression. One widely adopted system is the RCAS/TVA model, which employs the replication-competent avian sarcoma-leukosis virus long terminal repeat with splice acceptor (RCAS) to deliver genetic material. RCAS infects cells expressing the TVA receptor, which is naturally found only in avian cells but can be introduced into mammalian cells *via* genetic engineering [46]. GEMMs based on the RCAS/TVA system offer several advantages over Cre-loxP models, including limited viral transduction, which ensures that only a small subset of cells express the target gene. These models support sequential or concurrent gene delivery, enabling the study of multi-gene interactions in tumorigenesis. They also allow for flexible and rapid generation, making them suitable for investigating gene function, tumor initiation, and progression dynamics [47].

These models provide a powerful platform for dissecting the molecular mechanisms of brain cancer, evaluating therapeutic targets, and understanding tumor evolution in a controlled and biologically relevant context.

### *Patient-Derived Xenografts to Curate In Vivo Model*

The development of effective cancer therapeutics has often been hindered by the lack of preclinical models that accurately reflect the complexity and heterogeneity of human tumors. Among the most robust and translationally relevant tools in cancer research are patient-derived xenograft (PDX) models. These models have proven invaluable in co-clinical trials, personalized medicine, anti-cancer drug development, immunotherapy, gene therapy, and the establishment of PDX biobanks. To generate a PDX *in vivo* model, tumor tissue is extracted from patients with brain cancer and transplanted into immunodeficient host animals. Xenografts can be administered *via* three primary routes:

Orthotopic transplantation, where tumor cells are implanted into the original anatomical site (*e.g.*, brain). Heterotopic transplantation, typically subcutaneous, offering ease of monitoring and tumor volume measurement. Metastatic models, which simulate the spread of tumors. Among these, heterotopic transplantation is the most widely used due to its technical simplicity, cost-effectiveness, and ease of drug efficacy evaluation [48, 49]. However, this method introduces limitations, including the formation of tumors in a non-native microenvironment and the absence of a blood-brain barrier (BBB), which may compromise the model's ability to accurately predict therapeutic responses. In contrast, orthotopic transplantation better replicates the tumor microenvironment, including interactions with the BBB, and is recommended for brain tumor studies. Both subcutaneous and orthotopic models remain prevalent in research due to their complementary advantages. Recent advances in pediatric brain tumor PDX models have shown promise in representing subtypes such as glioblastoma (GBM), diffuse midline glioma, ependymoma, and medulloblastoma. These models retain the histological and molecular fidelity of the original patient tumors, making them highly valuable for subtype-specific investigations. For example, St. Jude Children's Research Hospital has successfully developed 37 orthotopic PDX models from pediatric brain tumor patients, including 22 medulloblastomas and 5 ependymomas, all of which maintain genetic and histopathological consistency with the source tumors [50].

To model diffuse intrinsic pontine glioma (DIPG) with the H3.3K27M mutation, Funato *et al.* engineered neural progenitor cells derived from human embryonic stem cells, incorporating a constitutively active PDGFRA, short hairpin RNA against p53, and the H3.3K27M mutation [51].

In another study focused on medulloblastoma, researchers utilized human neuroepithelial stem cells derived from induced pluripotent stem cells (iPSCs) of a patient with Gorlin syndrome, which carries a germline mutation in the sonic hedgehog (SHH) receptor PTCH1. These cells were transplanted into the cerebellum of mice, successfully recapitulating SHH-driven medulloblastoma and accelerating tumor growth [52]. Overall, PDX models offer a powerful and scalable platform for preclinical drug discovery, particularly in pediatric brain tumors, by enabling the study of tumor biology and therapeutic response within a system that preserves the intact blood-brain barrier and patient-specific tumor characteristics.

### *Chemically Induced Model*

In experimental neuro-oncology, rat brain tumor models have played a crucial role in advancing our understanding of tumor biology, biochemistry, imaging, and therapeutic strategies for brain cancers. Compared to mice, rats are more effective in generating chemically induced brain tumor models, with chemical carcinogens and oncogenic viruses commonly employed to initiate tumor formation. In particular, n-nitrosourea compounds have been used to induce various glioma subtypes in rats, including astrocytomas, oligodendrogliomas, and ependymal tumors [53, 54]. Chemical carcinogens can be administered *via* oral, intravenous, or local routes, with neonatal animals being more susceptible to tumor induction. In older animals, repeated administration may be required to enhance tumorigenic efficacy. From these chemically induced models, several glioma cell lines have been established, including C6, 9L, T9, F98, RG2, BTC, and CNS-1, all of which have become widely utilized in brain tumor research.

Among these, the C6 rat glioma model and 9L gliosarcoma are frequently employed to evaluate therapeutic efficacy, drug resistance, and drug transport mechanisms across the blood-brain barrier (BBB) and blood-tumor barrier (BTB). The 9L model is also extensively used in brain tumor imaging studies [55]. One key advantage of these models is their compatibility with syngeneic immunocompetent hosts, allowing for the engagement of both innate and adaptive immune responses in experimental treatments. This is particularly valuable for assessing immunotherapies. However, certain rat glioma cell lines such as C6, 9L, and T9 may exhibit immunogenicity, leading to spontaneous tumor rejection. This phenomenon can confound therapeutic assessments, especially when evaluating immune-stimulating agents, as tumor regression may be misinterpreted as treatment efficacy. Despite their utility, chemically induced tumor models present limitations. The tumor types, anatomical locations, and biological characteristics can vary significantly, and the resulting tumors may not fully replicate the histopathological and molecular features of human brain tumors. Therefore, while these models offer valuable insights, their translational relevance must be carefully considered in the context of preclinical drug development.

## HURDLES/LIMITATIONS ASSOCIATED WITH EFFECTIVE BRAIN TUMOR THERAPIES

Brain tumor treatment remains one of the most formidable challenges in oncology, largely due to the molecular complexity and biological resilience of these malignancies. These tumors often reside in inaccessible regions of the brain, making surgical resection difficult even for the most skilled neurosurgeons. Compounding this challenge is the presence of the blood-brain barrier (BBB), a highly selective network of tight junctions and transport proteins that protects the brain from systemic agents, including many chemotherapeutic drugs. As a result, systemic therapies often fail to reach therapeutic concentrations within brain tissues.

Moreover, brain tumors exhibit unique developmental, genetic, epigenetic, and microenvironmental features that render them resistant to both conventional and novel treatment modalities. The relative rarity of brain tumors further complicates these biological intricacies compared to other cancers, which has led to limited pharmaceutical investment, fragmented research efforts, and a small, dispersed scientific community. In response to these challenges, Cancer Research UK (CRUK) convened a global consortium of brain cancer experts, encompassing disciplines such as neurobiology, preclinical tumor modeling, genomics, drug discovery, neuropathology, neurosurgery, imaging, radiotherapy, and medical oncology. This collaborative initiative was tasked with identifying the most critical barriers to progress and outlining a strategic roadmap to improve outcomes for patients with brain tumors [56, 57]. The panel's findings are presented in this Position Paper, which highlights seven key priorities for future research and investment see Fig. (**2**). These priorities aim to catalyze innovation, foster interdisciplinary collaboration, and ultimately pave the way for curative therapies for all patients with brain tumors.

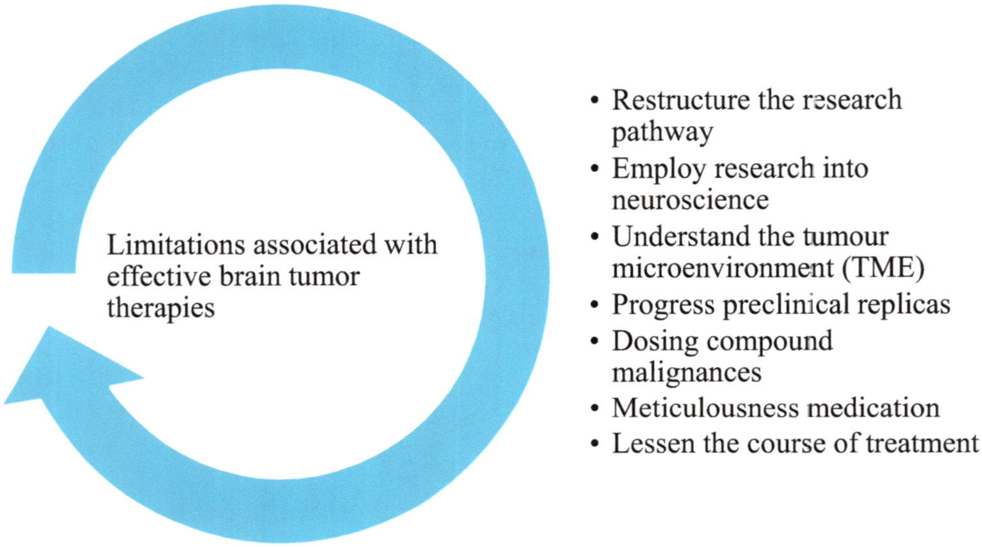


**Fig. (2).** Limitations associated with effective brain tumor therapies.

## CONCLUSION

Genomic, transcriptomic, and proteomic analyses of human brain tumor specimens have significantly enhanced our understanding of the signaling pathways that govern various cancer subtypes. These studies have consistently shown that cellular and genetic heterogeneity within tumors poses a significant challenge to achieving consistent therapeutic efficacy. This challenge is further compounded by the blood-brain barrier (BBB), one of the most complex physiological structures, which restricts the entry of therapeutic agents into the brain and limits effective drug delivery to the target site.

Despite these barriers, the use of *in vitro*, *in vivo*, and *ex vivo* models offers substantial advantages for high-throughput screening of potential therapeutic compounds. These platforms enable molecular and cellular-level investigations, facilitate replication of experimental results, and accelerate the identification of promising drug candidates. Notably, patient-derived tumor tissues used in glioblastoma (GBM) research have demonstrated considerable variability across cellular, genomic, proteomic, and extracellular domains, underscoring the need for multiscale analytical techniques to effectively characterize tumor heterogeneity. To deepen insights into the brain tumor microenvironment and develop novel therapeutic strategies, researchers have established a diverse array of experimental models. Among these, patient-derived xenografts (PDXs) represent one of the most technically advanced systems, closely replicating the human disease phenotype. PDXs are particularly valuable for modeling both common and rare pediatric brain tumors, including embryonal tumors, medulloblastoma (MB), anaplastic ependymoma, and atypical teratoid rhabdoid tumors. For preclinical drug testing and therapeutic validation, patient-derived orthotopic xenograft (PDOX) models offer a promising platform to expedite the development of effective treatments for pediatric brain cancers. Emerging technologies such as digital twins—virtual replicas of biological systems hold the potential to generate highly precise models of brain tumors, enabling personalized treatment protocols and improving clinical outcomes. Similarly, the integration of artificial intelligence (AI) into neuro-oncology is in its nascent stages. However, it offers exciting possibilities for accurately simulating the tumor microenvironment, predicting drug responses, and optimizing treatment strategies. In parallel, organoid models are gaining traction as tools for medium-throughput drug screening, capable of detecting heterogeneous treatment responses that often only emerge during clinical trials. By enabling combinatorial experiments in a controlled 3D configuration, organoids may help overcome the feasibility limitations of *in vivo* studies. As brain cancer research continues to evolve, refining experimental models remains central to advancing our understanding of tumor biology and improving targeted therapeutic approaches. Through interdisciplinary innovation and technological integration, researchers aim to mimic the complex tumor microenvironment better and enhance treatment outcomes for patients affected by this devastating disease.

## ABBREVIATION

| | |
|---|---|
| **AI** | Artificial Intelligence |
| **BBB** | Blood-Brain Barrier |
| **CDK** | Cyclin-Dependent Kinase |
| **GBM** | Glioblastoma |
| **GEMMs** | Genetically Engineered Mouse Models |
| **PDX** | Patient-Derived Xenograft |
| **PDO** | Patient-Derived Organoid |
| **WHO** | World Health Organization |

## AUTHORS' CONTRIBUTIONS

Tanmoy Kanp: Writing-original draft, conceptualize, and analysis; Anish Dhuri: Writing-original draft, conceptualize, and analysis; Bharath M: Formal Analysis and Data Curation; Ujala Gupta: Investigation, Formal Analysis, and Data Curation; Khushi Rode: Data Curation; Prashant Tiwari: Formal analysis and Review & Editing; Neha Singh: Writing – Review & and editing; Pankaj Kumar Singh: Conceptualization, Supervision, Resources, Review & editing, Visualization.

## REFERENCES

[1] Antonica F, Santomaso L, Pernici D, *et al.* A slow-cycling/quiescent cells subpopulation is involved in glioma invasiveness. Nat Commun 2022; 13(1): 4767.
[http://dx.doi.org/10.1038/s41467-022-32448-0] [PMID: 35970913]

[2] Aldoghachi AF, Aldoghachi AF, Breyne K, Ling KH, Cheah PS. Recent advances in the therapeutic strategies of glioblastoma multiforme. Neuroscience 2022; 491: 240-70.
[http://dx.doi.org/10.1016/j.neuroscience.2022.03.030] [PMID: 35395355]

[3] Alibek K, Kakpenova A, Baiken Y. Role of infectious agents in the carcinogenesis of brain and head and neck cancers. Infect Agent Cancer 2013; 8(1): 7.
[http://dx.doi.org/10.1186/1750-9378-8-7] [PMID: 23374258]

[4] Paolillo M, Comincini S, Schinelli S. *In Vitro* Glioblastoma Models: A Journey into the Third Dimension. Cancers (Basel) 2021; 13(10): 2449.
[http://dx.doi.org/10.3390/cancers13102449] [PMID: 34070023]

[5] Desroches J, Jermyn M, Pinto M, *et al.* A new method using Raman spectroscopy for *in vivo* targeted brain cancer tissue biopsy. Sci Rep 2018; 8(1): 1792.
[http://dx.doi.org/10.1038/s41598-018-20233-3] [PMID: 29379121]

[6] Ribatti D. The chick embryo chorioallantoic membrane as an experimental model to study *in vivo* angiogenesis in glioblastoma multiforme. Brain Res Bull 2022; 182: 26-9.
[http://dx.doi.org/10.1016/j.brainresbull.2022.02.005] [PMID: 35143927]

[7] Lapointe S, Perry A, Butowski NA. Primary brain tumours in adults. Lancet 2018; 392(10145): 432-46.
[http://dx.doi.org/10.1016/S0140-6736(18)30990-5] [PMID: 30060998]

[8] Lovely M. Symptom management of brain tumor patients. Semin Oncol Nurs 2004; 20(4): 273-83.
[http://dx.doi.org/10.1016/S0749-2081(04)00091-9] [PMID: 15612603]

[9] Available from: https://www.cancerresearchuk.org/about-cancer/brain-tumours/types/primary-secondary-tumours#:~:text=Tumours%20that%20start%20in%20the,primary%20and%20secondary%20brain%20tumours

[10] Amin J, Sharif M, Haldorai A, Mussarat Y, Nayak R. Brain tumor detection and classification using machine learning: a comprehensive survey. Complex & intelligent systems, 2022. 8(4): p. 3161-3183.
[http://dx.doi.org/10.1007/s40747-021-00563-y]

[11] Alentorn A, Hoang-Xuan K, Mikkelsen T. Presenting signs and symptoms in brain tumors. Handb Clin Neurol 2016; 134: 19-26.
[http://dx.doi.org/10.1016/B978-0-12-802997-8.00002-5] [PMID: 26948346]

[12] Snyder H, Robinson K, Shah D, Brennan R, Handrigan M. Signs and symptoms of patients with brain tumors presenting to the emergency department. J Emerg Med 1993; 11(3): 253-8.
[http://dx.doi.org/10.1016/0736-4679(93)90042-6] [PMID: 8340578]

[13] Omuro AMP, Leite CC, Mokhtari K, Delattre JY. Pitfalls in the diagnosis of brain tumours. Lancet Neurol 2006; 5(11): 937-48.
[http://dx.doi.org/10.1016/S1474-4422(06)70597-X] [PMID: 17052661]

[14] Weinberg BD, Kuruva M, Shim H, Mullins ME. Clinical applications of magnetic resonance spectroscopy in brain tumors: from diagnosis to treatment. Radiol Clin North Am 2021; 59(3): 349-62.
[http://dx.doi.org/10.1016/j.rcl.2021.01.004] [PMID: 33926682]

[15] Abd-Ellah MK, Awad AI, Khalaf AAM, Hamed HFA. A review on brain tumor diagnosis from MRI images: Practical implications, key achievements, and lessons learned. Magn Reson Imaging 2019; 61: 300-18.
[http://dx.doi.org/10.1016/j.mri.2019.05.028] [PMID: 31173851]

[16] Butowski NA. Epidemiology and diagnosis of brain tumors. Continuum (Minneap Minn) 2015; 21(2): 301-13.
[http://dx.doi.org/10.1212/01.CON.0000464171.50638.fa] [PMID: 25837897]

[17] Xu F, Na L, Li Y, Chen L. RETRACTED ARTICLE: Roles of the PI3K/AKT/mTOR signalling pathways in neurodegenerative diseases and tumours. Cell Biosci 2020; 10(1): 54.
[http://dx.doi.org/10.1186/s13578-020-00416-0] [PMID: 32266056]

[18] Von Achenbach C, Weller M, Kaulich K, *et al.* Synergistic growth inhibition mediated by dual PI3K/mTOR pathway targeting and genetic or direct pharmacological AKT inhibition in human glioblastoma models. J Neurochem 2020; 153(4): 510-24.
[http://dx.doi.org/10.1111/jnc.14899] [PMID: 31618458]

[19] Courtney KD, Corcoran RB, Engelman JA. The PI3K pathway as drug target in human cancer. J Clin Oncol 2010; 28(6): 1075-83.
[http://dx.doi.org/10.1200/JCO.2009.25.3641] [PMID: 20085938]

[20] Vivanco I, Sawyers CL. The phosphatidylinositol 3-Kinase–AKT pathway in human cancer. Nat Rev Cancer 2002; 2(7): 489-501.
[http://dx.doi.org/10.1038/nrc839] [PMID: 12094235]

[21] Chalhoub N, Zhu G, Zhu X, Baker SJ. Cell type specificity of PI3K signaling in Pdk1- and Pten-deficient brains. Genes Dev 2009; 23(14): 1619-24.
[http://dx.doi.org/10.1101/gad.1799609] [PMID: 19605683]

[22] Mlakar V, Morel E, Mlakar SJ, Ansari M, Gumy-Pause F. A review of the biological and clinical implications of RAS-MAPK pathway alterations in neuroblastoma. J Exp Clin Cancer Res 2021; 40(1): 189.
[http://dx.doi.org/10.1186/s13046-021-01967-x] [PMID: 34103089]

[23] Bahar ME, Kim HJ, Kim DR. Targeting the RAS/RAF/MAPK pathway for cancer therapy: from mechanism to clinical studies. Signal Transduct Target Ther 2023; 8(1): 455.
[http://dx.doi.org/10.1038/s41392-023-01705-z] [PMID: 38105263]

[24] Du W, Searle J. The rb pathway and cancer therapeutics. Curr Drug Targets 2009; 10(7): 581-9.
[http://dx.doi.org/10.2174/138945009788680392] [PMID: 19601762]

[25] Morris EJ, Ji JY, Yang F, *et al.* E2F1 represses β-catenin transcription and is antagonized by both pRB and CDK8. Nature 2008; 455(7212): 552-6.
[http://dx.doi.org/10.1038/nature07310] [PMID: 18794899]

[26] He J, Allen JR, Collins VP, *et al.* CDK4 amplification is an alternative mechanism to p16 gene homozygous deletion in glioma cell lines. Cancer Res 1994; 54(22): 5804-7.
[PMID: 7954404]

[27] Hegi ME, Rajakannu P, Weller M. Epidermal growth factor receptor. Curr Opin Neurol 2012; 25(6): 774-9.
[http://dx.doi.org/10.1097/WCO.0b013e328359b0bc] [PMID: 23007009]

[28] Pai SG, Carneiro BA, Mota JM, *et al.* Wnt/beta-catenin pathway: modulating anticancer immune response. J Hematol Oncol 2017; 10(1): 101.
[http://dx.doi.org/10.1186/s13045-017-0471-6] [PMID: 28476164]

[29] Silva-García O, Valdez-Alarcón JJ, Baizabal-Aguirre VM. Wnt/β-catenin signaling as a molecular target by pathogenic bacteria. Front Immunol 2019; 10: 2135.
[http://dx.doi.org/10.3389/fimmu.2019.02135] [PMID: 31611869]

[30] Jing J, Wu Z, Wang J, *et al.* Hedgehog signaling in tissue homeostasis, cancers and targeted therapies. Signal Transduct Target Ther 2023; 8(1): 315.
[http://dx.doi.org/10.1038/s41392-023-01559-5] [PMID: 37596267]

[31] Reichrath J, Reichrath S, Eds. Fasoulakis Z, Daskalakis G, Theodora M. The relevance of notch signaling in cancer progression. Adv Exp Med Biol 2021; 1287: 169-81.
[http://dx.doi.org/10.1007/978-3-030-55031-8_11]

[32] Dhandapani M, Goldman A. Preclinical cancer models and biomarkers for drug development: New technologies and emerging tools. J Mol Biomark Diagn 2017; 8(5): 356.
[http://dx.doi.org/10.4172/2155-9929.1000356] [PMID: 29285415]

[33] Imamura Y, Mukohara T, Shimono Y, *et al.* Comparison of 2D- and 3D-culture models as drug-testing platforms in breast cancer. Oncol Rep 2015; 33(4): 1837-43.
[http://dx.doi.org/10.3892/or.2015.3767] [PMID: 25634491]

[34] Caragher S, Chalmers AJ, Gomez-Roman N. Glioblastoma's next top model: novel culture systems for brain cancer radiotherapy research. Cancers (Basel) 2019; 11(1): 44.
[http://dx.doi.org/10.3390/cancers11010044] [PMID: 30621226]

[35] Slika H, Karimov Z, Alimonti P, *et al.* Preclinical models and technologies in glioblastoma research: evolution, current state, and future avenues. Int J Mol Sci 2023; 24(22): 16316.
[http://dx.doi.org/10.3390/ijms242216316] [PMID: 38003507]

[36] Ruiz-Garcia H, Alvarado-Estrada K, Schiapparelli P, Quinones-Hinojosa A, Trifiletti DM. Engineering three-dimensional tumor models to study glioma cancer stem cells and tumor microenvironment. Front Cell Neurosci 2020; 14: 558381.
[http://dx.doi.org/10.3389/fncel.2020.558381] [PMID: 33177991]

[37] Griffith LG, Swartz MA. Capturing complex 3D tissue physiology *in vitro*. Nat Rev Mol Cell Biol 2006; 7(3): 211-24.
[http://dx.doi.org/10.1038/nrm1858] [PMID: 16496023]

[38] Hubert CG, Rivera M, Spangler LC, *et al.* A three-dimensional organoid culture system derived from human glioblastomas recapitulates the hypoxic gradients and cancer stem cell heterogeneity of tumors found *in vivo*. Cancer Res 2016; 76(8): 2465-77.
[http://dx.doi.org/10.1158/0008-5472.CAN-15-2402] [PMID: 26896279]

[39] Gomez-Roman N, Stevenson K, Gilmour L, Hamilton G, Chalmers AJ. A novel 3D human glioblastoma cell culture system for modeling drug and radiation responses. Neuro-oncol 2016; 19(2): now164.
[http://dx.doi.org/10.1093/neuonc/now164] [PMID: 27576873]

[40] Dubey S, Suraj MR, Goni T, *et al.* 3D QSAR studies of 3, 16 and 17 position modifications in steroidal derivatives for CNS anticancer activity. Curr Res Chem. 2023;15(1):1–4.
[http://dx.doi.org/10.3923/crc.2023.1.4]

[41] Lee J, Jo DH, Kim JH, *et al.* Development of a patient-derived xenograft model of glioblastoma *via* intravitreal injection in mice. Exp Mol Med 2019; 51(4): 1-9.
[http://dx.doi.org/10.1038/s12276-019-0241-3] [PMID: 30992429]

[42] Liu SJ, Malatesta M, Lien BV, *et al.* CRISPRi-based radiation modifier screen identifies long non-coding RNA therapeutic targets in glioma. Genome Biol 2020; 21(1): 83.
[http://dx.doi.org/10.1186/s13059-020-01995-4] [PMID: 32234056]

[43] Chen CC, Li HW, Wang YL, *et al.* Patient-derived tumor organoids as a platform of precision treatment for malignant brain tumors. Sci Rep 2022; 12(1): 16399.
[http://dx.doi.org/10.1038/s41598-022-20487-y] [PMID: 36180511]

[44] Tejero R, Huang Y, Katsyv I, *et al.* Gene signatures of quiescent glioblastoma cells reveal mesenchymal shift and interactions with niche microenvironment. EBioMedicine 2019; 42: 252-69.
[http://dx.doi.org/10.1016/j.ebiom.2019.03.064] [PMID: 30952620]

[45] Serrano M, Lee HW, Chin L, Cordon-Cardo C, Beach D, DePinho RA. Role of the INK4a locus in tumor suppression and cell mortality. Cell 1996; 85(1): 27-37.
[http://dx.doi.org/10.1016/S0092-8674(00)81079-X] [PMID: 8620534]

[46] Orsulic S. An RCAS-TVA-based approach to designer mouse models. Mamm Genome 2002; 13(10): 543-7.
[http://dx.doi.org/10.1007/s00335-002-4003-4] [PMID: 12420130]

[47] Von Werder A, Seidler B, Schmid RM, Schneider G, Saur D. Production of avian retroviruses and tissue-specific somatic retroviral gene transfer *in vivo* using the RCAS/TVA system. Nat Protoc 2012; 7(6): 1167-83.
[http://dx.doi.org/10.1038/nprot.2012.060] [PMID: 22635109]

[48] Tracey AT, Murray KS, Coleman JA, Kim K. Patient-derived xenograft models in urological malignancies: Urothelial cell carcinoma and renal cell carcinoma. Cancers (Basel) 2020; 12(2): 439.
[http://dx.doi.org/10.3390/cancers12020439] [PMID: 32069881]

[49] Abdolahi S, Ghazvinian Z, Muhammadnejad S, Saleh M, Asadzadeh Aghdaei H, Baghaei K. Patient-derived xenograft (PDX) models, applications and challenges in cancer research. J Transl Med 2022; 20(1): 206.
[http://dx.doi.org/10.1186/s12967-022-03405-8] [PMID: 35538576]

[50] Smith KS, Xu K, Mercer KS, *et al.* Patient-derived orthotopic xenografts of pediatric brain tumors: a St. Jude resource. Acta Neuropathol 2020; 140(2): 209-25.
[http://dx.doi.org/10.1007/s00401-020-02171-5] [PMID: 32519082]

[51] Funato K, Major T, Lewis PW, Allis CD, Tabar V. Use of human embryonic stem cells to model pediatric gliomas with H3.3K27M histone mutation. Science 2014; 346(6216): 1529-33.
[http://dx.doi.org/10.1126/science.1253799] [PMID: 25525250]

[52] Susanto E, Marin Navarro A, Zhou L, *et al.* Modeling SHH-driven medulloblastoma with patient iPS cell-derived neural stem cells. Proc Natl Acad Sci USA 2020; 117(33): 20127-38.
[http://dx.doi.org/10.1073/pnas.1920521117] [PMID: 32747535]

[53] Huszthy PC, Daphu I, Niclou SP, *et al. In vivo* models of primary brain tumors: Pitfalls and perspectives. Neuro-oncol 2012; 14(8): 979-93.
[http://dx.doi.org/10.1093/neuonc/nos135] [PMID: 22679124]

[54] Day CP, Merlino G, Van Dyke T. Preclinical mouse cancer models: a maze of opportunities and challenges. Cell 2015; 163(1): 39-53.
[http://dx.doi.org/10.1016/j.cell.2015.08.068] [PMID: 26406370]

[55] Barth RF, Kaur B. Rat brain tumor models in experimental neuro-oncology: the C6, 9L, T9, RG2, F98, BT4C, RT-2 and CNS-1 gliomas. J Neurooncol 2009; 94(3): 299-312.
[http://dx.doi.org/10.1007/s11060-009-9875-7] [PMID: 19381449]

[56] Aldape K, Brindle KM, Chesler L, *et al.* Challenges to curing primary brain tumours. Nat Rev Clin Oncol 2019; 16(8): 509-20.
[http://dx.doi.org/10.1038/s41571-019-0177-5] [PMID: 30733593]

[57] Woodworth GF, Dunn GP, Nance EA, Hanes J, Brem H. Emerging insights into barriers to effective brain tumor therapeutics. Front Oncol 2014; 4: 126.
[http://dx.doi.org/10.3389/fonc.2014.00126] [PMID: 25101239]

# CHAPTER 3

# Natural Bioactive Acting on Brain Tumor Microenvironment

**Deependra Soni**[1], **Renjil Joshi**[2], **Pankaj Kumar Singh**[3] and **Anshita Gupta**[4,*]

[1] *Faculty of Pharmacy, MATS University Campus, Aarang, Raipur, Chhattisgarh, India*

[2] *Raigarh College of Pharmacy, Raigarh, Chhattisgarh, India*

[3] *Department of Pharmaceutics, National Institute of Pharmaceutical Education and Research (NIPER), Hyderabad-500037, India*

[4] *Department of Pharmacy, Guru Ghasidas Vishwavidyalaya (A Central University), Koni, Bilaspur-495009, Chhattisgarh, India*

**Abstract:** The brain tumor microenvironment (TME) is a complex and dynamic niche composed of tumor cells, immune cells, blood vessels, extracellular matrix (ECM), and signaling molecules that interact to support tumor progression, angiogenesis, and immune evasion. Targeting the TME has emerged as a promising strategy for combating brain tumors, particularly gliomas, which remain challenging to treat with conventional therapies. In this context, natural bioactive compounds have gained significant attention due to their multi-targeting abilities, biocompatibility, and reduced side effects compared to synthetic drugs. These abstract reviews key natural bioactive compounds that modulate the brain tumor microenvironment, focusing on their mechanisms of action and potential therapeutic benefits. The potential for modulating the microenvironment of brain tumors has been highlighted by recent studies on natural bioactives, which are compounds found in plants, fungi, and marine organisms. These bioactives can influence key pathways involved in tumor growth, immune evasion, angiogenesis, and metastasis. This review focuses on the mechanisms by which natural bioactives, such as curcumin, resveratrol, epigallocatechin gallate (EGCG), and cannabinoids, exert their effects on the brain tumor microenvironment. It has been demonstrated that these compounds inhibit tumor growth by modulating oxidative stress, inflammation, and signaling pathways, including MAPK, PI3K, and NF-κB. Additionally, the review discusses the challenges and future directions in translating these findings from preclinical models to clinical applications. Bioactives and their interactions with the brain tumor microenvironment provide the basis for developing new therapeutic strategies which can improve patient outcomes and reduce the side effects associated with standard treatment.

* **Corresponding author Anshita Gupta:** Department of Pharmacy, Guru Ghasidas Vishwavidyalaya (A Central University), Koni, Bilaspur-495009, Chhattisgarh, India; E-mail: anshita1912@gmail.com

**Prashant Tiwari, Pankaj Kumar Singh & Sunil Kumar Kadiri (Eds.)**

**Keywords:** Brain tumor, Curcumin, Cannabinoids, EGCG, Inflammation, Microenvironment, Natural bioactive, Oxidative stress, Resveratrol, Signaling pathways.

## INTRODUCTION

Brain tumors, encompassing both primary and metastatic neoplasms, present significant therapeutic challenges due to their intricate and resilient microenvironment. A diverse mixture of immune cells, stromal cells, tumor cells, and extracellular matrix constituents comprises the tumor microenvironment. It contributes greatly to tumor progression, therapeutic resistance, and poor prognoses. Conventional treatment modalities, including surgery, radiation, and chemotherapy, often fall short due to the tumor's adaptive mechanisms and the protective nature of the Blood-Brain Barrier (BBB). Recent advances in cancer research have shifted focus towards the potential use of natural bioactive compounds—substances derived from plants, fungi, and marine organisms—as therapeutic agents [1]. These compounds, known for their diverse pharmacological properties, offer a promising avenue for targeting the brain tumor microenvironment. Unlike synthetic drugs, natural bioactives often possess multi-targeted mechanisms of action, enabling them to modulate various signaling pathways simultaneously. Tumour cells, stromal cells, immune cells, and extracellular matrix constituents make up the diverse mixture that is the tumour microenvironment [2]. There are a number of natural bioactives that have garnered a considerable amount of interest, including curcumin, resveratrol, epigallocatechin gallate (EGCG), and cannabinoids. Tumorigenesis is influenced by oxidative stress, inflammation, angiogenesis, and cell proliferation, all of which are mediated by bioactive compounds [3]. Additionally, they have shown potential in modulating the immune response and enhancing the permeability of the BBB, thereby improving drug delivery to the tumor site [4].

## KEY COMPONENTS OF THE BRAIN TUMOR MICROENVIRONMENT

Brain tumor microenvironments are highly complex and dynamic ecosystems that significantly influence tumor development, progression, and resistance to therapy. The development of effective therapeutic strategies requires a thorough understanding of its key components (Fig. **1**). The primary elements of the brain tumor microenvironment include:

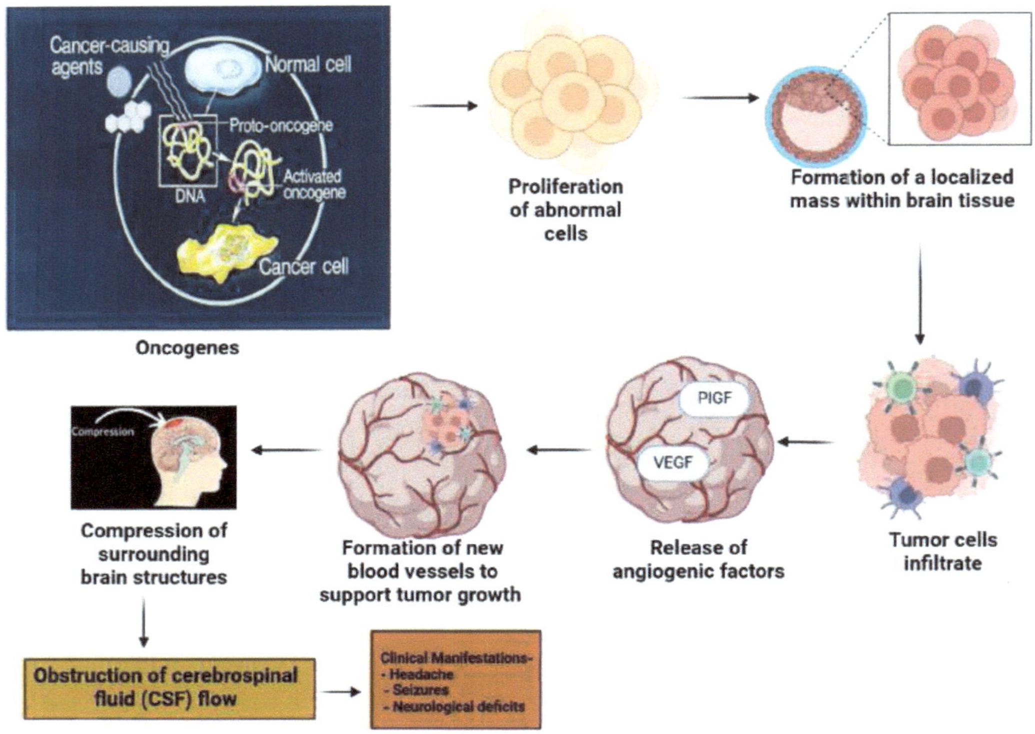


**Fig. (1).** Molecular pathology of progress of brain tumour for better understanding to develop effective therapeutic strategies.

## Tumor Cells

### *Heterogeneity*

Brain tumors, particularly gliomas, exhibit significant cellular heterogeneity, meaning they comprise diverse cell populations with varying genetic, molecular, and phenotypic characteristics. This heterogeneity contributes to the complexity and adaptability of the tumor, leading to varied responses to treatment and challenges in effectively targeting all tumor cells. It includes subpopulations such as cancer stem cells that possess high resistance to conventional therapies and are implicated in tumor recurrence. Treatment strategies for brain tumor patients must take into account and address this heterogeneity in order to be effective and personalized.

### *Stem-like Cells*

Within brain tumors, there is a subpopulation of cells known as tumor-initiating cells or cancer stem cells that exhibit stem-like characteristics and possess the ability to differentiate and self-renew. These cells are highly resilient and often resistant to conventional therapies such as chemotherapy and radiation, contributing to tumor recurrence and progression [5]. Their presence and unique properties make them a critical target for developing new therapeutic strategies aimed at achieving more durable and effective treatment outcomes for patients with brain tumors [6].

## Non-Tumor Cells

The brain tumor microenvironment comprises various non-tumor cells, including Glioma-Associated Microglia/Macrophages (GAMs), astrocytes, and endothelial cells. The growth of tumors is supported by the use of GAMs, which suppress the immune response [7]. Astrocytes, typically aiding neuronal function, can be recruited to facilitate tumor invasion. Angiogenesis is facilitated by endothelial cells, which create new blood vessels that provide the tumour with oxygen and nutrients. [8]. These non-tumor cells interact with tumor cells to create a supportive microenvironment, promoting tumor progression and resistance to therapy.

### *Astrocytes*

Neuronal function is supported and the blood-brain barrier is maintained by astrocytes, which play a significant role in the tumor's microenvironment. During tumor growth, astrocytes can cooperate with tumor invasion to facilitate the tumor's expansion [9]. Their secreted factors promote tumor cell survival, proliferation, and migration, thereby helping to create a protumorigenic niche. Additionally, astrocytes can modulate the immune response, contributing to an environment that protects tumor cells from immune surveillance and therapeutic interventions.

## Extracellular Matrix (ECM)

In the extracellular matrix (ECM), proteins, glycoproteins, and proteoglycans form a network that provides structural support to tissues. In brain tumors, the ECM is often remodeled to facilitate tumor growth and invasion [10]. Tumor cells interact with the ECM to migrate and invade surrounding tissues, aided by enzymes such as matrix metalloproteinases (MMPs) that degrade ECM components. This dynamic remodeling of the ECM not only supports tumor progression but also influences cellular signaling and the tumor microenvironment, making it a critical target for therapeutic strategies.

### *Composition*

The extracellular matrix (ECM), which provides cells with structural and metabolic support, is composed of a diverse array of proteins, glycoproteins, and proteoglycans. Key components include collagen, which provides tensile strength; fibronectin and laminin, which facilitate cell adhesion and migration; and proteoglycans, such as heparan sulfate, which modulate cell signaling. In brain tumors, the composition of the ECM is often altered to create a pro-tumorigenic environment, supporting tumor growth, invasion, and resistance to therapy [11].

### *Role in Invasion*

The extracellular matrix (ECM) plays a crucial role in tumor invasion by providing a scaffold that tumor cells can adhere to and migrate through. Tumor cells secrete enzymes like matrix metalloproteinases (MMPs) to degrade ECM components, facilitating their movement into surrounding tissues. This degradation process not only allows physical invasion but also releases bioactive molecules that promote tumor progression. By interacting with the ECM, tumor cells can navigate through the brain's complex structure, thereby enhancing their invasive capabilities and contributing to the tumor's spread.

## Signaling Molecules

A variety of signaling molecules play crucial roles in communication between tumor cells and non-tumor cells in the brain tumor microenvironment. These molecules include cytokines, chemokines, growth factors, and extracellular vesicles [12]. Cytokines and chemokines mediate immune responses, recruit immune cells, and promote inflammation or immune evasion. Growth factors, including Vascular Endothelial Growth Factor (VEGF) and Epidermal Growth Factor (EGF), promote the angiogenesis, proliferation, and survival of tumor cells [13]. Extracellular vesicles, including exosomes, transport bioactive molecules, such as proteins and nucleic acids, between cells, thereby influencing tumor behavior and the surrounding microenvironment. There is a complex interplay between these signaling molecules in the tumor microenvironment, which affects tumor growth, invasion, and response to therapy.

### *Cytokines and Chemokines*

Small signaling proteins known as chemokines and cytokines play pivotal roles in the microenvironment of brain tumors. Immune responses, inflammation, and cell proliferation within tumors are regulated by cytokines, including interleukins and Tumor Necrosis Factor (TNF). Based on the context, they may either promote or inhibit tumor growth. It is believed that chemokines, such as CXCL12 and CCL2, attract immune cells to the tumor site, thereby influencing immune cell infiltration and function. These molecules also facilitate cross-talk between tumor cells and stromal cells, contributing to tumor progression, angiogenesis, and therapeutic resistance [14]. Understanding the dynamic regulation of cytokines and chemokines in brain tumors is essential for developing targeted therapies that manipulate the tumor microenvironment to improve treatment outcomes [15].

### *Growth Factors*

Growth factors are important regulators that affect the angiogenesis, survival, and proliferation of tumor cells in the microenvironment of brain tumors. In order to support angiogenesis—the process of forming new blood vessels that supply oxygen and nutrients to tumors—Vascular Endothelial Growth Factor (VEGF) is essential [16]. Platelet-derived growth factor (PDGF) and epidermal growth factor (EGF) stimulate downstream signaling pathways linked to cell survival and proliferation, thereby promoting the development and migration of tumor cells. Tumour development and survival are also influenced by insulin-like growth factors, or IGFs [17]. Dysregulation of these growth factors can enhance tumor aggressiveness and resistance to therapies. Targeting these signaling pathways represents a promising approach for developing novel therapies to disrupt tumor growth and improve patient outcomes in brain tumors.

## Blood-Brain Barrier (BBB)

A specialist structure called the Blood-Brain Barrier (BBB) tightly regulates the movement of substances between the brain parenchyma and the circulation. It is composed of endothelial cells, pericytes, and astrocytes [18]. It protects the brain from toxins and pathogens while maintaining a stable environment for proper neuronal function. In brain tumors, particularly gliomas, the BBB is often disrupted, leading to increased permeability and leakage [19]. The distribution of therapeutic drugs to the tumour site is facilitated, and circulating tumour cells are allowed access into the body due to this altered blood-brain barrier. However, the BBB remains a significant challenge for drug delivery in brain tumor treatment, necessitating innovative strategies such as nanotechnology or focused ultrasound to enhance drug penetration while minimizing off-target effects. Understanding BBB dynamics is crucial for developing effective therapies that can effectively target brain tumors while preserving brain function [20].

### *Structure and Function*

By separating the circulation from the brain parenchyma, the highly specialised blood-brain barrier (BBB) protects the brain from outside agents and maintains the brain's homeostasis [21]. From a structural perspective, the capillaries lining the brain are surrounded by endothelial cells, which form the Blood-Brain Barrier (BBB). Tight junctions, which keep substances from going through, connect these cells. These endothelial cells are supported by astrocytes, which surround the blood vessels and contribute to the BBB's integrity through signaling and metabolic support [22]. Pericytes also play a role in regulating BBB permeability. Functionally, the BBB serves several critical roles:

### *Therapeutic Challenge*

Overcoming the Blood-Brain Barrier (BBB), which limits the transport of medications and therapeutic agents to the tumour site, is one of the main treatment hurdles in brain tumours [23]. Because the BBB hinders sufficient drug entry into the brain parenchyma, it reduces the effectiveness of targeted treatments and standard chemotherapy. To improve medication delivery while reducing systemic toxicity, techniques such as targeted ultrasound to temporarily break the blood-brain barrier and the development of BBB-penetrating medicines are being employed. Additionally, the heterogeneous nature of brain tumors, including cellular diversity and genetic mutations, contributes to treatment resistance and tumor recurrence, necessitating personalized treatment approaches. Overcoming these challenges requires innovative therapeutic strategies that effectively penetrate the BBB, target specific tumor subpopulations, and address resistance mechanisms to improve outcomes for patients with brain tumors.

## Hypoxia

A typical characteristic of the microenvironment of solid tumours, particularly brain tumours, is hypoxia, or low oxygen tension. Rapid tumour growth frequently outpaces the formation of new blood vessels in brain tumours, leaving certain areas of the tumour without enough oxygen supply. Hypoxia-Inducible Factors (HIFs) are transcription factors that control the expression of genes involved in low-oxygen adaption. Hypoxia activates HIFs.

In response to hypoxia, tumor cells undergo metabolic reprogramming to enhance survival and proliferation. This includes resistance to apoptosis, increased glycolysis, and angiogenesis (due to the overexpression of pro-angiogenic proteins like VEGF). Brain tumours that include hypoxic areas are more likely to be aggressive, invasive, and resistant to treatment [24]. Targeting hypoxia-induced pathways represents a promising therapeutic strategy to improve treatment outcomes by disrupting tumor adaptation mechanisms and enhancing the efficacy of conventional therapies in brain tumors [25].

### *Oxygen Deprivation*

Oxygen deprivation, or hypoxia, is a critical factor in the microenvironment of brain tumors where insufficient oxygen levels occur due to rapid tumor growth and inadequate blood supply. Hypoxia triggers a cascade of adaptive responses in tumor cells, primarily mediated by Hypoxia-Inducible Factors (HIFs). The expression of genes related to angiogenesis, glycolysis, pH control, and cell survival is regulated by these factors [26].

In brain tumors, regions of oxygen deprivation contribute to tumor aggressiveness, therapy resistance, and poor patient prognosis. Tumor cells adapt to low oxygen conditions by altering their metabolism to favor glycolysis over oxidative phosphorylation, which promotes the release of angiogenic factors, such as Vascular Endothelial Growth Factor (VEGF), and enhances their ability to invade surrounding tissues. Understanding and targeting hypoxia-related pathways are essential for developing effective therapies that can mitigate tumor progression and improve outcomes for patients with brain tumors [2].

### *Adaptive Responses*

Brain tumors, facing various challenges such as hypoxia, nutrient deprivation, and therapeutic pressures, exhibit several adaptive responses crucial for survival and progression. These responses include metabolic reprogramming, where tumor cells switch to glycolysis even in the presence of oxygen (the Warburg effect), ensuring energy production and biosynthesis under adverse conditions. Additionally, tumors promote angiogenesis through the upregulation of pro-angiogenic factors, such as VEGF, to sustain a nutrient and oxygen supply. They also enhance DNA repair mechanisms and activate anti-apoptotic pathways, conferring resistance to chemotherapy and radiation. Tumor cells also engage in interactions with the microenvironment in an effort to inhibit anti-tumor immune responses and evade immune surveillance, which further the process of immune evasion and treatment resistance. Comprehending these adaptive reactions is crucial for creating tailored treatments that can successfully impede tumor expansion and enhance patient outcomes in cases of brain tumors.

## NATURAL BIOACTIVES AND THEIR MECHANISMS OF ACTION

Natural bioactive compounds, such as curcumin, resveratrol, EGCG, quercetin, ginsenosides, and caffeic acid, exhibit promising mechanisms of action against brain tumors [27]. Curcumin, derived from turmeric, inhibits inflammatory and angiogenic pathways by targeting NF-kB, VEGF, and MMPs, thereby reducing tumor growth and invasion [28]. Resveratrol activates SIRT1 and inhibits NF-kB to provide neuroprotective and anti-inflammatory effects. It is present in red wine and grapes. EGCG from green tea functions as an anti-oxidant and anti-angiogenic agent, scavenging Reactive Oxygen Species (ROS) and inhibiting VEGF and MMPs [29] (Table **1**). Quercetin, present in many fruits and vegetables, also acts as an anti-oxidant and anti-inflammatory compound, targeting the PI3K/Akt pathway to prevent tumor progression [30]. Ginsenosides from ginseng exhibit anti-proliferative and pro-apoptotic effects by modulating the MAPK signaling pathway. Lastly, caffeic acid, found in coffee and various plants, enhances neuroprotection and modulates blood-brain barrier permeability, facilitating drug delivery and exerting anti-inflammatory effects [31] (Table **2**). These bioactives collectively offer multifaceted approaches to targeting the complex pathophysiology of brain tumors, making them valuable candidates for further research and potential therapeutic development (Table **3**).

**Table 1. List of Natural bioactive, describing their biological sources, and chemical constituents with their potential role as health benefits.**

| Bioactive Compound | Source | Biological Name | Family | Key Chemical Constituents | Mechanisms of Action | Health Benefits | Refs. |
|---|---|---|---|---|---|---|---|
| Curcumin | Turmeric | Curcuma longa | Zingiberaceae | Curcuminoids (Curcumin, Demethoxycurcumin, Bisdemethoxycurcumin) | - Inhibits NF-κB pathway. Scavenges free radicals Induces apoptosis Arrest cell cycle Inhibits PI3K/Akt pathway - Reduces VEGF | - Anti-inflammatory Anti-oxidant Anti-cancer - Neuroprotective | [32] |
| Resveratrol | Grapes, Red Wine | Vitis vinifera | Vitaceae | Stilbenoids (Resveratrol, Piceid) | - Activates SIRT1 Inhibits NF-κB Modulates AMPK Inhibits COX-2 Induces apoptosis | - Cardioprotective Anti-cancer - Anti-inflammatory Anti-aging | [33] |
| Epigallocatechin Gallate (EGCG) | Green Tea | Camellia sinensis | Theaceae | Catechins (EGCG, EGC, ECG, EC) | - Inhibits NF-κB Activates AMPK Induces apoptosis Inhibits VEGF Scavenges free radicals | - Anti-oxidant Anti-cancer Cardioprotective Anti-inflammatory | [34] |

*(Table 1) cont.....*

| Bioactive Compound | Source | Biological Name | Family | Key Chemical Constituents | Mechanisms of Action | Health Benefits | Refs. |
|---|---|---|---|---|---|---|---|
| Quercetin | Onions, Apples | Allium cepa | Amaryllidaceae | Flavonoids (Quercetin, Rutin) | - Inhibits PI3K/Akt pathway - Scavenges free radicals Induces apoptosis Inhibits VEGF Modulates MAPK pathway | - Anti-oxidant Anti-inflammatory Anti-cancer - Antiviral | [35] |
| Genistein | Soybeans | Glycine max | Fabaceae | Isoflavones (Genistein, Daidzein) | - Inhibits tyrosine kinases Modulates estrogen receptors Induces apoptosis Inhibits NF-κB Inhibits angiogenesis | - Anti-cancer - Cardioprotective Bone health | [36] |
| Lycopene | Tomatoes | Solanum lycopersicum | Solanaceae | Carotenoids (Lycopene, Phytoene, Phytofluene) | - Scavenges free radicals Inhibits NF-κB Modulates IGF signaling - Induces apoptosis | - Anti-oxidant Cardioprotective Anti-cancer - Skin protection | [37] |
| Berberine | Goldenseal, Barberry | Berberis vulgaris | Berberidaceae | Isoquinoline Alkaloids (Berberine, Palmatine) | - Activates AMPK Inhibits NF-κB Modulates insulin signaling Induces apoptosis Antimicrobial properties | - Antidiabetic Anti-cancer Antimicrobial Cardioprotective | [38] |
| Gingerol | Ginger | Zingiber officinale | Zingiberaceae | Phenolic Compounds (Gingerol, Shogaol, Paradol) | - Inhibits NF-κB Scavenges free radicals Modulates COX-2 and LOX Induces apoptosis Anti-nausea properties | - Anti-inflammatory Anti-oxidant Anti-cancer Digestive health | [39] |

*(Table 1) cont.....*

| Bioactive Compound | Source | Biological Name | Family | Key Chemical Constituents | Mechanisms of Action | Health Benefits | Refs. |
|---|---|---|---|---|---|---|---|
| Alicin | Garlic | Allium sativum | Amaryllidaceae | Organosulfur Compounds (Allicin, Diallyl Disulfide) | - Inhibits NF-κB Antimicrobial properties Scavenges free radicals Induces apoptosis Reduces blood pressure | - Antimicrobial Cardioprotective Anti-cancer Immune boosting | [40] |

**Table 2. Patents related to natural bioactives acting on the brain tumor microenvironment.**

| Patent Number | Title | Publication Date | Key Bioactives | Mechanism of Action | Inventors/Assignee |
|---|---|---|---|---|---|
| US20210056789 | Nanoparticles of curcumin and resveratrol for brain tumors | 2021-02-25 | Curcumin, Resveratrol | Modulates immune cells, inhibits tumor proliferation | John Smith *et al.* / XYZ Biotech |
| US20220234567 | Herbal extract combination for glioblastoma treatment | 2022-07-15 | Ginsenoside, Quercetin | Induces apoptosis, inhibits angiogenesis | Li Wei *et al.* / ABC Pharmaceuticals |
| WO2019034567 | Formulation of plant-derived compounds for brain tumor therapy | 2019-03-14 | Berberine, Epigallocatechin gallate | Modulates tumor microenvironment, inhibits invasion and metastasis | Maria Garcia *et al.* / Green Health Inc. |
| EP3567890 | Natural extract compositions for enhancing immune response in glioma | 2020-10-30 | Polysaccharides, Flavonoids | Enhances immune cell activity, reduces tumor immune evasion | Hans Muller *et al.* / Natural Cure GmbH |
| CN110345678 | Use of turmeric and ginger compounds in brain cancer treatment | 2019-11-05 | Curcumin, Gingerol | Reduces inflammation, modulates macrophage activity | Zhang Wei *et al.* / Beiji |

**Table 3. Recent Patents granted related to Natural bioactive acting on brain tumor microenvironment.**

| Patent Number | Year | Title | Natural Bioactive | Mechanism of Action | Application |
|---|---|---|---|---|---|
| US10940118B2 | 2021 | "Curcumin nanoformulations for brain cancer treatment" | Curcumin | Modulates inflammatory pathways, reduces oxidative stress, enhances apoptosis in tumor cells | Brain cancer therapy |
| US10688155B2 | 2020 | "Natural compounds for modulating tumor microenvironment" | Resveratrol, EGCG | Anti-angiogenic effects, immune modulation, inhibition of tumor cell proliferation | Brain metastases treatment |
| US10869834B2 | 2021 | "Alkaloid-based therapies for brain tumors" | Berberine, Matrine | Alters tumor-associated macrophages, reduces inflammatory cytokines, enhances immune response. | Glioblastoma treatment |
| US11241409B2 | 2022 | "Herbal extracts for brain cancer" | Ginsenosides, Luteolin | Modulates the TME by affecting macrophage polarization and reducing hypoxia. | Glioma therapy |
| US11344572B2 | 2023 | "Phytochemical formulations targeting brain tumors" | Quercetin, Kaempferol | Inhibits tumor growth by modulating signaling pathways, enhancing immune surveillance. | Brain tumor management |
| US11556911B2 | 2024 | "Bioactive peptides for brain tumor treatment" | Bovine lactoferrin-derived peptides | Disrupts tumor cell communication and reduces the immunosuppressive microenvironment. | Brain metastasis treatment |

**Table 4. Summary of recent clinical trials related to natural bioactive compounds acting on the brain tumor microenvironment.**

| Trial Title | Bioactive Compound | Clinical Phase | Study Focus | Year | Identifier |
|---|---|---|---|---|---|
| Efficacy of Curcumin in Glioblastoma Patients | Curcumin | Phase II | Evaluating curcumin's effect on glioblastoma multiforme progression | 2022 | NCT03015344 |

*(Table 4) cont.....*

| Trial Title | Bioactive Compound | Clinical Phase | Study Focus | Year | Identifier |
|---|---|---|---|---|---|
| Combination Therapy with Resveratrol for Brain Tumors | Resveratrol | Phase I/II | Assessing safety and efficacy of resveratrol combined with standard chemotherapy | 2020 | NCT02680085 |
| Anti-tumor Effects of EGCG in Brain Cancer | Epigallocatechin Gallate (EGCG) | Phase I | Studying EGCG's impact on brain tumor cells' viability and patient outcomes | 2021 | NCT04003195 |
| Evaluating Boswellia Serrata Extract in Glioma Patients | Boswellic acids | Phase II | Investigating anti-inflammatory and anti-cancer effects in glioma | 2019 | NCT02963684 |
| Clinical Trial of Quercetin in Brain Tumor Therapy | Quercetin | Phase I | Determining the safety and preliminary efficacy of quercetin in brain tumor treatment | 2018 | NCT03476330 |

**Table 5. Natural Bioactive containing Marketed preparation available for used in brain tumor.**

| Natural Bioactive Compound | Marketed Preparation | Application in Brain Tumor | Mechanism of Action | Clinical Status |
|---|---|---|---|---|
| Curcumin | Curcumin C3 Complex®, BCM-95® | Anti-inflammatory, Anti-oxidant, Anti-angiogenic | Inhibition of NF-kB, VEGF, MMPs | Research ongoing; clinical trials |
| Resveratrol | Resveratrol dietary supplements | Neuroprotective, Anti-inflammatory | Activation of SIRT1, NF-kB inhibition | Preclinical and clinical studies |
| EGCG (Epigallocatechin gallate) | Green tea extracts (*e.g.*, Polyphenon E®) | Anti-oxidant, Anti-angiogenic, Neuroprotective | ROS scavenging, Inhibition of VEGF, MMPs | Clinical trials ongoing |
| Quercetin | Quercetin supplements | Anti-oxidant, Anti-inflammatory, Anti-angiogenic | ROS scavenging, Inhibition of PI3K/Akt pathway | Preclinical studies |
| Ginsenosides | Ginseng extracts | Anti-proliferative, Apoptotic | Inhibition of MAPK signaling, Induction of apoptosis | Preclinical studies |
| Caffeic Acid | Coffee extracts, dietary supplements | Neuroprotective, Anti-inflammatory | Modulation of BBB permeability, Anti-oxidant | Preclinical studies |

Recent clinical investigations have increasingly concentrated on the potential of natural bioactive compounds in addressing the Brain Tumor Microenvironment (BTM). Natural bioactive compounds, sourced from plants and other natural origins, are being investigated for their potential to alter the BTM and enhance therapeutic effectiveness. A notable focus of this research is curcumin, the active ingredient found in turmeric. Clinical studies have demonstrated that curcumin can inhibit the growth of glioblastoma cells and improve the efficacy of chemotherapy. It does so by blocking various signaling pathways that contribute to tumor survival and promoting apoptosis. Research has indicated that resveratrol can inhibit the migration and invasion of glioma cells. Its mechanism appears to involve the modification of the tumor microenvironment by changing the composition of the extracellular matrix and hindering angiogenesis, the process of new blood vessel formation that is essential for tumor growth and dissemination (Table **4**). Furthermore, the flavonoid quercetin has attracted interest for its potential to improve the effectiveness of radiation therapy. Recent studies have shown that quercetin can sensitize tumor cells to radiation, thereby increasing cell death rates while simultaneously safeguarding normal cells. This dual effect is particularly advantageous in treating gliomas, where it is crucial to protect healthy tissue while effectively targeting malignant cells. Additionally, research has investigated the effects of omega-3 fatty acids, particularly docosahexaenoic acid (DHA), on brain tumor management. Clinical trials indicate that DHA may modify the inflammatory response within the brain tumor microenvironment, potentially slowing tumor progression. By influencing immune cell activity in this context, omega-3 fatty acids may offer supplementary benefits when combined with conventional therapies (Table **5**).

## CHALLENGES AND FUTURE DIRECTIONS

Recent developments in drug discovery have made it possible to develop tailored medicines for various illnesses, including brain tumors. The intricate landscape of brain tumors is a significant obstacle to the efficient targeting and management of these potent cancers. Tumour development, invasion, and resistance to treatment are facilitated by the protective niche created through the dynamic interactions between immune cells, the extracellular matrix, and tumour cells. Comprehending the complex molecular mechanisms underlying the interplay between tumor cells and the microenvironment is essential for the discovery of new bioactive substances that can specifically target tumor-promoting signals while preserving healthy brain tissue. In this context, overcoming the challenges in drug discovery for bioactives targeting the brain tumor microenvironment requires a multidisciplinary approach that integrates cutting-edge technologies with in-depth knowledge of tumor biology and drug development strategies. Several challenges hindering drug delivery have been discussed here.

Bioavailability, metabolism, and permeability of BBB are crucial factors in the clinical feasibility of natural compounds in the treatment of glioma. These properties vary significantly between substances and, in many cases, must be exceeded to achieve sufficient clinical and *in vivo* concentrations [41]. Oral bioavailability is a measure of the ability of a compound to reach the systemic circulation upon ingestion and is deficient in various natural compounds [42]. Curcumin, for example, has limited oral bioavailability due to its poor absorption and rapid first-pass metabolism. Therefore, although increasing evidence shows that curcumin is capable of crossing the BBB, its limited solubility in water, hydrophobic nature, and physico-chemical stability reduce its bioavailability [43]. Thus RES, despite its promising potential as a therapeutic agent, faces challenges in its administration due to various limitations. RES exhibits low bioavailability due to its rapid metabolism and clearance, short biological half-life, poor water solubility, chemical instability, and high photosensitivity. RES once reaches the bloodstream is quickly absorbed; it is metabolized in the liver and is rapidly eliminated [44]. Furthermore, the healthful effects of EGCG are hindered by its degradation in biological fluids, or during the processing and storage of tea. This degradation happens in two main reactions: auto-oxidation and epimerization. Additionally, it has been reported that the oral bioavailability of EGCG is low, both in animal models and in humans, due to its high molecular weight and metabolic alterations [45, 46]. Therefore, the low bioavailability hinders the accumulation of these compounds in target tissues at the concentrations necessary for therapeutic success. Consequently, because the success of a therapy for CNS disorders largely depends on the bioavailability of the therapeutic agent, new therapeutic approaches are necessary. Currently, considerable efforts are being devoted to the development of new delivery systems to overcome these limits, and at the same time to enhance the therapeutic efficacy of natural compounds. There are several types of materials that have been used for the encapsulation of natural compounds and the increasing of their activity [47]. Some of the proposed systems possess advantageous characteristics, including bioavailability, improved stability, and biocompatibility. For example, silica nanoparticles (NPs) significantly enhance the biological activity and carrying capacity of the RES [48]. Several publications have examined the effectiveness of RES delivery systems in treating glioblastoma. For example, Figueiro *et al.* investigated the anti-tumor effect of RES-loaded lipid core nano capsules against C6 glioma cells in both *in vitro* and *in vivo* models. *In vitro*, RES-loaded NPs significantly limited the viability of cancer cells compared to free RES, particularly upon induction of cell death by apoptosis. *In vivo*, treatment with the nano-complex resulted in a significant reduction in tumor size compared to free RES [49]. For example, when loaded into casein NPs, the oral availability of RES increased by up to 10 times [50]. Curcumin nano formulated with Poly-Lactic-co-Glycolic Acid (PLGA) in a mouse model showed three- to four-times greater longevity and concentration in CNS tissue at similar oral doses compared to standard curcumin preparations [51].

Liposomes have been used as drug delivery systems for decades. Liposomal curcumin enhanced the bioavailability of curcumin. Curcumin micelles are another form of higher bioavailability of curcumin, approximately 19-times greater than standard preparation curcumin [52]. Many studies have shown that the encapsulation of EGCG within a polymer matrix, such as chitosan-based nanocarriers, enhances its pharmacokinetics and pharmacodynamics [53]. Thus, the potential of EGCG in clinical applications is improved by these technologies.

## CONCLUSION

In summary, the development of natural bioactives targeting the brain tumor microenvironment presents a variety of challenges that require creative solutions. Developing effective treatment drugs is hindered by the complexity of the blood-brain barrier, the heterogeneity of brain tumors, and the dynamic nature of the tumor microenvironment. Despite advances in technologies such as high-throughput screening and computational modeling, translation to clinical applications remains limited. Another significant obstacle to medication development is the lack of suitable preclinical models that accurately replicate the tumor microenvironment. To overcome these obstacles in the future, researchers, physicians, and pharmaceutical corporations must work together. Modern methods, such as organoid cultures and genetic profiling, can be used to help create targeted medicines and provide important insights into the biology of brain tumors. Addressing these obstacles through interdisciplinary approaches and innovative strategies will be crucial in advancing drug discovery for natural bioactives targeting the brain tumor microenvironment.

In tackling the challenges faced in discovering natural bioactives that target the brain tumor microenvironment, it is imperative to address key obstacles, including identifying suitable drug targets, overcoming delivery limitations, and dealing with tumor heterogeneity. One potential solution lies in leveraging advanced technologies, such as genomics and proteomics, to identify specific molecular targets within the tumor microenvironment that can be exploited for drug development. Using nanotechnology to improve medication distribution across the blood-brain barrier, enhance effectiveness, and reduce systemic toxicity is another intriguing strategy. Additionally, incorporating patient-specific tumor profiling and personalized medicine strategies can help circumvent issues related to tumor heterogeneity and optimize treatment outcomes. By incorporating these innovative strategies and technologies, researchers can overcome the complexities of drug discovery for natural bioactives targeting the brain tumor microenvironment, paving the way for more effective therapeutic interventions.

## AUTHORS' CONTRIBUTIONS

DEEPENDRA SONI 1: Writing – original draft, Investigation, Resources. Formal analysis, RENJIL JOSHI 2: Data curation, Writing – review & PANKAJ KUMAR SINGH 3: Conceptualization, Supervision Resources; ANSHITA GUPTA SONI 4: Editing, Investigation Visualization.

## REFERENCES


[1] Shakya AK, Naik RR. The chemotherapeutic potentials of compounds isolated from the plant, marine, fungus, and microorganism: Their mechanism of action and prospects. 2022. [http://dx.doi.org/10.1155/2022/5919453]

[2] Barouki R, Kogevinas M, Audouze K, *et al.* The COVID-19 pandemic and global environmental change: Emerging research needs. Environ Int 2021; 146: 106272. [http://dx.doi.org/10.1016/j.envint.2020.106272] [PMID: 33238229]

[3] Adinew GM, Taka E, Mendonca P, Messeha SS, Soliman KF. The anticancer effects of flavonoids through miRNAs modulations in triple-negative breast cancer. Nutrients. 2021; 13(4): 1212.

[4] Zhou Y, Chen X, Cao J, Gao H. Overcoming the biological barriers in the tumor microenvironment for improving drug delivery and efficacy. J Mater Chem B Mater Biol Med 2020; 8(31): 6765-81. Available from: https://pubs.rsc.org/en/content/articlehtml/2020/tb/d0tb00649a [http://dx.doi.org/10.1039/D0TB00649A] [PMID: 32315375]

[5] Singh B, Dutt S. Role of cancer stem cells in maintenance of tumor heterogeneity in brain tumors. Cancer Stem Cells Signal Pathways. 2024 Jan 1; 165–84. [http://dx.doi.org/10.1016/B978-0-443-13212-4.00008-8]

[6] Bhargav AG, Mondal SK, Garcia CA, Green JJ, Quiñones-Hinojosa A. Nanomedicine revisited: Next generation therapies for brain cancer. Adv Ther (Weinh) 2020; 3(10): 2000118. Available from: https://onlinelibrary.wiley.com/doi/full/10.1002/adtp.202000118 [http://dx.doi.org/10.1002/adtp.202000118]

[7] Roberts N. Repurposing oxaliplatin for the treatment of glioblastoma. 2019 [cited 2024 Jun 29]; Available from: https://archive.hshsl.umaryland.edu/handle/10713/8989

[8] Vallon M, Chang J, Zhang H, Kuo CJ. Developmental and pathological angiogenesis in the central nervous system. Cell Mol Life Sci 2014; 71(18): 3489-506. Available from: https://link.springer.com/article/10.1007/s00018-014-1625-0 [http://dx.doi.org/10.1007/s00018-014-1625-0] [PMID: 24760128]

[9] Placone AL, Quiñones-Hinojosa A, Searson PC. The role of astrocytes in the progression of brain cancer: complicating the picture of the tumor microenvironment. Tumour Biol 2016; 37(1): 61-9. Available from: https://link.springer.com/article/10.1007/s13277-015-4242-0 [http://dx.doi.org/10.1007/s13277-015-4242-0] [PMID: 26493995]

[10] Zimmermann DR, Dours-Zimmermann MT. Extracellular matrix of the central nervous system: from neglect to challenge. Histochem Cell Biol 2008; 130(4): 635-53. [http://dx.doi.org/10.1007/s00418-008-0485-9] [PMID: 18696101]

[11] Faisal SM, Comba A, Varela ML, *et al.* The complex interactions between the cellular and non-cellular components of the brain tumor microenvironmental landscape and their therapeutic implications. Front Oncol 2022; 12: 1005069. [http://dx.doi.org/10.3389/fonc.2022.1005069] [PMID: 36276147]

[12] Pancholi S, Tripathi A, Bhan A, Acharya MM, Pillai P. Emerging Concepts on the role of extracellular vesicles and its cargo contents in glioblastoma-microglial cross-talk 2022. Available from: https://link.springer.com/article/10.1007/s12035-022-02752-3

[13] Chimal-Ramírez GK, Espinoza-Sánchez NA, Fuentes-Pananá EM. Protumor activities of the immune response: insights in the mechanisms of immunological shift, oncotraining, and oncopromotion. J Oncol 2013; 2013(1): 1-16. Available from: https://onlinelibrary.wiley.com/doi/full/10.1155/2013/835956 [http://dx.doi.org/10.1155/2013/835956] [PMID: 23577028]

[14] Singh AJ, Gray JW. Chemokine signaling in cancer-stroma communications. J Cell Commun Signal 2021; 15(3): 361-81. [http://dx.doi.org/10.1007/s12079-021-00621-7] [PMID: 34086259]

[15] Nengroo MA, Verma A, Datta D. Cytokine chemokine network in tumor microenvironment: Impact on CSC properties and therapeutic applications. Cytokine 2022; 156: 155916. [http://dx.doi.org/10.1016/j.cyto.2022.155916] [PMID: 35644058]

[16] Mittal K, Ebos J, Rini B. Angiogenesis and the tumor microenvironment: vascular endothelial growth factor and beyond. Semin Oncol 2014; 41(2): 235-51. [http://dx.doi.org/10.1053/j.seminoncol.2014.02.007] [PMID: 24787295]

[17] Deuel TF, Kawahara RS, Mustoe TA, Pierce GF. Growth factors and wound healing: Platelet-derived growth factor as a model cytokine. Annu Rev Med [Internet]. 1991 Feb 1 [cited 2024 Jun 29];42(Volume 42, 1991):567–84. Available from: https://www.annualreviews.org/content/journals/10.1146/annurev.me.42.020191.003031

[18] Correale J, Villa A. Cellular elements of the blood-brain barrier. Neurochem Res 2009; 34(12): 2067-77. Available from: https://link.springer.com/article/10.1007/s11064-009-0081-y [http://dx.doi.org/10.1007/s11064-009-0081-y] [PMID: 19856206]

[19] Upadhyay RK. Drug delivery systems, CNS protection, and the blood brain barrier. BioMed Res Int 2014; 2014(1): 1-37. Available from: https://onlinelibrary.wiley.com/doi/full/10.1155/2014/869269 [http://dx.doi.org/10.1155/2014/869269] [PMID: 25136634]

[20] Fortin D. The blood-brain barrier: its influence in the treatment of brain tumors metastases. Curr Cancer Drug Targets 2012; 12(3): 247-59. Available from: https://www.eurekaselect.com/article/21402 [http://dx.doi.org/10.2174/156800912799277511] [PMID: 22229251]

[21] Alahmari A. Blood-brain barrier overview: Structural and functional correlation. Neural Plast 2021; 2021(1): 1-10. Available from: https://onlinelibrary.wiley.com/doi/full/10.1155/2021/6564585 [http://dx.doi.org/10.1155/2021/6564585] [PMID: 34912450]

[22] Cardoso FL, Brites D, Brito MA. Looking at the blood–brain barrier: Molecular anatomy and possible investigation approaches. Brain Res Brain Res Rev 2010; 64(2): 328-63. [http://dx.doi.org/10.1016/j.brainresrev.2010.05.003] [PMID: 20685221]

[23] Parrish KE, Sarkaria JN, Elmquist WF. Improving drug delivery to primary and metastatic brain tumors: Strategies to overcome the blood–brain barrier. Clin Pharmacol Ther 2015; 97(4): 336-46. Available from: https://onlinelibrary.wiley.com/doi/full/10.1002/cpt.71 [http://dx.doi.org/10.1002/cpt.71] [PMID: 25669487]

[24] Roy S, Kumaravel S, Sharma A, Duran CL, Bayless KJ, Chakraborty S. Hypoxic tumor microenvironment: Implications for cancer therapy. https://doi.org/101177/1535370220934038 [Internet]. 2020 Jun 27 [cited 2024 Jun 29]; 245(13): 1073–86. Available from: https://journals.sagepub.com/doi/10.1177/1535370220934038

[25] Chen Z, Han F, Du Y, Shi H, Zhou W. Hypoxic microenvironment in cancer: molecular mechanisms and therapeutic interventions. [cited 2024 Jun 29]; [http://dx.doi.org/10.1038/s41392-023-01332-8]

[26] Brahimi-Horn MC, Pouysségur J. Hypoxia in cancer cell metabolism and pH regulation. Essays Biochem [Internet]. 2007 Aug 10 [cited 2024 Jun 29];43:165–78. Available from: /essaysbiochem/article/doi/10.1042/bse0430165/88112/Hypoxia-in-cancer-cell-metabolism-and-pH

[27] Marino A, Battaglini M, Moles N, Ciofani G. Natural antioxidant compounds as potential pharmaceutical tools against neurodegenerative diseases. ACS Omega 2022; 7(30): 25974-90. Available from: https://pubs.acs.org/doi/full/10.1021/acsomega.2c03291 [http://dx.doi.org/10.1021/acsomega.2c03291] [PMID: 35936442]

[28] Mohankumar K, Francis AP, Pajaniradje S, Rajagopalan R. Synthetic curcumin analog: inhibiting the invasion, angiogenesis, and metastasis in human laryngeal carcinoma cells *via* NF-kB pathway. Mol Biol Rep 2021; 48(8): 6065-74. Available from: https://link.springer.com/article/10.1007/s11033-02--06610-8 [http://dx.doi.org/10.1007/s11033-021-06610-8] [PMID: 34355287]

[29] Sunkaria A, Yadav A, Sharma SK, Sandhir R. Molecular mechanisms involved in the neuroprotective action of phytochemicals. Neuroprotective Eff Phytochem Neurol Disord [Internet]. 2017 Jan 2 [cited 2024 Jun 29];515–59. Available from: https://onlinelibrary.wiley.com/doi/full/10.1002/9781119155195.ch27 [http://dx.doi.org/10.1002/9781119155195.ch27]

[30] Zughaibi TA, Suhail M, Tarique M, Tabrez S. Targeting PI3K/Akt/mTOR pathway by different flavonoids: A cancer chemopreventive approach. Int J Mol Sci 2021, Vol 22, Page 12455 [Internet]. 2021 Nov 18 [cited 2024 Jun 29];22(22):12455. Available from: https://www.mdpi.com/1422-0067/22/22/12455/htm

[31] Ciaurelli S, Origlia C, Adamo A, Giannattasio S, González M, Quinto L. Paradigm Shift: an Overview on Nutrigenetics and Nutrigenomics. 2021 Jul 6 [cited 2024 Jun 29]; Available from: https://www.preprints.org/manuscript/202107.0137/v1 [http://dx.doi.org/10.20944/preprints202107.0137.v1]

[32] Shehzad A, Lee J, Lee YS. Curcumin in various cancers. Biofactors 2013; 39(1): 56-68. Available from: https://onlinelibrary.wiley.com/doi/full/10.1002/biof.1068 [http://dx.doi.org/10.1002/biof.1068] [PMID: 23303705]

[33] Bejenaru LE, Biţă A, Belu I, Segneanu AE, Radu A, Dumitru A. Resveratrol: A review on the biological activity and applications. Appl Sci 2024, Vol 14, Page 4534 [Internet]. 2024 May 25 [cited 2024 Jun 29];14(11):4534. Available from: https://www.mdpi.com/2076-3417/14/11/4534/htm [http://dx.doi.org/10.3390/app14114534]

[34] Samanta S. Potential bioactive components and health promotional benefits of tea (Camellia sinensis). J Am Nutr Assoc [Internet]. 2022 [cited 2024 Jun 29]; Available from: https://www.tandfonline.com/doi/abs/10.1080/07315724.2020.1827082

[35] Moneim AA, Abdel-Moneim A, Magdy A. Review on medicinal plants as potential sources of cancer prevention and treatment. 2016 [cited 2024 Jun 29]; Available from: www.ejbps.com

[36] Katanić Stanković JS, Mihailović N, Mihailović V. Genistein: Advances on resources, biosynthesis pathway, bioavailability, bioactivity, and pharmacology. Handb Diet Flavonoids [Internet]. 2023 [cited 2024 Jun 29];1–40. Available from: https://link.springer.com/referenceworkentry/10.1007/978-3-0-0-94753-8_45-1

[37] Friedman M. Anticarcinogenic, cardioprotective, and other health benefits of tomato compounds lycopene, α-tomatine, and tomatidine in pure form and in fresh and processed tomatoes. J Agric Food Chem 2013; 61(40): 9534-50. Available from: https://pubs.acs.org/doi/abs/10.1021/jf402654e [http://dx.doi.org/10.1021/jf402654e] [PMID: 24079774]

[38] Kumar A, Ekavali , Chopra K, Mukherjee M, Pottabathini R, Dhull DK. Current knowledge and pharmacological profile of berberine: An update. Eur J Pharmacol 2015; 761: 288-97. [http://dx.doi.org/10.1016/j.ejphar.2015.05.068] [PMID: 26092760]

[39] Ayustaningwarno F, Anjani G, Ayu AM, Fogliano V. A critical review of Ginger's (*Zingiber officinale*) antioxidant, anti-inflammatory, and immunomodulatory activities. Front Nutr 2024; 11: 1364836. [http://dx.doi.org/10.3389/fnut.2024.1364836] [PMID: 38903613]

[40] Goncharov N V., Belinskaia DA, Ukolov AI, Jenkins RO, Avdonin P V. Organosulfur compounds as nutraceuticals. Nutraceuticals Effic Saf Toxic. 2021 Jan 1;911–24. [http://dx.doi.org/10.1016/B978-0-12-821038-3.00054-9]

[41] Zhai K, Siddiqui M, Abdellatif B, Liskova A, Kubatka P, Büsselberg D. Natural compounds in glioblastoma therapy: Preclinical insights, mechanistic pathways, and outlook. Cancers 2021, Vol 13, Page 2317 [Internet]. 2021 May 12 [cited 2025 Mar 23];13(10):2317. Available from: https://www.mdpi.com/2072-6694/13/10/2317/htm

[42] Sadgrove NJ, Jones GL. From petri dish to patient: Bioavailability estimation and mechanism of action for antimicrobial and immunomodulatory natural products. Front Microbiol 2019; 10(OCT): 2470. Available from: www.frontiersin.org [http://dx.doi.org/10.3389/fmicb.2019.02470] [PMID: 31736910]

[43] Nelson KM, Dahlin JL, Bisson J, Graham J, Pauli GF, Walters MA. The essential medicinal chemistry of curcumin. J Med Chem 2017; 60(5): 1620-37. Available from: https://pubs.acs.org/doi/full/10.1021/acs.jmedchem.6b00975 [http://dx.doi.org/10.1021/acs.jmedchem.6b00975] [PMID: 28074653]

[44] Wang S, Wang Z, Yang S, *et al.* Tissue distribution of *trans*-resveratrol and its metabolites after oral administration in human eyes. J Ophthalmol 2017; 2017(1): 4052094. Available from: https://onlinelibrary.wiley.com/doi/full/10.1155/2017/4052094 [PMID: 28409021]

[45] Krupkova O, Ferguson SJ, Wuertz-Kozak K. Stability of (−)-epigallocatechin gallate and its activity in liquid formulations and delivery systems. J Nutr Biochem 2016; 37: 1-12. [http://dx.doi.org/10.1016/j.jnutbio.2016.01.002] [PMID: 27770867]

[46] Fernández VA, Toledano LA, Lozano NP, Tapia EN, Roig MDG, Fornell RDLT. Bioavailability of epigallocatechin gallate administered with different nutritional strategies in healthy volunteers. Antioxidants 2020, Vol 9, Page 440 [Internet]. 2020 May 19 [cited 2025 Mar 23]; 9(5): 440. Available from: https://www.mdpi.com/2076-3921/9/5/440/htm

[47] Martínez-Ballesta Mc, Gil-Izquierdo Á, García-Viguera C, Domínguez-Perles R. Nanoparticles and controlled delivery for bioactive compounds: Outlining challenges for new "smart-foods" for health. Foods 2018, Vol 7, Page 72 [Internet]. 2018 May 7 [cited 2025 Mar 23]; 7(5): 72. Available from: https://www.mdpi.com/2304-8158/7/5/72/htm

[48] Ahmadi Z, Mohammadinejad R, Ashrafizadeh M. Drug delivery systems for resveratrol, a non-flavonoid polyphenol: Emerging evidence in last decades. J Drug Deliv Sci Technol 2019; 51: 591-604. [http://dx.doi.org/10.1016/j.jddst.2019.03.017]

[49] Figueiró F, Bernardi A, Frozza RL, *et al.* Resveratrol-loaded lipid-core nanocapsules treatment reduces *in vitro* and *in vivo* glioma growth. J Biomed Nanotechnol 2013; 9(3): 516-26. [http://dx.doi.org/10.1166/jbn.2013.1547] [PMID: 23621009]

[50] Peñalva R, Morales J, González-Navarro CJ, Larrañeta E, Quincoces G, Peñuelas I. Increased oral bioavailability of resveratrol by its encapsulation in casein nanoparticles. Int J Mol Sci 2018, Vol 19, Page 2816 [Internet]. 2018 Sep 18 [cited 2025 Mar 23]; 19(9): 2816. Available from: https://www.mdpi.com/1422-0067/19/9/2816/htm [http://dx.doi.org/10.3390/ijms19092816]

[51] Dende C, Meena J, Nagarajan P, Nagaraj VA, Panda AK, Padmanaban G. Nanocurcumin is superior to native curcumin in preventing degenerative changes in Experimental Cerebral Malaria. Sci Reports 2017 71 [Internet]. 2017 Aug 30 [cited 2025 Mar 23]; 7(1): 1–12. Available from: https://www.nature.com/articles/s41598-017-10672-9 [http://dx.doi.org/10.1038/s41598-017-10672-9]

[52] Zhang Q, Polyakov NE, Chistyachenko YS, *et al.* Preparation of curcumin self-micelle solid dispersion with enhanced bioavailability and cytotoxic activity by mechanochemistry. Drug Deliv

2018; 25(1): 198-209. Available from: https://www.tandfonline.com/doi/abs/10.1080/10717544.2017.1422298 [http://dx.doi.org/10.1080/10717544.2017.1422298] [PMID: 29302995]

[53] Hu B, Ting Y, Zeng X, Huang Q. Cellular uptake and cytotoxicity of chitosan–caseinophosphopeptides nanocomplexes loaded with epigallocatechin gallate. Carbohydr Polym 2012; 89(2): 362-70. [http://dx.doi.org/10.1016/j.carbpol.2012.03.015] [PMID: 24750731]

# CHAPTER 4

# Advancements in Brain Tumor Treatment: Harnessing Nanotechnology for Promising Solutions

**Dhaval J. Patel**[1,*], **Vaibhavkumar B. Patel**[2] and **Unnati V. Patel**[3]

[1] *Department of Pharmaceutics, Saraswati Institute of Pharmaceutical Sciences, At & Po-Dhanap, Gandhinagar, Gujarat, India*

[2] *Department of Pharmacology and Pharmacy Practice, Saraswati Institute of Pharmaceutical Sciences, At & Po-Dhanap, Gandhinagar, Gujarat, India*

[3] *Quality control Specialist, Analytical Method and development and validation team, Avanti Polar Lipids, Part of Croda International Plc, Alabaster, AL-35007, USA*

**Abstract:** Brain tumors, including brain metastases and gliomas, are among the worst cancers in the world due to the ineffectiveness of existing treatment techniques. Currently, many treatments are used to relieve pain and prolong survival time, including chemotherapy, radiotherapy, and surgery, although all are destructive and prolong the patient's lifespan by more than one year, and relapse is common even after effective treatment. Numerous factors contribute to the failure of cancer treatments, including physiological barriers like the Blood-Brain Tumor Barrier (BTB) and the Blood Brain Barrier (BBB), which are difficult for existing macromoleculararantitumor medications to cross. These failure factors, together with the long-term success of treatment, necessitate new developments in brain tumor treatments. Nanomedicine has emerged as one of the most promising options for advancing or improving brain tumor care. Targeted drug delivery with nanoparticles has the potential to significantly reduce dosage, improve release characteristics, increase specificity and bioavailability, lengthen the shelf life, and lessen toxicity and side effects. Some nanodrugs can cross the BBB and BTB, which are major impediments to treating brain cancers. In this chapter, we will examine the current state of the art, as well as the most unique and exceptional innovations in treatment options, including a concise summary of preclinical and clinical research on nanodrugs in brain tumor therapy.

**Keywords:** Brain tumor, Blood-brain barrier, Nanomedicine, Nanoparticles, Targeted drug delivery.

* **Corresponding author Dhaval J. Patel:** Department of Pharmaceutics, Saraswati Institute of Pharmaceutical Sciences, At & Po-Dhanap, Gandhinagar, Gujarat, India; E-mail: dhaval6668@gmail.com

**Prashant Tiwari, Pankaj Kumar Singh & Sunil Kumar Kadiri (Eds.)**

## INTRODUCTION

One of the most aggressive forms of cancer that has ever been identified is brain tumors. Even with improvements in modern therapeutic methods, these illnesses usually have disastrous consequences. The mechanism and properties of tumor cell development are extensively explored in the current research. A door to new treatments may be opened with the discovery of novel targets or drugs that work better than ones that already exist. A normal cell can become cancerous through a series of genetic changes that occur during the multistep process of cancer growth [1]. Six physiological factors have been found by Hanahan and Weinberg [2] to be capable of inducing cancer like self-sufficiency in generating signals for cell development, namely (i) resistance to cell proliferation inhibitors, (ii) insensitivity to inhibitors of cell proliferation, (iii) avoiding the processes of apoptosis, (iv) uncontrolled and unchecked replication/duplication process, (v) capacity to stimulate angiogenesis which support the tumor cell growth, and (vi) capacity to infiltrate biological materials. Two more characteristics were also identified in cancer development, including alterations in metabolic processes within the cell that may facilitate the formation of malignant cells and the capacity to evade the immunological reaction. By understanding the mechanisms underlying carcinogenesis, one can prevent the development of tumors that are linked to unchecked cellular replication brought on by either a decrease in tumor suppressor genes, which inhibit cancer cell proliferation, or an increase in protooncogenic factors, which promote normal cell proliferation.

Central Nervous System (CNS) malignancies sometimes exhibit significant histological variety. As a result, in 1957, the WHO proposed a tumor classification in order to harmonize detection and, eventually, develop a universal classification [3]. Based on the degree of Malignancy, the WHO currently divides brain tumors into four categories, Grades I to IV, as determined by histological analysis like cellular pleomorphism, endothelial hyperplasia, nuclear atypia, cell division activity, necrosis process, angiogenesis, tumor suppression gene factors, protooncogenic factors, clinical features of disease and responses to the medical treatments. The following are the four types of grades:

- Grade I, also known as pilocytic astrocytoma, is characterized by moderate growth, the absence of malignancy, and extended survival. It is treated surgically.
- Grade II, also known as low-grade astrocytoma, is characterized by relatively moderate growth; malignancy may be present or not, and recurrence often occurs when the tumor progresses to a higher grade.

- Grade III, also known as anaplastic astrocytoma, is characterized by malignant, a tendency to progress to a higher grade, and high mitotic activity. Most patients are treated with chemotherapy and radiotherapy.
- Grade IV, also known as glioblastoma mutiforme, is characterized by fast growth, aggressive malignancy, extensive necrosis, and a very low survival rate.

As per the Central Brain Tumor Registry of the United States (CBTRUS) reports from 2005 to 2009, 20.6 cases of CNS tumors were reported in the US for every 100,000 individuals annually, of which 7.3 cases had malignant tumors and 13.3 cases had low-grade tumors per 100,000 individuals [4]. 50% of the primary grade brain tumors convert into glioblastoma multiforme, and its 5-year survival rate is less than 10% [5].

The global survey, which included personal health data from 37.5 million cancer patients diagnosed between 2000 and 2014, revealed that overall cancer survival rates are increasing—particularly for highly malignant diseases [6]. Cancer is responsible for approximately one in every six deaths worldwide. According to Global Cancer Statistics 2020, 19.3 million new cancer cases resulted in 10 million deaths [7]. More than 70% of these deaths occur in low- and middle-income countries. In adults, brain metastases account for nearly 150,000 cases of intracranial malignancies in the United States alone. Between 8% and 10% of adult cancer patients develop brain metastases; however, the incidence varies significantly depending on the type of primary tumor. Approximately 70% of brain metastases originate from breast and lung cancers. Other cancers that may metastasize to the brain include colorectal, renal cell carcinoma, and melanoma [8, 9].

Glioblastoma and other high-grade brain tumors are often fatal due to their resistance to radiation and chemotherapy, as well as their highly invasive nature [10]. The most common malignant brain tumor is Glioblastoma Multiforme (GBM), which exists in two forms: primary and secondary. Primary GBM arises from normal glial cells, while secondary GBM develops from anaplastic astrocytoma [11 - 14]. Despite the advent of advanced targeted therapies and immunotherapies, the five-year survival rate for GBM remains low, estimated between 4% and 7% [15 - 18]. GBM is typically diagnosed in patients in their mid-60s, with a higher incidence in men—approximately 1.6 times that of women. Studies report an occurrence rate of 1.1 to 5.0 cases per 100,000 individuals annually, with Caucasian populations exhibiting higher rates than Asian and African populations [17 - 22]. Recent data indicate a consistent and statistically significant increase in GBM across all age groups [19]. Early clinical signs of GBM are often non-specific and may include headaches, nausea, personality changes, memory loss, and seizures. As the tumor grows, patients may rapidly progress to unconsciousness.

## CURRENT TREATMENT OF BRAIN TUMOR

Five key dietary and lifestyle factors: high Body Mass Index (BMI), smoking, physical inactivity, inadequate consumption of fruits and vegetables, and excessive alcohol intake, account for nearly one-third of cancer-related deaths [23]. Cancer treatment remains challenging due to its complex and heterogeneous pathogenesis. Moreover, the emergence of multidrug-resistant cancer cells is a major contributor to therapeutic failure. Consequently, no single treatment regimen is universally effective, and recurrence is common as cancer often develops resistance to previously successful therapies [24]. The primary modalities for cancer treatment include chemotherapy, radiation therapy, and surgical intervention [25]. These approaches have been the cornerstone of cancer management for decades. However, they are associated with significant limitations, including poor selectivity for tumor cells, systemic toxicity, multidrug resistance, and severe adverse effects on overall health. Despite their widespread use, these treatments have not achieved complete eradication of the disease [26, 27]. The likelihood of treatment failure and metastasis increases substantially in the absence of early diagnosis. In the early stages of brain cancer, accurate assessment of tumor volume and margins is particularly challenging due to the accumulation of extracellular fluid around cancer cells. To achieve optimal clinical outcomes, treatment strategies must integrate both diagnostic and therapeutic approaches tailored to individual patients. Monoclonal antibody-based therapies targeting Tumor-Associated Antigens (TAAs) and Tumor-Specific Antigens (TSAs) have been explored. Antibody-Drug Conjugates (ADCs), a class of biopharmaceuticals, are designed to selectively target cancer cells while sparing healthy tissue [28]. This approach has successfully transitioned from laboratory research to clinical application. Although these methods represent significant advancements, each has its own set of advantages and limitations.

Researchers have investigated a wide range of disciplines including pharmacology, oncology, supramolecular chemistry, organic chemistry, and nanotechnology to overcome these challenges [29]. Nanotechnology, particularly in the field of nanotheranostics, has shown promise in enhancing both therapeutic and diagnostic capabilities. Due to their small size and the poor vascularization of tumors, nanosystems can preferentially accumulate in tumor cells. Additionally, Nanoparticles (NPs) can be engineered to deliver drugs with high precision [30, 31]. Nanotherapeutics, derived from the rapidly evolving science of nanotechnology, offer significant potential in cancer treatment. Nanotechnology involves the manipulation of materials at the nanoscale, enabling high drug-loading capacity due to the large surface area of nanomaterials [32]. Their small size allows for active interaction with cellular components, making them suitable for both *in vitro* and *in vivo* biomedical applications.

Compared to conventional chemotherapy, nanoparticle-based drug delivery systems offer several advantages, including protection of drugs from degradation within the body, enhanced uptake by tumor cells, and reduced interaction with healthy cells [33]. Ideal characteristics for nanoparticle drug carriers include long circulation time, non-immunogenic and non-inflammatory properties, biocompatibility and biodegradability, physiological stability in blood, avoidance of aggregation and thrombogenicity, compatibility with nucleic acids, peptides, proteins, and small molecules, as well as cost-effectiveness and ease of manufacturing. Nanocarriers can be composed of various organic and inorganic materials, including polymers, lipids, proteins, and viral components. Among them, dendrimers (< 10 nm) are the smallest, while liposomes (80–200 nm), micelles (20–60 nm), and polymeric nanoparticles (40–100 nm) are larger [34]. Numerous studies have documented the use of nanocarriers for delivering anticancer drugs, some of which have undergone clinical trials and received FDA approval [35]. Nanomedicine is emerging as one of the most promising strategies for targeting Glioblastoma Multiforme (GBM), offering improved treatment selectivity and reduced side effects. It aims to enhance clinical outcomes through disease diagnosis, monitoring, therapy, prevention, and understanding of pathophysiology [36]. This field utilizes nanomaterials ranging from 10 to 500 nm for various applications, including brain tumor treatment. Nanomedicines, such as polymeric nanoparticles and liposomes, offer benefits like improved drug biodistribution and pharmacokinetics, increased drug accumulation in target tissues, reduced dose-dependent adverse effects, minimized drug accumulation in non-target tissues, capability to co-deliver multiple drugs, and surface modification to enhance circulation and target specificity.

Given these advantages, nanomedicine has been extensively explored to address major challenges in brain tumor treatment, including complex brain anatomy, diffuse and heterogeneous tumor nature, difficulty in delineating tumor boundaries, poor therapeutic agent accumulation at the tumor site, and risk of chemotherapy-induced resistance. A critical barrier to effective brain tumor treatment is the presence of three distinct Central Nervous System (CNS) barriers, particularly the Blood–Brain Barrier (BBB). The BBB's complex structure restricts the entry of harmful substances, including toxins, nutrients, ions, and small molecules. It also features an efficient efflux system that removes waste and toxic compounds. The second barrier, the brain–blood barrier, further limits drug penetration. Tumor growth compromises BBB integrity, leading to the formation of the Blood–Brain Tumor Barrier (BBTB). While the BBTB may exhibit increased permeability in high-grade gliomas, it also presents high interstitial pressure that hinders drug absorption. Variability in endothelial cell spacing within tumor capillaries further impairs uniform drug distribution, reducing therapeutic efficacy [37].

Invasive methods such as intracerebral or intraventricular drug administration, or BBB disruption *via* chemical, physical, or biological stimuli, are considered unsafe and costly for brain tumor treatment. Although certain drugs like paclitaxel, doxorubicin, and Temozolomide (TMZ) can cross the BBB, their efficacy is limited and often requires high doses [38].

A promising solution lies in nano-drug delivery systems with tunable properties such as size, shape, and surface chemistry [39, 40]. By optimizing pharmacokinetics, tumor-targeting ability, and BBB permeability, these systems can enhance therapeutic outcomes. Traditional chemotherapy drugs damage DNA indiscriminately, affecting both cancerous and healthy cells and increasing the risk of severe side effects. Therefore, innovative approaches are needed to improve BBB penetration, selectively target tumor cells, preserve healthy brain function, and minimize immune-related adverse effects in nanodrug delivery systems.

## METHODS FOR CROSSING THE BBB

- Surface Alteration for Crossing BBB.

Substances traveling across the BBB can do so by a variety of transport mechanisms, including receptor-mediated transport, adsorption-mediated transport, transporter-mediated transport, paracellular routes, and cell-piercing peptides [41]. Surface modifications of nanodrug carriers, such as BBB Receptor-targeted ligands, BBB Transporter-targeted Ligands, and substances that trigger adsorption of the BBB, can be used to easily control the corresponding BBB crossing ability and improve the crossing mechanism.

- BBB-Crossing Biomimetic Nano Drug Delivery System (DDS).

Biometric Drug Delivery Systems (Biometric DDS) represent a groundbreaking advancement in therapeutic technology, offering distinct advantages over conventional drug delivery methods. As a result, they are increasingly being explored for the treatment of various chronic metabolic disorders. In Biometric DDS, therapeutic molecules are delivered directly into biological structures or processes, enhancing precision and efficacy.

Nanoparticles derived from biological sources, such as cell membranes, viral vectors, whole cells, endogenous proteins, and extracellular vesicles, are referred to as nanobiomimetic drug carriers. These biologically inspired vectors mimic native molecules, thereby minimizing undesired immune responses and prolonging systemic circulation. By inheriting both the functional and structural complexity of their source materials, these carriers offer superior performance in drug delivery. Consequently, bionic DDS may provide several key benefits, including enhanced ability to cross the Blood–Brain Barrier (BBB), high biocompatibility, optimal accumulation at target sites, and protection of therapeutic agents from degradation. These attributes make Biometric DDS a promising platform for overcoming major challenges in drug delivery, particularly in the context of neurological and metabolic diseases [42].

- Modify the Administration Mode to Avoid the BBB.

The medicine and drug carrier are often administered intravenously or orally. Tailored nanodrug carriers have the potential to target and penetrate the BBB; nevertheless, these medications must first transit through the systemic circulation. Once in other organs, the drug will inevitably cause toxicity, particularly in cases where patients are very sick and need high dosages [43]. Therefore, it is possible to transport medications directly into the brain and avoid the BBB by altering the drug delivery approach, such as *via* local administration.

- Destroy the BBB Using Physicochemical Techniques.

The blood-brain barrier can be temporarily disrupted and its permeability altered by various physical and chemical techniques. The BBB can be irreversibly disrupted by intra-arterial injection of hyperosmotic drugs like mannitol or alkylglycerol [38]. Relevant studies demonstrate that the BBB's permeability is strongly temperature-dependent, enabling noticeably faster material transfer at temperatures above 38–39 °C [44]. Magnetic nanoparticles laden with drugs produce heat by an alternating magnetic field, which has the potential to break down the BBB and improve through-BBB mobility [45]. Furthermore, the combination of concentrated ultrasound and microbubbles can temporarily open the BBB. Microbubbles in conjunction with focused ultrasound are a frequently utilized technology.

## TECHNIQUES FOR BETTER TARGETING OF BRAIN TUMOR CELLS

Brain nanodrug delivery systems should be more specifically targeted to brain tumor locations in order to reduce the harm that nanomedicines cause to the immune system and normal brain function. At the moment, active targeting strategies are the primary approach to brain tumor targeting, while passive targeting tactics improve medication accumulation at tumor locations. Recently, a number of novel approaches have emerged, including tumor microenvironment-triggered DDS and magnetically focused nano-DDS [46, 47]. Here we focus on magnetic targeting, passive targeting, and active targeting.

- Passive Targeting.

To develop brain cancer-targeted nanomedicines, passive targeting strategies have been employed due to the highly permeable vasculature surrounding tumor tissues [48]. The primary mechanism underlying passive targeting is the Enhanced Permeability And Retention (EPR) effect. Pathological conditions, such as edema, swelling, and elevated intracranial pressure, can compromise the integrity of the Blood–Brain Barrier (BBB), further facilitating nanoparticle penetration [38]. The EPR effect arises from the disruption of blood vessels during solid tumor formation, allowing nanoparticles of specific sizes to infiltrate tumor tissues through gaps in the endothelial lining [49]. This leads to increased accumulation of nanoparticles at the tumor site. To optimize drug delivery *via* the EPR effect, researchers have engineered nanoparticles with tailored physicochemical properties, such as surface charge, size, geometry, and hydrophilicity, to enhance tumor-specific aggregation [50]. Studies suggest that the optimal nanoparticle size range for exploiting the EPR effect is between 10 and 200 nm. Particles larger than 200 nm are unable to fully penetrate the tumor interstitium, while those smaller than 10 nm are rapidly cleared by renal filtration, preventing sufficient accumulation at the tumor site [48, 49].

Additionally, nanoparticle lipophilicity and stability significantly influence the EPR effect. Drugs are often encapsulated in liposomes or polymeric nanoparticles to enhance EPR-mediated delivery. PEGylation—a surface modification technique involving polyethylene glycol—is commonly used to prolong systemic circulation and improve nanoparticle stability. Biomimetic Drug Delivery Systems (DDS) also demonstrate enhanced EPR effects due to their extended circulation times and biological compatibility [42]. Despite these advantages, a significant proportion of administered nanoparticles (>95%) tend to accumulate in non-target organs, such as the lungs, spleen, and liver [51]. Nevertheless, numerous studies have documented the successful clinical application of passive targeting strategies. For example, PEGylated liposomal doxorubicin was the first FDA-approved nanomedicine, used to treat HIV-associated Kaposi's sarcoma and ovarian cancer. Encapsulation in lipid carriers significantly reduced doxorubicin's cellular and systemic toxicity, improved pharmacokinetics, and enabled controlled release and biodistribution [52].

More recently, the FDA approved CPX-351 (Vyxeos™), a liposomal formulation combining daunorubicin and cytarabine, which demonstrated an overall survival rate of 9.6 months in newly diagnosed acute myeloid leukemia patients compared to 6.0 months with conventional therapy [53]. These approvals have opened new avenues for cancer treatment using nano-formulations. Additional targeting layers can be incorporated into these systems to further enhance biodistribution and reduce immune clearance. However, the therapeutic efficacy of passive targeting is limited by the heterogeneity of EPR effects across and within tumors. Variability in endothelial gaps due to rapid tumor growth can result in inconsistent nanoparticle extravasation [54]. Some studies challenge the assumption that the hypoxic core of tumors exhibits higher nanoparticle permeability than the peripheral regions [55]. This spatial heterogeneity complicates passive targeting. Moreover, extravasation is influenced not only by vascular permeability but also by blood flow dynamics around the tumor site, which vary both spatially and temporally [56].

Even when nanoparticles successfully extravasate, the dense interstitial matrix of tumors can hinder deeper penetration. In such cases, nanoparticle size and size-dependent properties become critical. To address this, Wong *et al.* developed a multi-stage nanosystem comprising 100 nm gelatin particles that degrade into 10 nm fragments upon exposure to tumor-associated matrix metalloproteinases, thereby enhancing tissue penetration [57]. Unfortunately, the lack of robust human tumor models that accurately replicate EPR effects limits our understanding of passive targeting. Most current data are derived from rapidly growing subcutaneous xenograft models, which exhibit exaggerated EPR effects. Consequently, findings from these models may overestimate the efficacy of passively targeted nanomedicines [58]. Furthermore, there is a notable scarcity of patient-derived experimental evidence on the EPR phenomenon. Therefore, deeper investigation into tumor biology and EPR dynamics across various tumor types is essential. Such insights will enable the rational design of nanomaterials for improved therapeutic outcomes in personalized cancer treatment.

- Active Targeting.

Active targeting refers to the interaction between drug carriers and target cells based on ligand–receptor specificity [51]. Unlike passive targeting, active targeting offers greater precision in delivering therapeutic or diagnostic agents to brain tumors, resulting in enhanced efficacy and reduced cytotoxicity [48]. Tumor progression is closely associated with the expression of specific receptors that are either absent or minimally expressed in healthy tissues but are highly upregulated in brain tumor cells or the vasculature surrounding brain tumors. These receptors can be exploited to direct drug carriers to accumulate at tumor sites or penetrate tumor cells by modifying the carrier surface with appropriate ligands. Receptor-mediated targeting strategies are particularly relevant for crossing the Blood–Brain Barrier (BBB) and enhancing drug delivery to brain tissues. However, these strategies may not exclusively target tumor cells, potentially leading to off-target effects and neurotoxicity [59]. To mitigate this risk, receptor-mediated endocytosis pathways, such as those involving interleukin receptors, integrin receptors, and neuropilin-1 (NRP-1) receptors, are utilized. These receptors exhibit low expression in the BBB but are abundantly expressed in brain tumor regions, making them suitable for selective tumor targeting [60 - 62].

In one study, liposomes functionalized with anti-HER2 ligands demonstrated increased nanoparticle uptake in HER2-expressing cancer cells. However, both non-targeted liposomes and anti-HER2-modified liposomes accumulated primarily in the perivascular and stromal regions of the tumor. Rapid clearance by macrophages resulted in limited internalization by tumor cells, thereby reducing therapeutic efficacy [63]. In another investigation, Zhou *et al.* conjugated the anthracycline doxorubicin to magnetic Iron Oxide Nanoparticles (IONPs) functionalized with Insulin-like Growth Factor 1 (IGF1) [64]. Using a Patient-Derived Xenograft (PDX) model of pancreatic cancer, they achieved enhanced tumor targeting and deeper tissue penetration following intravenous administration. The targeted theranostic IGF1–IONP–doxorubicin system significantly suppressed tumor growth, demonstrating promising therapeutic potential.

To further improve specificity and safety, dual-targeted drug delivery systems have been developed. These systems incorporate two distinct ligands on the surface of the drug carrier—one targeting the BBB and the other targeting brain tumor cells. This dual-ligand approach enhances treatment efficacy while minimizing systemic toxicity and adverse effects [57]. Studies have shown that peptides, such as RGD (arginine–glycine–aspartic acid) and TGN, exhibit excellent tissue penetration and high selectivity for brain tumors [65, 66].

- Magnetic Targeting.

Magnetic targeting is an advanced drug delivery method that utilizes external magnetic fields to guide magnetic drug carriers toward tumor sites [67]. This approach relies on the magnetic attraction between nanoparticles and an externally applied magnetic field, requiring the incorporation of magnetic nanoparticles into nanocarriers. Magnetic nanoparticles are widely employed as both therapeutic carriers and diagnostic imaging agents. Common examples include Superparamagnetic Iron Oxide Nanoparticles (SPIONs), fluorescent Magnetic

Nanoparticles (MNPs), and Iron Oxide Nanoparticles (IONPs). In magnetic targeting, specific ligands can be conjugated to magnetic nanoparticles to enhance selectivity and facilitate targeted delivery to brain tumors. Following intravenous administration, magnetic nanoparticles are exposed to an external magnetic field directed at the tumor site, accelerating their transport from systemic circulation to the brain tumor [68]. This technique enables localized drug accumulation by applying a magnetic field to the cranial region.

Cui *et al.* [69] investigated a dual-targeting strategy that incorporated magnetic guidance. Their findings demonstrated that dual-targeted nanoparticles exhibited significantly enhanced brain penetration and cellular uptake compared to non-targeted nanoparticles. The approach also improved drug delivery efficiency, reduced systemic side effects, and increased survival rates in murine models exposed to magnetic fields. Despite its advantages, magnetic targeting faces limitations in selectivity. While the external magnetic field can concentrate drug carriers near the tumor site, it may not precisely localize them within the tumor tissue. Nevertheless, numerous studies have highlighted the benefits of magnetic targeting, including enhanced therapeutic efficacy, reduced dosage requirements, and minimized adverse effects [69, 70]. In conclusion, the cellular uptake, therapeutic efficacy, and biodistribution of nanoparticle-based materials are strongly influenced by their physicochemical properties, such as shape, size, surface chemistry, and surface charge. Additionally, a wide array of other factors significantly affect the absorption and dispersion of nanomaterials within cells. These include nanoparticle purity and chemical composition, particle stability and crystal structure, surface-to-volume ratio and aggregation state, protein–nanoparticle interactions, incubation conditions, tumor cell type, and Disease-specific variables. A comprehensive understanding of these parameters is essential for optimizing nanoparticle design and improving therapeutic outcomes in brain cancer treatment.

## NANOMATERIALS AS DRUG DELIVERY AGENTS FOR THE TREATMENT OF BRAIN TUMOR

Rapid advancements in biological nanotechnology have significantly influenced the field of nanomedicine. Nanotechnology facilitates the precise manipulation and production of materials at the nanoscale, enabling the development of innovative tools for monitoring, controlling, diagnosing, and treating biological systems. Nanomaterials are defined as substances with structural dimensions ranging from 1 to 100 nanometers, either externally or internally [71]. Due to their ability to transport drug molecules directly to target sites and efficiently cross the Blood-Brain Barrier (BBB), nanostructured materials offer a promising strategy for brain-targeted drug delivery [72]. Various nanomaterials including carbon nanotubes, liposomes, dendrimers, gold nanoparticles, and micelles have been extensively investigated for their potential as Drug Delivery Systems (DDS) in neurological applications. Nanoparticles (NPs) can bypass or mask the restrictive nature of the BBB, thereby enhancing the delivery of therapeutic agents to the brain. To achieve therapeutic efficacy, it is essential that drug delivery systems enable the administration of lower doses compared to conventional free drug formulations, while maintaining or improving effectiveness [73]. Compared to their bulk counterparts, nanoparticles exhibit unique physicochemical properties, such as high drug-loading capacity, increased surface area for molecular interaction, compatibility with both hydrophobic and hydrophilic molecules, and enhanced stability across various physiological conditions [74]. In addition to composition, the size and shape of nanoparticles critically influence their behavior and performance. Controlling these parameters is essential to minimize aggregation and produce monodispersed nanoparticles that can be efficiently internalized by cells [75]. The high surface-to-mass ratio of nanoparticles allows them to bind, absorb, or carry other molecules, which underpins many of their biomedical applications. Furthermore, nanoparticles can be synthesized or modified using composite materials to enhance their physical and functional properties [76].

- Liposomes.

Liposomes are spherical vesicles composed of aqueous compartments enclosed by biodegradable natural or synthetic phospholipid bilayers. Their amphiphilic nature enables spontaneous self-assembly into nanospheres [77]. Based on synthesis techniques and post-formation processing, liposomes are classified as Unilamellar Vesicles (ULVs) or Multilamellar Vesicles (MLVs). ULVs consist of a single lipid bilayer surrounding an aqueous core, capable of encapsulating hydrophilic molecules. In contrast, MLVs contain multiple lipid bilayers (typically 1–5 µm in diameter) and are suited for encapsulating lipophilic compounds. Although MLVs offer a larger encapsulation volume, ULVs tend to release their contents more rapidly due to their hydrodynamic properties [78].

Liposomes can release drugs by fusing with tumor cells or entering the extracellular matrix *via* endocytosis [79]. Both passive and active targeting strategies can be employed to direct liposomes to tumor sites. While active targeting particularly toward blood tumors, vasculature, and micrometastases offers advantages, it is not universally superior to passive targeting [80]. Polyethylene Glycol (PEG) coating enhances liposome biocompatibility, prolongs circulation time, and reduces toxicity and side effects [81]. Additionally, macromolecules, such as polymers, peptides, and polysaccharides, can be conjugated to the liposome surface to improve circulation and facilitate brain-specific drug delivery.

Liposome half-life in circulation is significantly influenced by particle size. Nanostructured liposomes around 100 nm can efficiently penetrate tumor tissues, whereas larger liposomes are more readily cleared by the phagocytic system [79]. Liposomes have been extensively utilized as nanocarriers in the treatment of neurological disorders and various cancers. Lakkadwala *et al.* developed dual-functional liposomal nanotherapeutics by modifying the liposome surface with Penetratin (Pen) peptide to facilitate receptor-mediated transcytosis and Transferrin (Tf) to enhance cellular uptake in the brain [82]. When erlotinib and doxorubicin were co-encapsulated in Tf–Pen liposomes, the formulation demonstrated approximately 15% greater BBB translocation compared to free drugs and showed significant efficacy against glioblastoma *in vivo*. The nano-liposomal system eliminated nearly 90% of the tumor in murine models without inducing adverse effects.

In another study, Muthu *et al.* formulated docetaxel-loaded liposomes using TPGS (an FDA-approved PEGylated form of vitamin E), with particle sizes ranging from 126 to 191 nm. These liposomes exhibited enhanced cellular uptake and cytotoxicity in C6 glioma cells compared to free drug molecules [83]. Paclitaxel, a microtubule-targeting anticancer agent, is effective against brain, ovarian, and lung cancers. However, its commercial formulations are unable to cross the BBB, limiting its efficacy against gliomas [84, 85]. Artemether, known for its cytotoxic activity against various cancer cells, downregulates matrix metalloproteinase-9, vascular endothelial growth factor, and other tumor-associated proteins [86]. It has been proposed as a nanotherapeutic agent capable of eliminating brain stem cells, facilitating drug translocation across the BBB, and disrupting vasculogenic mimicry channels. A novel liposomal system co-loaded with artemether and paclitaxel demonstrated enhanced apoptosis in brain cancer cells by activating pro-apoptotic enzymes and proteins while inhibiting anti-apoptotic pathways [87]. Ravi *et al.* [88] explored liposome-mediated radiation therapy, which enhances therapeutic efficacy while reducing the required radiation dose. Liposomes can confine radiation effects to tumor interiors, minimizing damage to surrounding healthy tissues. In a Phase I study of recurrent Glioblastoma Multiforme (GBM), Rhenium-186 Nanoliposomes (186RNL) successfully delivered β-emitting radiation selectively to tumors [89, 90].

Convection-Enhanced Delivery (CED) offers an alternative to intravenous administration by directly transporting drugs to the central nervous system, proving especially useful in GBM treatment [91]. Co-flow contrast agents can be incorporated into liposomes to enable real-time MRI monitoring [92]. For example, the topoisomerase I inhibitor CPT-11 has been successfully delivered to GBM *via* CED of nanoliposomes, allowing for targeted therapy [93]. However, CED requires advanced technology, and further research is needed to address its limitations and explore unresolved challenges. Multiple studies have identified Histone Deacetylases (HDACs) as promising targets for brain tumor therapy. However, HDAC inhibitors suffer from poor pharmacokinetics, including short half-life, rapid clearance, low solubility, and limited permeability. PEGylated liposomes loaded with HDAC inhibitors have demonstrated improved drug efficacy *in vivo*, offering a viable strategy to overcome these limitations.

- Nano-Micelles.

Micelles are a unique class of amphiphilic spherical nanostructures formed through the self-aggregation of amphiphilic molecules in aqueous environments at a specific threshold known as the Critical Micelle Concentration (CMC). These structures possess both hydrophilic and hydrophobic domains: the hydrophilic segments form the outer shell, while the hydrophobic segments constitute the core, which effectively encapsulates lipophilic bioactive compounds [94]. Micelles have garnered significant attention as nanocarriers due to their high drug-loading capacity, particularly for chemotherapeutics targeting ovarian cancer at specific sites [95]. With sizes typically ranging from 10 to 100 nm, micelles facilitate efficient penetration and endocytosis by ovarian cancer cells while minimizing non-specific uptake by healthy cells [96]. Compared to other drug delivery systems, nano-micelles exhibit several advantageous properties, including prolonged plasma circulation, enhanced biocompatibility, superior thermodynamic and kinetic stability, preferential accumulation in inflamed and tumor tissues, and capability to encapsulate hydrophobic drugs within their core due to their dynamic structural properties [94]. Recently, micelles have emerged as promising nanocarriers for targeted drug delivery in brain tumor therapy. Their small size renders them less detectable by the phagocytic system, while their hydrophilic shells contribute to improved permeability and retention [97].

Agrawal *et al.* [98] developed bio-adhesive micelles encapsulating docetaxel for brain tumor treatment. Chitosan was incorporated during micelle formation to enhance mucoadhesive properties, and transferrin was added to facilitate receptor-mediated transcytosis across the Blood–Brain Barrier (BBB). This nano-theragnostic system demonstrated superior drug delivery efficiency in glioma cells compared to free docetaxel. In a subsequent study, Agrawal *et al.* [99] formulated TPGS-based micelles containing docetaxel and transferrin, achieving sustained drug release over 24 hours, with approximately 50% of the drug released during that period. Dual-drug targeting strategies have recently gained traction for brain tumor therapy due to their enhanced targeting precision and rapid BBB penetration. Building on this approach, micelles co-loaded with paclitaxel and artemether were developed, demonstrating targeted delivery to brain tumor cells and potent anti-cancer activity. Mechanistic analysis revealed that this dual-drug micellar system significantly increased caspase-3 expression and induced apoptosis in U87 glioma cell lines, indicating adequate enrichment within the glioma microenvironment and successful BBB traversal [100]. In another study, Niu *et al.* [101] designed a dual-targeting docetaxel-integrated micelle (GF-DOX) for brain tumor therapy in murine models. The micelles exhibited enhanced uptake by Brain Microvascular Endothelial Cells (BMVECs) *via* clathrin- and caveolae-mediated endocytosis. Additionally, GF-DOX demonstrated improved bioavailability of docetaxel compared to its free form. These findings suggest that dual-targeting nanotherapeutics may significantly enhance the efficacy of anticancer drugs in both *in vitro* and *in vivo* brain tumor models.

• Dendrimers.

Dendrimers are a class of well-defined, hyperbranched nanoscale polymers characterized by their highly organized three-dimensional architecture. Over the past few decades, researchers have developed a wide variety of dendrimers and continue to design new generations with enhanced functionalities [102]. Due to their extensive surface functional groups and internal cavities, dendrimers are considered highly promising carriers for drug delivery. Therapeutic agents can either be conjugated to the dendrimer's surface or encapsulated within its interior voids. The abundance of functional groups on the dendrimer surface allows for versatile conjugation strategies, accommodating a wide range of medicinal compounds.

Compared to other drug delivery systems, dendrimers offer several advantageous properties, including nanoscale size with tunable surface functionalities, high structural uniformity, rapid cellular uptake, reduced recognition and clearance by macrophages, targeted delivery capabilities, and efficient transcytosis across cellular membranes [103].

Surface modifications, such as PEGylation, amino acid functionalization, glycosylation, and acetylation, can significantly enhance the biocompatibility of dendrimer-based polymers. Moreover, dendrimers functionalized with ligands that selectively bind to tumor-specific receptors can facilitate targeted delivery to brain tumor cells while sparing healthy tissues [104].

Among dendritic polymers, poly(amidoamine) (PAMAM) dendrimers have gained substantial attention due to their role in targeted gene and drug delivery in cancer therapy [105]. Leveraging their brain-targeting capabilities, Liu *et al.* [106] developed PAMAM dendrimers conjugated with angiopep-2 and Epidermal Growth Factor (EGF), along with Doxorubicin (DOX). This formulation demonstrated reduced toxicity to normal cells and tissues, while enhancing anticancer efficacy through improved BBB penetration. The DOX-loaded dendrimer nanoformulation exhibited strong trans-epithelial transport across cellular barriers, long-term physicochemical stability, high biocompatibility and, enhanced therapeutic potency and safety [107]. These findings underscore the potential of surface-modified dendrimers as nanocarriers capable of crossing the BBB *via* receptor-mediated transcytosis, offering a targeted and effective approach for brain tumor therapy.

- Carbon Nanotubes and Carbon Dots.

Carbon Nanotubes (CNTs) are cylindrical nanostructures that have garnered significant interest in pharmaceutical research due to their unique mechanical, surface, and electrical properties. Their high surface reactivity enhances biocompatibility by enabling the attachment of various guest molecules. CNTs can be functionalized with carbohydrates, polymers, organic compounds, and peptides, allowing for tailored solutions to meet specific therapeutic needs. Their primary applications lie in cancer therapy and targeted drug delivery to tumor cells [108]. Harsha *et al.* [109] synthesized PEGylated carbon nanotubes conjugated with mangiferin and evaluated their efficacy using U-87 glioblastoma cell lines. The formulation demonstrated enhanced drug release, potent anticancer activity, improved safety profile, and increased bioavailability compared to the free drug [110].

A recent advancement in carbon-based nanostructures is the development of Carbon Dots (CDs). These zero-dimensional, ultra-small (< 10 nm), spherical nanoparticles are non-toxic, biocompatible, and possess excellent photoluminescence and electron-transfer capabilities. Their versatile surface chemistry and nanoscale dimensions make them highly attractive for pharmaceutical and biomedical applications [111]. Fluorescent carbon dots, in particular, have emerged as promising drug nanocarriers due to their efficient cargo-loading capacity, tunable surface functionalities, low cytotoxicity, and strong biocompatibility [112]. Zheng *et al.* [113] developed a novel nanoparticle formulation that combines fluorescent carbon dots with oxaliplatin, demonstrating targeted delivery to brain tumor cells and significantly enhancing the anticancer efficacy of oxaliplatin, while maintaining favorable biocompatibility. More recently, a triple-conjugated nano-formulation was engineered using a transferrin ligand and two chemotherapeutic agents epirubicin and temozolomide. This formulation exhibited superior cytotoxicity against brain tumor cells compared to either free drugs or transferrin-conjugated monotherapy, highlighting the potential of multifunctional carbon-based nanocarriers in the treatment of brain cancer [114].

• Silver and Gold Nanoparticles.

Gold Nanoparticles (AuNPs) are a highly versatile nanoplatform with unique physicochemical properties that make them suitable for a wide range of theragnostic applications. Their inherent biocompatibility, tunable size and shape, and compatibility with various functional moieties have positioned AuNPs as promising candidates for cancer diagnosis and therapy [115]. Like other inorganic nanostructures, AuNPs are susceptible to oxidative stress-induced cytotoxicity. However, they also possess several advanced features, including monodispersity, adjustable core size, ease of synthesis, high surface area, low toxicity, light-scattering capabilities, and strong affinity for biomolecular conjugation [116].

Therapeutic agents can be loaded onto AuNPs *via* covalent bonding or electrostatic interactions [117]. One of the most significant advantages of AuNPs is their size variability, which facilitates efficient circulation and tumor targeting *via* passive or active mechanisms [118]. Alle *et al.* [119] synthesized Carboxymethyl Xanthan Gum (CMXG)-coated AuNPs using microwave irradiation. This green synthesis approach offers several benefits, including environmental sustainability, non-toxicity, cost-effectiveness, and the dual role of xanthan gum as a capping and reducing agent. Doxorubicin (DOX) was loaded onto these crystalline, spherical AuNPs *via* electrostatic interactions. The formulation exhibited pH-sensitive drug release, with minimal release at physiological pH and enhanced release in acidic environments. Compared to free DOX, the CMXG-coated AuNPs demonstrated superior anticancer efficacy due to their enhanced uptake and pH-dependent release profile.

A PEGylated biogenic AuNP-based multifunctional dual nanocarrier was also developed to deliver transferrin antibodies and folate to glioma cells. Curcin, a ribosome-inactivating protein, was conjugated to AuNPs to enable pH-controlled drug release in the acidic tumor microenvironment while maintaining cellular compatibility. Curcin effectively inhibited glioma cell proliferation and migration, demonstrating potent therapeutic activity. Mechanistic studies revealed cytoskeletal disruption, mitochondrial instability, and the generation of Reactive Oxygen Species (ROS), indicating potential cytotoxic mechanisms. Additionally,

AuNPs exhibit exceptional photothermal ablation properties, enabling precise destruction of cancerous tissues upon light activation [120].

Silver nanoparticles (AgNPs) also exhibit a broad spectrum of biological activities, which vary depending on their size, shape, concentration, target tissue, and interaction mechanisms. Their unique chemical, physical, and biological properties—combined with potent antibacterial and anticancer effects—have attracted considerable research interest [121]. The primary mechanism underlying their therapeutic action involves the generation of ROS following AgNP disintegration, which releases silver ions into cells. These ROS can damage cellular components, such as lipids, proteins, and DNA, ultimately leading to cell death [122]. Salazar-Garcia *et al.* [123] reported that AgNPs reduced the viability of glioma cells and arrested cell cycle progression at the G2/M phase. A multifunctional nanoplatform was developed by conjugating silver nanoparticles and the anticancer drug alisertib with chlorotoxin—a 36-amino-acid peptide that selectively binds to the MMP-2 receptor expressed in brain tumor cells. This silver/alisertib@PNPs–chlorotoxin system significantly inhibited tumor growth *in vivo* [124]. AgNPs offer high drug-loading capacity, reduced systemic toxicity, and enhanced biocompatibility, contributing to improved tumor-specific delivery. These properties have enabled the successful encapsulation of Histone Deacetylase (HDAC) inhibitors in AgNPs, thereby increasing their therapeutic specificity and efficacy in brain tumor models.

• Zinc Oxide NPs.

Zinc Oxide Nanoparticles (ZnO-NPs) are emerging as a cost-effective, versatile, and aesthetically appealing alternative to gold and silver nanoparticles for a wide range of applications, including piezoelectric devices, solar cells, gas sensors, light-emitting systems, food packaging, optics, agriculture, cosmetics, and anticancer therapeutics [125]. ZnO-NPs exhibit rapid electron-transfer kinetics, high adsorption capacity, excellent biocompatibility, and strong catalytic efficiency [126]. Various morphologies of ZnO nanostructures—such as nanorods, nanowires, nanobridges, nanotubes, nanobelts, nanonails, and nanoribbons—have been synthesized using techniques like thermal evaporation, hydrothermal synthesis, and ultrasonication [127]. Notably, ZnO-NPs have demonstrated the ability to reach the brain following oral administration, either *via* neural transport mechanisms or by crossing the Blood–Brain Barrier (BBB). Shim *et al.* [128] reported cytotoxic effects in both brain and blood tissues resulting from interactions between ZnO-NP, plasma, and neural components. Among the proteins identified on the surface of ZnO-NPs, apolipoprotein E was found to play a key role in facilitating BBB penetration and targeting brain pathologies. ZnO-NPs have been shown to induce oxidative stress and energy depletion in microglial cells, ultimately leading to cell death [129]. Compared to nanoparticles of similar size made from $Al_2O_3$, $TiO_2$, $CrO_3$, and $Fe_3O_4$, ZnO-NPs exhibited greater cytotoxicity in murine brain cancer cells [130]. Wahab *et al.* [131] fabricated ZnO nanostructures under varying experimental conditions to evaluate their antitumor activity. The results demonstrated significant inhibition of cell proliferation and induction of apoptosis in HeLa and U87 glioma cell lines. A marked increase in micronuclei formation was observed in U87 cells, suggesting that ZnO-NPs may interfere with DNA strand rejoining, thereby enhancing cell death. However, prolonged exposure to ZnO-NPs over a seven-day period resulted in oxidative stress in brain tissues, as evidenced by elevated malondialdehyde levels and reduced antioxidant activity. Additionally, both low and high doses of ZnO-NPs led to DNA fragmentation and increased expression of stress-related biomarkers, including heat shock protein-70, caspase-3, Fas, and pro-inflammatory cytokines [132].

## LIMITATIONS

Animal studies investigating Nanoparticle (NP)-based therapies for brain tumors have yielded promising results. However, several critical limitations must be addressed before these technologies can be translated into clinical use. One major challenge is the restricted ability of NPs to penetrate the Blood–Brain Barrier (BBB) and reach brain tissue, as many are sequestered or degraded before reaching their target. Following intravenous administration, NPs may be opsonized by plasma proteins, such as immunoglobulins, complement components, and serum albumin. This opsonization facilitates recognition by phagocytes, particularly macrophages, which internalize and degrade the particles *via* lysosomal pathways. Additionally, NPs that evade internalization may still be opsonized, obscuring targeting ligands and reducing BBB permeability [133].

The physical characteristics of NPs, including charge, size, and composition, significantly influence their circulation time and biodistribution, and warrant further investigation. Chemotherapeutic-loaded NPs may also cause off-target toxicity in organs such as the spleen and liver due to sequestration by the Reticuloendothelial System (RES), thereby reducing tumor uptake [134]. Techniques such as PEGylation are commonly employed to extend serum half-life and reduce degradation. Moreover, NPs can be engineered to release their payload in response to specific stimuli [135]. Despite their therapeutic potential, NPs pose a risk of neurotoxicity. When directed toward the brain, NPs may activate Central Nervous System (CNS) immune cells—particularly microglia, which function similarly to peripheral macrophages—thereby exacerbating neuroinflammation [136]. Activated microglia release Reactive Oxygen Species (ROS) and nitric oxide, leading to excitotoxicity and neuronal damage [137]. Concerns have been raised regarding chronic or routine exposure to NPs used in biomedical and industrial applications, as even non-brain-targeted NPs have been shown to accumulate in brain cells, causing DNA damage, apoptosis, and oxidative stress [138]. Metallic NPs, in particular, have demonstrated the potential to induce neuroinflammation upon accumulation in brain tissue [139]. Silica nanoparticles have been implicated in neurotoxicity and damage to dopaminergic neurons in the striatum [140].

Interestingly, the same mechanisms that contribute to neuroinflammation may be harnessed therapeutically. Targeting activated microglia with NPs may help reduce neuroinflammation under controlled conditions [141]. While *in vivo* studies generally support the safety of NPs designed for brain tumor imaging and therapy, careful consideration of NP dose, composition, and physicochemical properties remains essential [142]. Notably, cationic NPs have demonstrated higher toxicity compared to anionic counterparts [143]. Furthermore, NPs may inadvertently increase BBB permeability, facilitating the entry of harmful substances [144]. NPs can be tailored to respond to specific stimuli within the tumor microenvironment, such as low pH, hypoxia, and protein biomarkers. They may also be activated by external stimuli, including magnetic fields (for magnetic hyperthermia) and ultrasound (for photoacoustic therapy) [145]. In complex tumor environments, where multiple biomarkers and genetic alterations coexist, NPs can be designed to respond to multiple inputs simultaneously using molecular logic gates. These logic-gated systems enhance specificity and reduce side effects by enabling drug release and imaging only under defined conditions [146, 147]. Functional moieties on NPs can undergo conformational changes in response to external stimuli, enabling targeted delivery to cancer-specific biomarkers. Quantum Dots (QDs), for example, have been used to modulate fluorescence intensity in response to environmental molecules [148]. Badeu *et al.* demonstrated that stimulus-responsive synthetic cross-linkers in hydrogels allow precise control over degradation and drug release in response to reducing agents, enzymes, and light. Aptamer-based “DNA nanorobots” have also been developed to adapt their structure in response to stimuli, enabling targeted delivery [148]. Ma *et al.* [149] reported that DNA nanorobots can selectively target breast cancer cells and promote the degradation of membrane protein, illustrating the potential of molecular logic gating in cancer therapy.

NPs offer a promising strategy for enhancing drug delivery across the BBB, improving imaging and treatment of malignant brain tumors. However, the persistent challenges in treating aggressive brain cancers underscore the need for continued research. Ideal NP characteristics, such as size, charge, and surface functionalization, must be optimized in laboratory settings. Comparative studies between NP-based delivery and free drug administration are essential to evaluate differential efficacy and toxicity. Given the tendency of NPs to accumulate in the RES, thorough biodistribution studies are necessary to ensure selective tumor targeting. Auxiliary targeting mechanisms, such as magnetic field guidance or Focused Ultrasound (FUS) to transiently open the BBB, should also be explored. Alternative delivery routes, including intranasal administration and Convection-Enhanced Delivery (CED), may offer additional advantages. Following the successful demonstration of safety and efficacy in animal models, clinical trials are needed to determine therapeutic windows, pharmacokinetics, and overall effectiveness of NP-based therapies for brain cancer. NPs may also be used in combination with conventional treatments, such as surgery, radiation, and chemotherapy, to enhance therapeutic outcomes. Given the vast array of possible NP formulations and functional combinations, rigorous research is essential to identify the most suitable candidates for clinical application.

## CLINICAL TRIALS

Despite extensive *in vitro* and *in vivo* research confirming the efficacy and therapeutic potential of nanotechnology-based therapies for *glioblastoma multiforme* (GBM), relatively few clinical trials have been conducted to evaluate nanotherapeutics in human patients. Among the most studied formulations, Pegylated Liposomal Doxorubicin (PEG-Dox) and Temozolomide (TAZ) have been widely used as post-surgical treatments for patients with high-grade recurrent GBM and for newly diagnosed cases following chemoradiotherapy over the past two decades. Clinical trials have demonstrated the tolerability and feasibility of PEG-Dox, with the primary adverse effects being myelosuppression and palmar-plantar erythrodysesthesia [150, 151]. Targeting Epidermal Growth Factor Receptor (EGFR), which is overexpressed in approximately 50% of GBM patients, the EnGenelC Delivery Vehicle (EDV)—a nanocellular minicell platform—has been developed to encapsulate chemotherapeutic agents, such as doxorubicin. In a Phase I trial involving 14 GBM patients, EDV-doxorubicin was well tolerated, with only mild side effects, such as fever, chills, or rigors. Notably, patients did not experience myelosuppression or palmar-plantar erythrodysesthesia, which are common with conventional regimens [150 - 152]. However, despite their safety profiles, none of these nanoformulations have demonstrated significant improvements in clinical outcomes.

Magnetic hyperthermia is another promising approach that enhances the cytotoxic effects of radiation therapy through localized heating. Magnetic Iron Oxide Nanoparticles (MNPs) are injected directly into the tumor and activated by an alternating magnetic field to generate heat. This strategy has shown efficacy in preclinical animal models [154] and has progressed to clinical trials [153]. In

Europe, an MNP-based compound has been approved for use in patients with recurrent GBM undergoing radiation therapy in combination with magnetic hyperthermia [152]. Nevertheless, several challenges hinder its broader clinical adoption, including difficulty in precisely controlling intratumoral temperature, limitations for patients with pacemakers or defibrillators, requirement to remove dental crowns, implants, and fillings prior to treatment [153].

Several novel nanoformulations have demonstrated safety and preliminary efficacy in GBM treatment, including Semliki Forest virus vector encapsulating the IL-12 gene in cationic liposomes, 5-fluorouracil-releasing microspheres, and EDV-doxorubicin [152, 155]. However, further clinical trials are needed to validate their therapeutic impact. Notably, a Phase II trial is currently recruiting patients with recurrent GBM to evaluate a combination of TAZ and targeted P53 gene therapy (SGT-53), which has shown promise in other solid tumors [156].

## CONCLUSION

Glioma remains one of the most formidable brain malignancies due to its aggressive nature and poor prognosis, making complete treatment highly challenging. Chemotherapy continues to be a cornerstone in glioma management; however, the Blood–Brain Barrier (BBB) and Blood–Brain Tumor Barrier (BBTB) present significant obstacles to effective drug delivery. Nanotechnology-based drug delivery systems offer a promising solution by enhancing therapeutic efficacy while minimizing the systemic toxicity associated with conventional chemotherapy. The integration of contrast agents into nanocarrier systems enables real-time visualization of drug distribution, facilitating precise monitoring of treatment progress. Nonetheless, therapeutic failure remains a concern, as glioma cells exhibit resistance to many chemotherapeutic agents. Additionally, certain drug delivery mechanisms may compete with endogenous molecules, reducing the efficiency of drug transport. To mitigate carrier-induced toxicity, rapidly degradable materials are often employed in nanoparticle design; however, premature degradation may prevent the drug from reaching its intended site of action. Consequently, current research trends emphasize the development of multifunctional and multi-targeted nanocarrier systems that can overcome these limitations. This review highlights the state-of-the-art in nanomaterial-based drug delivery for brain cancer therapy, encompassing platforms such as liposomes, carbon dots, carbon nanotubes, dendrimers, nano-micelles, and metallic nanoparticles including silver, gold, and zinc oxide. These nanostructured materials possess the unique ability to traverse the BBB and deliver therapeutic agents directly to tumor sites, often at significantly lower doses than traditional formulations, thereby enhancing safety and efficacy. Nanotechnology also holds promise in augmenting immunotherapy, offering targeted modulation of immune responses within the tumor microenvironment. Moreover, while radiotherapy remains a standard approach for tumor size reduction, it is associated with collateral damage to healthy tissues. The incorporation of nanomaterials into radiotherapeutic strategies may enhance tumor specificity and reduce off-target effects, potentially improving clinical outcomes. Before nanomaterials can be widely adopted in clinical practice, several critical attributes must be rigorously evaluated, including biodegradability, biocompatibility, water solubility, shelf stability, and extended circulation half-life. Equally important is the comprehensive assessment of nanocarrier toxicity, which remains a key barrier to clinical translation. Future research should focus on optimizing nanoparticle design to balance therapeutic efficacy with safety, thereby paving the way for next-generation nanomedicine in brain cancer treatment.

## AUTHORS' CONTRIBUTIONS

Dr. Dhaval J. Patel: Conceptualization, writing-review and editing, investigation, resources, formal analysis, visualization. Dr. Vaibhavkumar B. Patel: Writing-original draft, investigation, formal analysis. Dr. Unnati V. Patel: Writing-review and editing, investigation, resources, formal analysis.

## REFERENCES

[1] Lima FR, Kahn SA, Soletti RC, *et al.* Glioblastoma: therapeutic challenges, what lies ahead. Biochim Biophys Acta 2012; 1826(2): 338-49.
[PMID: 22677165]

[2] Hanahan D, Weinberg RA. Hallmarks of cancer: the next generation. Cell 2011; 144(5): 646-74.
[http://dx.doi.org/10.1016/j.cell.2011.02.013] [PMID: 21376230]

[3] Louis DN, Ohgaki H, Wiestler OD, *et al.* The 2007 WHO classification of tumours of the central nervous system. Acta Neuropathol 2007; 114(2): 97-109.
[http://dx.doi.org/10.1007/s00401-007-0243-4] [PMID: 17618441]

[4] Dolecek TA, Propp JM, Stroup NE, Kruchko C. CBTRUS statistical report: primary brain and central nervous system tumors diagnosed in the United States in 2005-2009. Neuro-oncol 2012; 14(Suppl 5) (Suppl. 5): v1-v49.
[http://dx.doi.org/10.1093/neuonc/nos218] [PMID: 23095881]

[5] Harter DH, Wilson TA, Karajannis MA. Glioblastoma multiforme: State of the art and future therapeutics. Surg Neurol Int 2014; 5(1): 64.
[http://dx.doi.org/10.4103/2152-7806.132138] [PMID: 24991467]

[6] Chen W, Zheng R, Baade PD, *et al.* Cancer statistics in China, 2015. CA Cancer J Clin 2016; 66(2): 115-32.
[http://dx.doi.org/10.3322/caac.21338] [PMID: 26808342]

[7] Sung H, Ferlay J, Siegel RL, *et al.* Global cancer statistics 2020: GLOBOCAN estimates of incidence and mortality worldwide for 36 cancers in 185 countries. CA Cancer J Clin 2021; 71(3): 209-49.
[http://dx.doi.org/10.3322/caac.21660] [PMID: 33538338]

[8] Fokas E, Steinbach JP, Rödel C. Biology of brain metastases and novel targeted therapies: time to translate the research. Biochim Biophys Acta 2013; 1835(1): 61-75.
[PMID: 23142311]

[9] Rivkin M, Kanoff RB. Metastatic brain tumors: current therapeutic options and historical perspective. J Am Osteopath Assoc 2013; 113(5): 418-23. [PMID: 23667195]

[10] Liu C, Zong H. Developmental origins of brain tumors. Curr Opin Neurobiol 2012; 22(5): 844-9. [http://dx.doi.org/10.1016/j.conb.2012.04.012] [PMID: 22560511]

[11] Ostrom QT, Gittleman H, Truitt G, Boscia A, Kruchko C, Barnholtz-Sloan JS. CBTRUS statistical report: Primary brain and other central nervous system tumors diagnosed in the United States in 2011–2015. Neuro-oncol 2018; 20(4) (Suppl. 4): iv1-iv86. [http://dx.doi.org/10.1093/neuonc/noy131] [PMID: 30445539]

[12] Ohgaki H, Kleihues P. The definition of primary and secondary glioblastoma. Clin Cancer Res 2013; 19(4): 764-72. [http://dx.doi.org/10.1158/1078-0432.CCR-12-3002] [PMID: 23209033]

[13] Karcher S, Steiner HH, Ahmadi R, *et al.* Different angiogenic phenotypes in primary and secondary glioblastomas. Int J Cancer 2006; 118(9): 2182-9. [http://dx.doi.org/10.1002/ijc.21648] [PMID: 16331629]

[14] Renault IZ, Golgher D. Molecular genetics of glioblastomas: defining subtypes and understanding the biology. Neuroimaging Clin N Am 2015; 25(1): 97-103. [http://dx.doi.org/10.1016/j.nic.2014.09.007] [PMID: 25476515]

[15] Zhang X, Zhang W, Cao WD, Cheng G, Zhang YQ. Glioblastoma multiforme: Molecular characterization and current treatment strategy (Review). Exp Ther Med 2012; 3(1): 9-14. [http://dx.doi.org/10.3892/etm.2011.367] [PMID: 22969836]

[16] Le Rhun E, Preusser M, Roth P, *et al.* Molecular targeted therapy of glioblastoma. Cancer Treat Rev 2019; 80: 101896. [http://dx.doi.org/10.1016/j.ctrv.2019.101896] [PMID: 31541850]

[17] Chakrabarti I, Cockburn M, Cozen W, Wang YP, Preston-Martin S. A population-based description of glioblastoma multiforme in Los Angeles County, 1974–1999. Cancer 2005; 104(12): 2798-806. [http://dx.doi.org/10.1002/cncr.21539] [PMID: 16288487]

[18] Tian M, Ma W, Chen Y, *et al.* Impact of gender on the survival of patients with glioblastoma. Biosci Rep 2018; 38(6): BSR20180752. [http://dx.doi.org/10.1042/BSR20180752] [PMID: 30305382]

[19] Korja M, Raj R, Seppä K, *et al.* Glioblastoma survival is improving despite increasing incidence rates: a nationwide study between 2000 and 2013 in Finland. Neuro-oncol 2019; 21(3): 370-9. [http://dx.doi.org/10.1093/neuonc/noy164] [PMID: 30312433]

[20] Philips A, Henshaw DL, Lamburn G, O'Carroll MJ. Brain tumors: Rise in glioblastoma multiforme incidence in England 1995–2015 suggests an adverse environmental or lifestyle factor. J Environ Public Health 2018; 2018: 7910754. [PMID: 30034480]

[21] Thakkar JP, Dolecek TA, Horbinski C, *et al.* Epidemiologic and molecular prognostic review of glioblastoma. Cancer Epidemiol Biomarkers Prev 2014; 23(10): 1985-96. [http://dx.doi.org/10.1158/1055-9965.EPI-14-0275] [PMID: 25053711]

[22] Miranda-Filho A, Piñeros M, Soerjomataram I, Deltour I, Bray F. Cancers of the brain and CNS: global patterns and trends in incidence. Neuro-oncol 2017; 19(2): 270-80. [PMID: 27571887]

[23] Stanaway JD, Afshin A, Gakidou E, *et al.* Global, regional, and national comparative risk assessment of 84 behavioural, environmental and occupational, and metabolic risks or clusters of risks for 195 countries and territories, 1990–2017: a systematic analysis for the Global Burden of Disease Study 2017. Lancet 2018; 392(10159): 1923-94. [http://dx.doi.org/10.1016/S0140-6736(18)32225-6] [PMID: 30496105]

[24] Bukhari SZ, Zeth K, Iftikhar M, *et al.* Supramolecular lipid nanoparticles as delivery carriers for non-invasive cancer theranostics. Current Research in Pharmacology and Drug Discovery 2021; 2: 100067. [http://dx.doi.org/10.1016/j.crphar.2021.100067] [PMID: 34909685]

[25] Hanna NH, Einhorn LH. Testicular cancer--discoveries and updates. N Engl J Med 2014; 371(21): 2005-16. [http://dx.doi.org/10.1056/NEJMra1407550] [PMID: 25409373]

[26] Rios Velazquez E, Parmar C, Liu Y, *et al.* Somatic mutations drive distinct imaging phenotypes in lung cancer. Cancer Res 2017; 77(14): 3922-30. [http://dx.doi.org/10.1158/0008-5472.CAN-17-0122] [PMID: 28566328]

[27] Birgisson H, Påhlman L, Gunnarsson U, Glimelius B. Late adverse effects of radiation therapy for rectal cancer – a systematic overview. Acta Oncol 2007; 46(4): 504-16. [http://dx.doi.org/10.1080/02841860701348670] [PMID: 17497318]

[28] Mukherjee A, Waters AK, Babic I, *et al.* Antibody drug conjugates: Progress, pitfalls, and promises. Hum Antibodies 2018; 27(1): 53-62. [http://dx.doi.org/10.3233/HAB-180348] [PMID: 30223393]

[29] Norouzi M, Amerian M, Amerian M, Atyabi F. Clinical applications of nanomedicine in cancer therapy. Drug Discov Today 2020; 25(1): 107-25. [http://dx.doi.org/10.1016/j.drudis.2019.09.017] [PMID: 31586642]

[30] Mukherjee A, Waters AK, Kalyan P, Achrol AS, Kesari S, Yenugonda VM. Lipid–polymer hybrid nanoparticles as a next-generation drug delivery platform: state of the art, emerging technologies, and perspectives. Int J Nanomedicine 2019; 14: 1937-52. [http://dx.doi.org/10.2147/IJN.S198353] [PMID: 30936695]

[31] Yue X, Dai Z. Liposomal nanotechnology for cancer theranostics. Curr Med Chem 2018; 25(12): 1397-408. [http://dx.doi.org/10.2174/0929867324666170306105350] [PMID: 28266269]

[32] Alexis F, Rhee JW, Richie JP, Radovic-Moreno AF, Langer R, Farokhzad OC. New frontiers in nanotechnology for cancer treatment. Urol Oncol 2008; 26(1): 74-85. [http://dx.doi.org/10.1016/j.urolonc.2007.03.017] [PMID: 18190835]

[33] Bhowmik A, Khan R, Ghosh MK. Blood brain barrier: a challenge for effectual therapy of brain tumors. BioMed Res Int 2015; 2015: 1-20. [http://dx.doi.org/10.1155/2015/320941] [PMID: 25866775]

[34] Dostalova S, Heger Z, Kudr J, *et al.* Apoferritin: protein nanocarrier for targeted delivery. In: Naik J, Ed. Nano Based Drug Delivery Zagerb. IAPC Publishing 2015; pp. 217-33. [http://dx.doi.org/10.5599/obp.8.8]

[35] Ediriwickrema A, Saltzman WM. Nanotherapy for cancer: targeting and multifunctionality in the future of cancer therapies. ACS Biomater Sci Eng 2015; 1(2): 64-78. [http://dx.doi.org/10.1021/ab500084g] [PMID: 25984571]

[36] Van der Meel R, Vehmeijer LJC, Kok RJ, Storm G, van Gaal EVB. Ligand-targeted particulate nanomedicines undergoing clinical evaluation: Current status. Adv Drug Deliv Rev 2013; 65(10): 1284-98. [http://dx.doi.org/10.1016/j.addr.2013.08.012] [PMID: 24018362]

[37] Lesniak MS, Upadhyay U, Goodwin R, Tyler B, Brem H. Local delivery of doxorubicin for the treatment of malignant brain tumors in rats. Anticancer Res 2005; 25(6B): 3825-31. [PMID: 16312042]

[38] Sun C, Ding Y, Zhou L, *et al.* Noninvasive nanoparticle strategies for brain tumor targeting. Nanomedicine 2017; 13(8): 2605-21. [http://dx.doi.org/10.1016/j.nano.2017.07.009] [PMID: 28756093]

[39] Tsou YH, Zhang XQ, Zhu H, Syed S, Xu X. Drug Delivery to the Brain across the Blood–Brain Barrier Using Nanomaterials. Small 2017; 13(43): 1701921. [http://dx.doi.org/10.1002/smll.201701921] [PMID: 29045030]

[40] Jena L, McErlean E, McCarthy H. Delivery across the blood-brain barrier: nanomedicine for glioblastoma multiforme. Drug Deliv Transl Res 2020; 10(2): 304-18. [http://dx.doi.org/10.1007/s13346-019-00679-2] [PMID: 31728942]

[41] Patel MM, Patel BM. Crossing the blood–brain barrier: Recent advances in drug delivery to the brain. CNS Drugs 2017; 31(2): 109-33. [http://dx.doi.org/10.1007/s40263-016-0405-9] [PMID: 28101766]

[42] Chen Y, Wei C, Lyu Y, Chen H, Jiang G, Gao X. Biomimetic drug-delivery systems for the management of brain diseases. Biomater Sci 2020; 8(4): 1073-88. [http://dx.doi.org/10.1039/C9BM01395D] [PMID: 31728485]

[43] Saka R, Sathe P, Khan W. Brain local delivery strategy. Brain Targeted Drug Delivery System. Amsterdam, The Netherlands: Elsevier 2019; pp. 241-86. [http://dx.doi.org/10.1016/B978-0-12-814001-7.00011-1]

[44] Walter EJ, Carraretto M. The neurological and cognitive consequences of hyperthermia. Crit Care 2016; 20(1): 199. [http://dx.doi.org/10.1186/s13054-016-1376-4] [PMID: 27411704]

[45] Skandalakis GP, Rivera DR, Rizea CD, *et al.* Hyperthermia treatment advances for brain tumors. Int J Hyperthermia 2020; 37(2): 3-19. [http://dx.doi.org/10.1080/02656736.2020.1772512] [PMID: 32672123]

[46] Wu M, Chen W, Chen Y, *et al.* Focused ultrasound-augmented delivery of biodegradable multifunctional nanoplatforms for imaging-guided brain tumor treatment. Adv Sci (Weinh) 2018; 5(4): 1700474. [http://dx.doi.org/10.1002/advs.201700474] [PMID: 29721406]

[47] Lu YJ, Lin PY, Huang PH, *et al.* Magnetic graphene oxide for dual targeted delivery of doxorubicin and photothermal therapy. Nanomaterials (Basel) 2018; 8(4): 193. [http://dx.doi.org/10.3390/nano8040193] [PMID: 29584656]

[48] Kang YJ, Cutler EG, Cho H. Therapeutic nanoplatforms and delivery strategies for neurological disorders. Nano Converg 2018; 5(1): 35. [http://dx.doi.org/10.1186/s40580-018-0168-8] [PMID: 30499047]

[49] Hwang HH, Lee DY. Protein-based drug delivery in brain tumor therapy. Adv Exp Med Biol 2020; 1249: 203-21. [http://dx.doi.org/10.1007/978-981-15-3258-0_13] [PMID: 32602099]

[50] Gao W, Zhang L. Coating nanoparticles with cell membranes for targeted drug delivery. J Drug Target 2015; 23(7-8): 619-26. [http://dx.doi.org/10.3109/1061186X.2015.1052074] [PMID: 26453159]

[51] Bae YH, Park K. Targeted drug delivery to tumors: Myths, reality and possibility. J Control Release 2011; 153(3): 198-205. [http://dx.doi.org/10.1016/j.jconrel.2011.06.001] [PMID: 21663778]

[52] Shi J, Kantoff PW, Wooster R, Farokhzad OC. Cancer nanomedicine: progress, challenges and opportunities. Nat Rev Cancer 2017; 17(1): 20-37. [http://dx.doi.org/10.1038/nrc.2016.108] [PMID: 27834398]

[53] Lancet JE, Uy GL, Cortes JE, *et al.* Final results of a phase III randomized trial of CPX-351 *versus* 7+3 in older patients with newly diagnosed high risk (secondary) AML. J Clin Oncol 2016; 34(15_suppl): 7000. [http://dx.doi.org/10.1200/JCO.2016.34.15_suppl.7000]

[54] Chauhan VP, Jain RK. Strategies for advancing cancer nanomedicine. Nat Mater 2013; 12(11): 958-62.
[http://dx.doi.org/10.1038/nmat3792] [PMID: 24150413]

[55] Lee H, Hoang B, Fonge H, Reilly RM, Allen C. *In vivo* distribution of polymeric nanoparticles at the whole-body, tumor, and cellular levels. Pharm Res 2010; 27(11): 2343-55.
[http://dx.doi.org/10.1007/s11095-010-0068-z] [PMID: 20195708]

[56] Ernsting MJ, Murakami M, Roy A, Li SD. Factors controlling the pharmacokinetics, biodistribution and intratumoral penetration of nanoparticles. J Control Release 2013; 172(3): 782-94.
[http://dx.doi.org/10.1016/j.jconrel.2013.09.013] [PMID: 24075927]

[57] Wong C, Stylianopoulos T, Cui J, *et al.* Multistage nanoparticle delivery system for deep penetration into tumor tissue. Proc Natl Acad Sci USA 2011; 108(6): 2426-31.
[http://dx.doi.org/10.1073/pnas.1018382108] [PMID: 21245339]

[58] Bogart LK, Pourrov G, Murphy CJ, *et al.* Nanoparticles for imaging, sensing, and therapeutic intervention ACS Publications, Washington DC, 2014;8(4): 3107-3122.

[59] Gao H. Perspectives on dual targeting delivery systems for brain tumors. J Neuroimmune Pharmacol 2017; 12(1): 6-16.
[http://dx.doi.org/10.1007/s11481-016-9687-4] [PMID: 27270720]

[60] Kuang J, Song W, Yin J, *et al.* iRGD modified chemo-immunotherapeutic nanoparticles for enhanced immunotherapy against glioblastoma. Adv Funct Mater 2018; 28(17): 1800025.
[http://dx.doi.org/10.1002/adfm.201800025]

[61] Zhu Q, Ling X, Yang Y, *et al.* Embryonic stem cells-derived exosomes endowed with targeting properties as chemotherapeutics delivery vehicles for glioblastoma therapy. Adv Sci (Weinh) 2019; 6(6): 1801899.
[http://dx.doi.org/10.1002/advs.201801899] [PMID: 30937268]

[62] Wang B, Lv L, Wang Z, *et al.* Nanoparticles functionalized with Pep-1 as potential glioma targeting delivery system *via* interleukin 13 receptor α2-mediated endocytosis. Biomaterials 2014; 35(22): 5897-907.
[http://dx.doi.org/10.1016/j.biomaterials.2014.03.068] [PMID: 24743033]

[63] Kirpotin DB, Drummond DC, Shao Y, *et al.* Antibody targeting of long-circulating lipidic nanoparticles does not increase tumor localization but does increase internalization in animal models. Cancer Res 2006; 66(13): 6732-40.
[http://dx.doi.org/10.1158/0008-5472.CAN-05-4199] [PMID: 16818648]

[64] Zhou H, Qian W, Uckun FM, *et al.* IGF1 receptor targeted theranostic nanoparticles for targeted and image-guided therapy of pancreatic cancer. ACS Nano 2015; 9(8): 7976-91.
[http://dx.doi.org/10.1021/acsnano.5b01288] [PMID: 26242412]

[65] Zhong Y, Su T, Shi Q, *et al.* Co-administration of iRGD enhances tumor-targeted delivery and anti-tumor effects of paclitaxel-loaded PLGA nanoparticles for colorectal cancer treatment. Int J Nanomedicine 2019; 14: 8543-60.
[http://dx.doi.org/10.2147/IJN.S219820] [PMID: 31802868]

[66] Shi X, Ma R, Lu Y, *et al.* iRGD and TGN co-modified PAMAM for multi-targeted delivery of ATO to gliomas. Biochem Biophys Res Commun 2020; 527(1): 117-23.
[http://dx.doi.org/10.1016/j.bbrc.2020.04.064] [PMID: 32446354]

[67] Liu C, Zhao Z, Gao H, *et al.* Enhanced blood-brain-barrier penetrability and tumor-targeting efficiency by peptide-functionalized poly(amidoamine) dendrimer for the therapy of gliomas. Nanotheranostics 2019; 3(4): 311-30.
[http://dx.doi.org/10.7150/ntno.38954] [PMID: 31687320]

[68] Gandhi H, Sharma AK, Mahant S, Kapoor DN. Recent advancements in brain tumor targeting using magnetic nanoparticles. Ther Deliv 2020; 11(2): 97-112.

[http://dx.doi.org/10.4155/tde-2019-0077] [PMID: 31914859]

[69] Cui Y, Zhang M, Zeng F, Jin H, Xu Q, Huang Y. Dual-targeting magnetic plga nanoparticles for codelivery of paclitaxel and curcumin for brain tumor therapy. ACS Appl Mater Interfaces 2016; 8(47): 32159-69.
[http://dx.doi.org/10.1021/acsami.6b10175] [PMID: 27808492]

[70] Lu YJ, Chuang EY, Cheng YH, Anilkumar TS, Chen HA, Chen JP. Thermosensitive magnetic liposomes for alternating magnetic field-inducible drug delivery in dual targeted brain tumor chemotherapy. Chem Eng J 2019; 373: 720-33.
[http://dx.doi.org/10.1016/j.cej.2019.05.055]

[71] Sengul AB, Asmatulu E. Toxicity of metal and metal oxide nanoparticles: a review. Environ Chem Lett 2020; 18(5): 1659-83.
[http://dx.doi.org/10.1007/s10311-020-01033-6]

[72] Quader S, Kataoka K. Nanomaterial-enabled cancer therapy. Mol Ther 2017; 25(7): 1501-13.
[http://dx.doi.org/10.1016/j.ymthe.2017.04.026] [PMID: 28532763]

[73] Jain KK. Nanobiotechnology-based drug delivery to the central nervous system. Neurodegener Dis 2007; 4(4): 287-91.
[http://dx.doi.org/10.1159/000101884] [PMID: 17627131]

[74] Teleanu DM, Chircov C, Grumezescu AM, Teleanu RI. Neurotoxicity of nanomaterials: An up-to-date overview. Nanomaterials (Basel) 2019; 9(1): 96.
[http://dx.doi.org/10.3390/nano9010096] [PMID: 30642104]

[75] Frank D, Tyagi C, Tomar L, *et al.* Overview of the role of nanotechnological innovations in the detection and treatment of solid tumors. Int J Nanomedicine 2014; 9: 589-613.
[PMID: 24489467]

[76] Gatoo MA, Naseem S, Arfat MY, Mahmood Dar A, Qasim K, Zubair S. Physicochemical properties of nanomaterials: implication in associated toxic manifestations. BioMed Res Int 2014; 2014: 1-8.
[http://dx.doi.org/10.1155/2014/498420] [PMID: 25165707]

[77] Malam Y, Loizidou M, Seifalian AM. Liposomes and nanoparticles: nanosized vehicles for drug delivery in cancer. Trends Pharmacol Sci 2009; 30(11): 592-9.
[http://dx.doi.org/10.1016/j.tips.2009.08.004] [PMID: 19837467]

[78] Torchilin VP. Recent advances with liposomes as pharmaceutical carriers. Nat Rev Drug Discov 2005; 4(2): 145-60.
[http://dx.doi.org/10.1038/nrd1632] [PMID: 15688077]

[79] Bozzuto G, Molinari A. Liposomes as nanomedical devices. Int J Nanomedicine 2015; 10: 975-99.
[http://dx.doi.org/10.2147/IJN.S68861] [PMID: 25678787]

[80] Allen TM, Cullis PR. Liposomal drug delivery systems: From concept to clinical applications. Adv Drug Deliv Rev 2013; 65(1): 36-48.
[http://dx.doi.org/10.1016/j.addr.2012.09.037] [PMID: 23036225]

[81] Immordino ML, Dosio F, Cattel L. Stealth liposomes: review of the basic science, rationale, and clinical applications, existing and potential. Int J Nanomedicine 2006; 1(3): 297-315.
[PMID: 17717971]

[82] Lakkadwala S, dos Santos Rodrigues B, Sun C, Singh J. Dual functionalized liposomes for efficient co-delivery of anti-cancer chemotherapeutics for the treatment of glioblastoma. J Control Release 2019; 307: 247-60.
[http://dx.doi.org/10.1016/j.jconrel.2019.06.033] [PMID: 31252036]

[83] Muthu MS, Kulkarni SA, Xiong J, Feng SS. Vitamin E TPGS coated liposomes enhanced cellular uptake and cytotoxicity of docetaxel in brain cancer cells. Int J Pharm 2011; 421(2): 332-40.
[http://dx.doi.org/10.1016/j.ijpharm.2011.09.045] [PMID: 22001537]

[84] Zhan C, Gu B, Xie C, Li J, Liu Y, Lu W. Cyclic RGD conjugated poly(ethylene glycol)-co-poly(lactic acid) micelle enhances paclitaxel anti-glioblastoma effect. J Control Release 2010; 143(1): 136-42. [http://dx.doi.org/10.1016/j.jconrel.2009.12.020] [PMID: 20056123]

[85] Postma TJ, Heimans JJ, Luykx SA, *et al.* A phase II study of paclitaxel in chemonaïve patients with recurrent high-grade glioma. Ann Oncol 2000; 11(4): 409-13. [http://dx.doi.org/10.1023/A:1008376123066] [PMID: 10847458]

[86] Anfosso L, Efferth T, Albini A, Pfeffer U. Microarray expression profiles of angiogenesis-related genes predict tumor cell response to artemisinins. Pharmacogenomics J 2006; 6(4): 269-78. [http://dx.doi.org/10.1038/sj.tpj.6500371] [PMID: 16432535]

[87] Qiao Y, Wan J, Zhou L, *et al.* Stimuli-responsive nanotherapeutics for precision drug delivery and cancer therapy. Wiley Interdiscip Rev Nanomed Nanobiotechnol 2019; 11(1): e1527. [http://dx.doi.org/10.1002/wnan.1527] [PMID: 29726115]

[88] Ravi V, Abraham T, Madhankumar A, Slagle-Webb B, Pandya B, Connor J. Exth-63. Liposomal delivery of ferritin heavy chain (fth1) sirna results in increased radiation sensitivity in patient derived glioma initiating cells. Neuro-oncol 2019; 21 (Suppl. 6): vi95. [http://dx.doi.org/10.1093/neuonc/noz175.393]

[89] Floyd J, Phillips W, Goins B, Bao A, Brenner A. ATNT-03 first in human study of rhenium nanoliposomes for intratumoral therapy of glioma. Neuro-oncol 2015; 17 (Suppl. 5): v10.3. [http://dx.doi.org/10.1093/neuonc/nov205.03]

[90] Brenner AJ, Bao A, Phillips W, *et al.* Safety and feasibility of rhenium-186 nanoliposome ( $^{186}$ RNL) in recurrent glioma: The ReSPECT phase 1 trial. J Clin Oncol 2021; 39(15_suppl): 2061. [http://dx.doi.org/10.1200/JCO.2021.39.15_suppl.2061]

[91] Zhang C, Nance EA, Mastorakos P, *et al.* Convection enhanced delivery of cisplatin-loaded brain penetrating nanoparticles cures malignant glioma in rats. J Control Release 2017; 263: 112-9. [http://dx.doi.org/10.1016/j.jconrel.2017.03.007] [PMID: 28279797]

[92] Kumar K, Butowski N, Aghi M, *et al.* SCIDOT-02 a phase I study of convection-enhanced delivery of liposomal-irinotecan (ONIVYDE) using real-time imaging with gadolinium in patients with recurrent high grade gliomas: results thus far. Neuro-oncol 2019; 21 (Suppl. 6): vi272. [http://dx.doi.org/10.1093/neuonc/noz175.1143]

[93] Mendiburu-Eliçabe M, Gil-Ranedo J. Combination therapy of intraperitoneal rapamycin and convection-enhanced delivery of Nanoliposomal CPT-11 in rodent Orthotopic brain tumor xenografts. Curr Cancer Drug Targets 2015; 15(4): 352-62. [http://dx.doi.org/10.2174/1568009615666150225123120] [PMID: 25714699]

[94] Pantshwa JM, Kondiah PPD, Choonara YE, Marimuthu T, Pillay V. Nanodrug delivery systems for the treatment of ovarian cancer. Cancers (Basel) 2020; 12(1): 213. [http://dx.doi.org/10.3390/cancers12010213] [PMID: 31952210]

[95] Fathi M, Majidi S, Zangabad PS, Barar J, Erfan-Niya H, Omidi Y. Chitosan-based multifunctional nanomedicines and theranostics for targeted therapy of cancer. Med Res Rev 2018; 38(6): 2110-36. [http://dx.doi.org/10.1002/med.21506] [PMID: 29846948]

[96] Feng ST, Li J, Luo Y, *et al.* pH-sensitive nanomicelles for controlled and efficient drug delivery to human colorectal carcinoma LoVo cells. PLoS One 2014; 9(6): e100732. [http://dx.doi.org/10.1371/journal.pone.0100732] [PMID: 24964012]

[97] Kim S, Shi Y, Kim JY, Park K, Cheng JX. Overcoming the barriers in micellar drug delivery: loading efficiency, *in vivo* stability, and micelle–cell interaction. Expert Opin Drug Deliv 2010; 7(1): 49-62. [http://dx.doi.org/10.1517/17425240903380446] [PMID: 20017660]

[98] Agrawal P, Sonali , Singh RP, *et al.* Bioadhesive micelles of d -α-tocopherol polyethylene glycol succinate 1000: Synergism of chitosan and transferrin in targeted drug delivery. Colloids Surf B Biointerfaces 2017; 152: 277-88.

[http://dx.doi.org/10.1016/j.colsurfb.2017.01.021] [PMID: 28122295]

[99] Sonali , Agrawal P, Singh RP, *et al.* Transferrin receptor-targeted vitamin E TPGS micelles for brain cancer therapy: preparation, characterization and brain distribution in rats. Drug Deliv 2016; 23(5): 1788-98.
[http://dx.doi.org/10.3109/10717544.2015.1094681] [PMID: 26431064]

[100] Mittal S, Ashhar MU, Qizilbash FF, *et al.* Ligand conjugated targeted nanotherapeutics for treatment of neurological disorders. Curr Pharm Des 2020; 26(19): 2291-305.
[http://dx.doi.org/10.2174/1381612826666200417141600] [PMID: 32303160]

[101] Niu J, Wang L, Yuan M, Zhang J, Chen H, Zhang Y. Dual-targeting nanocarrier based on glucose and folic acid functionalized pluronic P105 polymeric micelles for enhanced brain distribution. J Drug Deliv Sci Technol 2020; 57: 101343.
[http://dx.doi.org/10.1016/j.jddst.2019.101343]

[102] Noriega-Luna B, Godínez LA, Rodríguez FJ, *et al.* Applications of dendrimers in drug delivery agents, diagnosis, therapy, and detection. J Nanomater 2014; 2014(1): 507273.
[http://dx.doi.org/10.1155/2014/507273]

[103] Mishra V, Kesharwani P. Dendrimer technologies for brain tumor. Drug Discov Today 2016; 21(5): 766-78.
[http://dx.doi.org/10.1016/j.drudis.2016.02.006] [PMID: 26891979]

[104] Ambekar RS, Choudhary M, Kandasubramanian B. Recent advances in dendrimer-based nanoplatform for cancer treatment: A review. Eur Polym J 2020; 126: 109546.
[http://dx.doi.org/10.1016/j.eurpolymj.2020.109546]

[105] Yan H, Wang L, Wang J, *et al.* Two-order targeted brain tumor imaging by using an optical/paramagnetic nanoprobe across the blood brain barrier. ACS Nano 2012; 6(1): 410-20.
[http://dx.doi.org/10.1021/nn203749v] [PMID: 22148835]

[106] Liu YL, Chen D, Shang P, Yin DC. A review of magnet systems for targeted drug delivery. J Control Release 2019; 302: 90-104.
[http://dx.doi.org/10.1016/j.jconrel.2019.03.031] [PMID: 30946854]

[107] Muniswamy VJ, Raval N, Gondaliya P, Tambe V, Kalia K, Tekade RK. 'Dendrimer-Cationize--Albumin' encrusted polymeric nanoparticle improves BBB penetration and anticancer activity of doxorubicin. Int J Pharm 2019; 555: 77-99.
[http://dx.doi.org/10.1016/j.ijpharm.2018.11.035] [PMID: 30448308]

[108] Mahajan S, Patharkar A, Kuche K, *et al.* Functionalized carbon nanotubes as emerging delivery system for the treatment of cancer. Int J Pharm 2018; 548(1): 540-58.
[http://dx.doi.org/10.1016/j.ijpharm.2018.07.027] [PMID: 29997043]

[109] Harsha PJ, Thotakura N, Kumar M, *et al.* A novel PEGylated carbon nanotube conjugated mangiferin: An explorative nanomedicine for brain cancer cells. J Drug Deliv Sci Technol 2019; 53: 101186.
[http://dx.doi.org/10.1016/j.jddst.2019.101186]

[110] You Y, Wang N, He L, *et al.* Designing dual-functionalized carbon nanotubes with high blood–brain-barrier permeability for precise orthotopic glioma therapy. Dalton Trans 2019; 48(5): 1569-73.
[http://dx.doi.org/10.1039/C8DT03948H] [PMID: 30499579]

[111] Mishra V, Patil A, Thakur S, Kesharwani P. Carbon dots: emerging theranostic nanoarchitectures. Drug Discov Today 2018; 23(6): 1219-32.
[http://dx.doi.org/10.1016/j.drudis.2018.01.006] [PMID: 29366761]

[112] Zuo J, Jiang T, Zhao X, Xiong X, Xiao S, Zhu Z. Preparation and application of fluorescent carbon dots. J Nanomater 2015; 2015(1): 787862.
[http://dx.doi.org/10.1155/2015/787862]

[113] Zheng M, Liu S, Li J, *et al.* Integrating oxaliplatin with highly luminescent carbon dots: an unprecedented theranostic agent for personalized medicine. Adv Mater 2014; 26(21): 3554-60.

[http://dx.doi.org/10.1002/adma.201306192] [PMID: 24634004]

[114] Hettiarachchi SD, Graham RM, Mintz KJ, *et al.* Triple conjugated carbon dots as a nano-drug delivery model for glioblastoma brain tumors. Nanoscale 2019; 11(13): 6192-205. [http://dx.doi.org/10.1039/C8NR08970A] [PMID: 30874284]

[115] Sonali MKV, Viswanadh MK, Singh RP, *et al.* Nanotheranostics: Emerging strategies for early diagnosis and therapy of brain cancer. Nanotheranostics 2018; 2(1): 70-86. [http://dx.doi.org/10.7150/ntno.21638] [PMID: 29291164]

[116] Fan Z, Fu PP, Yu H, Ray PC. Theranostic nanomedicine for cancer detection and treatment. Yao Wu Shi Pin Fen Xi 2014; 22(1): 3-17. [PMID: 24673900]

[117] Norden AD, Drappatz J, Wen PY. Novel anti-angiogenic therapies for malignant gliomas. Lancet Neurol 2008; 7(12): 1152-60. [http://dx.doi.org/10.1016/S1474-4422(08)70260-6] [PMID: 19007739]

[118] Pandey S, Oza G, Mewada A, Shah R, Thakur M, Sharon M. Folic acid mediated synaphic delivery of doxorubicin using biogenic gold nanoparticles anchored to biological linkers. J Mater Chem B Mater Biol Med 2013; 1(9): 1361-70. [http://dx.doi.org/10.1039/c2tb00168c] [PMID: 32260809]

[119] Alle M, G B, Kim TH, Park SH, Lee SH, Kim JC. Doxorubicin-carboxymethyl xanthan gum capped gold nanoparticles: Microwave synthesis, characterization, and anti-cancer activity. Carbohydr Polym 2020; 229: 115511. [http://dx.doi.org/10.1016/j.carbpol.2019.115511] [PMID: 31826400]

[120] Mohamed MS, Veeranarayanan S, Poulose AC, *et al.* Type 1 ribotoxin-curcin conjugated biogenic gold nanoparticles for a multimodal therapeutic approach towards brain cancer. Biochim Biophys Acta, Gen Subj 2014; 1840(6): 1657-69. [http://dx.doi.org/10.1016/j.bbagen.2013.12.020] [PMID: 24361614]

[121] Le Ouay B, Stellacci F. Antibacterial activity of silver nanoparticles: A surface science insight. Nano Today 2015; 10(3): 339-54. [http://dx.doi.org/10.1016/j.nantod.2015.04.002]

[122] Ebrahimzadeh Z, Salehzadeh A, Naeemi AS, Jalali A. Silver nanoparticles biosynthesized by Anabaena flos-aquae enhance the apoptosis in breast cancer cell line. Bull Mater Sci 2020; 43(1): 92. [http://dx.doi.org/10.1007/s12034-020-2064-1]

[123] Salazar-García S, García-Rodrigo JF, Martínez-Castañón GA, Ruiz-Rodríguez VM, Portales-Pérez DP, Gonzalez C. Silver nanoparticles (AgNPs) and zinc chloride (ZnCl2) exposure order determines the toxicity in C6 rat glioma cells. J Nanopart Res 2020; 22(9): 253. [http://dx.doi.org/10.1007/s11051-020-04984-7]

[124] Locatelli E, Naddaka M, Uboldi C, *et al.* Targeted delivery of silver nanoparticles and alisertib: *In vitro* and *in vivo* synergistic effect against glioblastoma. Nanomedicine (Lond) 2014; 9(6): 839-49. [http://dx.doi.org/10.2217/nnm.14.1] [PMID: 24433240]

[125] Rasmussen JW, Martinez E, Louka P, Wingett DG. Zinc oxide nanoparticles for selective destruction of tumor cells and potential for drug delivery applications. Expert Opin Drug Deliv 2010; 7(9): 1063-77. [http://dx.doi.org/10.1517/17425247.2010.502560] [PMID: 20716019]

[126] Jin BJ, Bae SH, Lee SY, Im S. Effects of native defects on optical and electrical properties of ZnO prepared by pulsed laser deposition. Mater Sci Eng B 2000; 71(1-3): 301-5. [http://dx.doi.org/10.1016/S0921-5107(99)00395-5]

[127] Ostrovsky S, Kazimirsky G, Gedanken A, Brodie C. Selective cytotoxic effect of ZnO nanoparticles on glioma cells. Nano Res 2009; 2(11): 882-90. [http://dx.doi.org/10.1007/s12274-009-9089-5]

[128] Shim KH, Hulme J, Maeng EH, Kim MK, An SSA. Analysis of zinc oxide nanoparticles binding proteins in rat blood and brain homogenate. Int J Nanomedicine 2014; 9(Suppl 2) (Suppl. 2): 217-24. [PMID: 25565839]

[129] Wei L, Wang J, Chen A, Liu J, Feng X, Shao L. Involvement of PINK1/parkin-mediated mitophagy in ZnO nanoparticle-induced toxicity in BV-2 cells. Int J Nanomedicine 2017; 12: 1891-903. [http://dx.doi.org/10.2147/IJN.S129375] [PMID: 28331313]

[130] Jeng HA, Swanson J. Toxicity of metal oxide nanoparticles in mammalian cells. J Environ Sci Health Part A Tox Hazard Subst Environ Eng 2006; 41(12): 2699-711. [http://dx.doi.org/10.1080/10934520600966177] [PMID: 17114101]

[131] Wahab R, Kaushik NK, Verma AK, *et al.* Fabrication and growth mechanism of ZnO nanostructures and their cytotoxic effect on human brain tumor U87, cervical cancer HeLa, and normal HEK cells. J Biol Inorg Chem 2011; 16(3): 431-42. [http://dx.doi.org/10.1007/s00775-010-0740-0] [PMID: 21140179]

[132] Attia H, Nounou H, Shalaby M. Zinc oxide nanoparticles induced oxidative DNA damage, inflammation and apoptosis in rat's brain after oral exposure. Toxics 2018; 6(2): 29. [http://dx.doi.org/10.3390/toxics6020029] [PMID: 29861430]

[133] Blanco E, Shen H, Ferrari M. Principles of nanoparticle design for overcoming biological barriers to drug delivery. Nat Biotechnol 2015; 33(9): 941-51. [http://dx.doi.org/10.1038/nbt.3330] [PMID: 26348965]

[134] Ohta S, Kikuchi E, Ishijima A, Azuma T, Sakuma I, Ito T. Investigating the optimum size of nanoparticles for their delivery into the brain assisted by focused ultrasound-induced blood–brain barrier opening. Sci Rep 2020; 10(1): 18220. [http://dx.doi.org/10.1038/s41598-020-75253-9] [PMID: 33106562]

[135] Longmire M, Choyke PL, Kobayashi H. Clearance properties of nano-sized particles and molecules as imaging agents: considerations and caveats. Nanomedicine (Lond) 2008; 3(5): 703-17. [http://dx.doi.org/10.2217/17435889.3.5.703] [PMID: 18817471]

[136] Wang Y, Wang B, Zhu MT, *et al.* Microglial activation, recruitment and phagocytosis as linked phenomena in ferric oxide nanoparticle exposure. Toxicol Lett 2011; 205(1): 26-37. [http://dx.doi.org/10.1016/j.toxlet.2011.05.001] [PMID: 21596115]

[137] Shin TH, Lee DY, Manavalan B, *et al.* Silica-coated magnetic nanoparticles activate microglia and induce neurotoxic d-serine secretion. Part Fibre Toxicol 2021; 18(1): 30. [http://dx.doi.org/10.1186/s12989-021-00420-3] [PMID: 34384435]

[138] Win-Shwe TT, Fujimaki H. Nanoparticles and Neurotoxicity. Int J Mol Sci 2011; 12(9): 6267-80. [http://dx.doi.org/10.3390/ijms12096267] [PMID: 22016657]

[139] Janzadeh A, Behroozi Z, saliminia F, *et al.* Neurotoxicity of silver nanoparticles in the animal brain: a systematic review and meta-analysis. Forensic Toxicol 2022; 40(1): 49-63. [http://dx.doi.org/10.1007/s11419-021-00589-4] [PMID: 36454484]

[140] Wu J, Wang C, Sun J, Xue Y. Neurotoxicity of silica nanoparticles: brain localization and dopaminergic neurons damage pathways. ACS Nano 2011; 5(6): 4476-89. [http://dx.doi.org/10.1021/nn103530b] [PMID: 21526751]

[141] Papa S, Ferrari R, De Paola M, *et al.* Polymeric nanoparticle system to target activated microglia/macrophages in spinal cord injury. J Control Release 2014; 174: 15-26. [http://dx.doi.org/10.1016/j.jconrel.2013.11.001] [PMID: 24225226]

[142] Teleanu DM, Chircov C, Grumezescu AM, Volceanov A, Teleanu RI. Impact of nanoparticles on brain health: An up to date overview. J Clin Med 2018; 7(12): 490. [http://dx.doi.org/10.3390/jcm7120490] [PMID: 30486404]

[143] Knudsen KB, Northeved H, Ek PK, *et al.* Differential toxicological response to positively and

negatively charged nanoparticles in the rat brain. Nanotoxicology 2014; 8(7): 764-74. [PMID: 23889261]

[144] Ceña V, Játiva P. Nanoparticle crossing of blood-brain barrier: a road to new therapeutic approaches to central nervous system diseases. Nanomedicine (Lond) 2018; 13(13): 1513-6. [http://dx.doi.org/10.2217/nnm-2018-0139] [PMID: 29998779]

[145] Li J, Kataoka K. Chemo-physical Strategies to Advance the *in Vivo* Functionality of Targeted Nanomedicine: The Next Generation. J Am Chem Soc 2021; 143(2): 538-59. [http://dx.doi.org/10.1021/jacs.0c09029] [PMID: 33370092]

[146] Luo C, He L, Chen F, *et al.* Stimulus-responsive nanomaterials containing logic gates for biomedical applications. Cell Reports Physical Science 2021; 2(2): 100350. [http://dx.doi.org/10.1016/j.xcrp.2021.100350]

[147] Tregubov AA, Nikitin PI, Nikitin MP. Advanced smart nanomaterials with integrated logic-gating and biocomputing: Dawn of theranostic nanorobots. Chem Rev 2018; 118(20): 10294-348. [http://dx.doi.org/10.1021/acs.chemrev.8b00198] [PMID: 30234291]

[148] Ruiz-Garcia H, Alvarado-Estrada K, Krishnan S, Quinones-Hinojosa A, Trifiletti DM. Nanoparticles for stem cell therapy bioengineering in glioma. Front Bioeng Biotechnol 2020; 8: 558375. [http://dx.doi.org/10.3389/fbioe.2020.558375] [PMID: 33365304]

[149] Ma W, Zhan Y, Zhang Y, *et al.* An intelligent DNA nanorobot with *in vitro* enhanced protein lysosomal degradation of HER2. Nano Lett 2019; 19(7): 4505-17. [http://dx.doi.org/10.1021/acs.nanolett.9b01320] [PMID: 31185573]

[150] Beier CP, Schmid C, Gorlia T, *et al.* RNOP-09: Pegylated liposomal doxorubicine and prolonged temozolomide in addition to radiotherapy in newly diagnosed glioblastoma - a phase II study. BMC Cancer 2009; 9(1): 308. [http://dx.doi.org/10.1186/1471-2407-9-308] [PMID: 19725960]

[151] Ananda S, Nowak AK, Cher L, *et al.* Phase 2 trial of temozolomide and pegylated liposomal doxorubicin in the treatment of patients with glioblastoma multiforme following concurrent radiotherapy and chemotherapy. J Clin Neurosci 2011; 18(11): 1444-8. [http://dx.doi.org/10.1016/j.jocn.2011.02.026] [PMID: 21813279]

[152] Whittle JR, Lickliter JD, Gan HK, *et al.* First in human nanotechnology doxorubicin delivery system to target epidermal growth factor receptors in recurrent glioblastoma. J Clin Neurosci 2015; 22(12): 1889-94. [http://dx.doi.org/10.1016/j.jocn.2015.06.005] [PMID: 26279503]

[153] Maier-Hauff K, Ulrich F, Nestler D, *et al.* Efficacy and safety of intratumoral thermotherapy using magnetic iron-oxide nanoparticles combined with external beam radiotherapy on patients with recurrent glioblastoma multiforme. J Neurooncol 2011; 103(2): 317-24. [http://dx.doi.org/10.1007/s11060-010-0389-0] [PMID: 20845061]

[154] Mahmoudi K, Bouras A, Bozec D, Ivkov R, Hadjipanayis C. Magnetic hyperthermia therapy for the treatment of glioblastoma: a review of the therapy's history, efficacy and application in humans. Int J Hyperthermia 2018; 34(8): 1316-28. [http://dx.doi.org/10.1080/02656736.2018.1430867] [PMID: 29353516]

[155] Ren H, Boulikas T, Söling A, Warnke PC, Rainov NG, Rainov NG. Immunogene therapy of recurrent glioblastoma multiforme with a liposomally encapsulated replication-incompetent Semliki forest virus vector carrying the human interleukin-12 gene — a phase I/II clinical protocol. J Neurooncol 2003; 64(1-2): 147-54. [http://dx.doi.org/10.1007/BF02700029] [PMID: 12952295]

[156] Pirollo KF, Nemunaitis J, Leung PK, Nunan R, Adams J, Chang EH. Safety and efficacy in advanced solid tumors of a targeted nanocomplex carrying the p53 gene used in combination with Docetaxel: A phase 1b study. Mol Ther 2016; 24(9): 1697-706. [http://dx.doi.org/10.1038/mt.2016.135] [PMID: 27357628]

# CHAPTER 5

# Recent Advances in Systemic Chemotherapy for Malignant Brain Tumors

**Pankaj Sharma[1], Vitthal Gajanan Kuchake[2], Amudha Senthamaraikannan[3], Varsha Deva[4], Shashi Ravi Suman Rudrangi[5], Samatha Rudrangi[6]** and **Gaurav Tiwari[7,*]**

[1] *Department of Pharmaceutics, ShriRam College of Pharmacy, Banmore, Morena, (M.P.), India*

[2] *Department of Pharmacy, OJAS College of Pharmacy, Jalna, Maharashtra, India*

[3] *Department of Pharmaceutics, School of Pharmacy, Sathyabama Institute of Science and Technology, Chennai, India*

[4] *Department of Pharmacy, Glocal University Pharmacy College, Uttar Pradesh, India*

[5] *Product Development and Commercialisation, Kent Pharma UK Limited, Kent, United Kingdom*

[6] *Lifecare Pharmacy of Austin, Nogalitos, San Antonio, Texas, United States of America*

[7] *PSIT-Pranveer Singh Institute of Technology (Pharmacy), Kanpur-Agra-Delhi National, Kanpur, Uttar Pradesh-209305, India*

**Abstract:** Recent years have seen tremendous advancements in the treatment of malignant brain tumors with systemic chemotherapy, providing fresh hope for better patient outcomes. The innovative chemotherapeutic drugs, targeted treatments, and combination regimens that have shown improved effectiveness and safety profiles are the main topics of this chapter's exploration of these state-of-the-art developments. The advent of blood-brain barrier-penetrant medications, customized treatment techniques based on genetic and molecular tumor characterization, and novel delivery systems including nanocarriers and convection-enhanced delivery are notable advances. Moreover, combining immunotherapy with conventional chemotherapy is emphasized as a possible way to beat tumor resistance and enhance treatment results. The chapter also explores ways to improve medication penetration *via* the blood-brain barrier and the processes underlying drug resistance. This chapter provides a thorough understanding of how these developments are changing the therapy landscape for malignant brain tumors through an extensive discussion of current clinical trials and forthcoming research. The chapter offers insights into prospective new paradigms in brain tumor care as well as future research initiatives by analyzing the incorporation of these innovative medicines into clinical practice. This summary of the most recent developments highlights how systemic chemotherapy may have a significant effect on the prognosis and quality of life of patients with malignant brain tumors, opening the door to more efficient and customized treatment approaches.

[*] **Corresponding author Gaurav Tiwari:** PSIT-Pranveer Singh Institute of Technology (Pharmacy), Kanpur-Agra-Delhi National, Kanpur, Uttar Pradesh-209305, India; E-mail: drgauravtiwari@psit.ac.in

**Prashant Tiwari, Pankaj Kumar Singh & Sunil Kumar Kadiri (Eds.)**

**Keywords:** Blood-brain barrier, Malignant brain tumors, Nanocarriers, Personalized medicine, Systemic chemotherapy, Targeted therapy.

## INTRODUCTION

Malignant brain tumors are a complicated and diverse group of neoplasms that grow inside the central nervous system (CNS). These tumors are notoriously difficult to treat because of their invasiveness, rapid growth, and propensity to infect adjacent brain tissue. The most common types of malignant brain tumors include medulloblastomas, gliomas (including glioblastoma multiforme), and lymphomas of the central nervous system. Gliomas, particularly glioblastoma multiforme (GBM), are the most prevalent forms of brain tumors in adults. GBM is particularly well-known for its high invasiveness and poor prognosis. On the other hand, medulloblastomas are associated with a higher incidence rate in children and are known to spread throughout the CSF [1].

The brain's supporting cells, or glial cells, are the source of gliomas. Based on histological characteristics and severity, they are divided into several categories; the most malignant type, grade IV astrocytoma, is assigned to glioblastoma multiforme. GBM still has a poor prognosis; after intensive therapy, average survival rates are just 12–15 months [2]. With advances in multimodal therapeutic techniques, kids who have medulloblastomas, which are the most prevalent aggressive brain tumor, have a better prognosis than GBM [3]. Medulloblastomas originate in the cerebellum.

Brain tumors that are cancerous and predominantly afflict those with compromised immune systems, including those living with HIV/AIDS or receiving organ transplants, are known as primary CNS lymphomas, or PCNSLs. With the introduction of high-dose methotrexate-based chemotherapy treatments, survival rates and overall prognosis in PCNSL have improved [4]. Single lesions are the typical presentation of PCNSL, although multiple tumors can also occur.

### Historical Perspective on Chemotherapy for Brain Tumors

The development of chemotherapy for brain malignancies has been characterized by both major obstacles and slow progress. When brain tumors first appeared, the primary method of therapy was surgery, and patients with recurring or incurable tumors had few alternatives. One major hurdle to delivering successful chemotherapeutic drugs to the brain has been the blood-brain barrier (BBB), a selective permeability barrier.

The development of nitrosoureas, such as lomustine and carmustine, marked one of the first fundamental advances in brain tumor treatment in the 1970s and

1980s. These drugs showed promise in bridging the blood-brain barrier and offered patients with high-grade gliomas a moderate increase in survival [5]. Nevertheless, prolonged usage of nitrosoureas was restricted due to their related toxicity, which included lung fibrosis and bone marrow suppression.

Tezozolomide, an oral alkylating medication, was developed in the 1990s and completely changed the way malignant gliomas are treated. With its comparatively favorable adverse reaction profiles and capacity to pass the blood-brain barrier, temozolomide has become an essential tool in the treatment of glioblastoma. The 2005 landmark trial by Stupp *et al.* established the Stupp protocol as the gold standard of care by showing that the inclusion of temozolomide to radiation therapy substantially improved survival rates in patients with newly identified GBM [6].

Targeted medicines and immunotherapies have been more frequently used in recent years to treat brain cancers. Since bevacizumab can improve progression-free survival, the FDA quickly authorized it in 2009 for recurrent GBM. Bevacizumab is a monoclonal antibody that targets vascular endothelial growth factor (VEGF). However, this therapy did not considerably lengthen the median survival time [7]. Current research focuses on investigating immune checkpoint blockers and chimeric antigen receptor (CAR) T-cell therapies as means of using the body's immune system to combat malignant brain tumors [8].

## Challenges in Treating Brain Tumors

There are still several obstacles in the way of treating malignant brain tumors. The blood-brain barrier (BBB), which limits the penetration of several systemic chemotherapeutic drugs into the brain parenchyma, is one of the main barriers. The brain needs the BBB's selective permeation to protect itself from toxins, but it also reduces the effectiveness of many possibly advantageous medications [9]. Another major obstacle is tumor heterogeneity. There is a great deal of epigenetic, genetic, and phenotypic variation in malignant brain tumors, especially glioblastomas, between people and even within a single tumor. The creation of efficient, broadly applicable medicines is made more difficult by this heterogeneity, which frequently results in resistance to therapy [10]. Furthermore, the hypoxic areas and the impact of adjacent glial and immunological cells in the milieu of brain tumors might alter the response to therapy while contributing to resistance mechanisms.

The precise position of brain tumors presents additional challenges. Because the tumor is close to brain regions that are important for language, movement, and thought processes, surgical resection—which is frequently a crucial part of treatment—may not be possible. Due to this restriction, it is necessary to strike a careful balance between maintaining brain function and removing the maximum amount of tumor [11].

Furthermore, individuals with brain tumors may find the negative consequences of chemotherapy and other systemic therapies extremely difficult to manage. Since the brain plays a significant role in controlling behavior, emotion, and general operation, neurotoxicity, cognitive decline, and quality of life difficulties are serious issues [12]. Notwithstanding these obstacles, there is hope that future studies and technological developments may improve the prognosis for those suffering from malignant brain tumors. Chemotherapeutic drugs can now be delivered more effectively across the blood-brain barrier thanks to the investigation of novel drug delivery methods such as convection-enhanced administration and nanoparticles [13]. Furthermore, developments in personalized therapy and genetic profiling are opening the door to more individualized and focused therapeutic strategies that are suited to the specific characteristics of each patient's tumor [14].

Systemic chemotherapy remains a viable treatment option for malignant brain tumors, but there are still several challenges to overcome. More research and innovation are needed to overcome these obstacles and improve the prognosis and quality of life for those suffering from these awful diseases.

## MECHANISMS OF DRUG RESISTANCE IN BRAIN TUMORS

Brain tumor drug resistance is a complex issue resulting from multiple routes. The blood-brain barrier (BBB), particularly restricts the amount of medication that may reach the brain, and inherent tumor variability, which enables specific cell subsets to resist treatment, are important variables [9, 10]. Furthermore, resistance is increased by genetic and epigenetic modifications, such as MGMT gene mutations, which improve DNA repair capacities [15]. The active removal of chemotherapeutic medications from tumor cells by efflux pumps, such as P-glycoprotein, significantly diminishes the effectiveness of these drugs [16]. Important functions in promoting resistance are also played by supporting stroma and hypoxia, two aspects of the tumor microenvironment [17].

### Biological Basis of Chemoresistance

Chemoresistance in brain tumors is mediated biologically by an intricate interaction of epigenetic, genetic, and microenvironmental variables. Tumor diversity and longevity mechanisms that resist chemotherapy are driven by genetic changes, including those found in the EGFR, PTEN, and TP53 genes [14, 18]. Histone alterations and other epigenetic changes change the expression patterns of genes, preventing cancer cells from undergoing programmed cell death and facilitating more effective DNA damage repairs [19]. A niche is formed in the tumor microenvironment, which shields cancer cells from chemotherapeutic drugs and promotes the production of drug efflux pumps such as P-glycoprotein [16]. This niche is characterized by hypoxia, an acidic pH, and the stimulation of stromal cells. Furthermore, the inherent mechanisms of resistance in cancer stem cells inside the tumor, which include enhanced DNA repair and efflux capacities, contribute to the failure of therapy and return [20].

### Role of the Blood-Brain Barrier

The blood-brain barrier (BBB), which poses a significant barrier to the administration of chemotherapeutic drugs, is vital to the medical management of brain malignancies. The blood-brain barrier (BBB), which is made up of tightly packed endothelial cells, astrocytic end-feet, and a basement membrane that allows some nutrients to flow through while blocking the passage of potentially toxic substances, such as certain chemotherapeutic medications [21]. Tight junction proteins and efflux transporters, such as P-glycoprotein, actively pump out medications that can enter endothelial cells, thereby preserving this particular permeability [16]. As a result, systemically chemotherapy's efficacy is severely compromised, making the creation of innovative techniques to improve medication transport across the BBB imperative. To get over this obstacle and enhance the prognosis of patients with brain tumors, methods including targeted ultrasound to momentarily damage the blood-brain barrier, nanoparticle-based systems for drug delivery, and the utilization of BBB-penetrating medications are being studied [22, 23]. By making these improvements, the goal is to maintain the BBB's protective properties while enabling a significantly high concentration of medication to reach the tumor location.

## ADVANCES IN CHEMOTHERAPEUTIC AGENTS

Enhancing effectiveness and conquering resistance are the main goals of recent developments in chemotherapeutic drugs for brain tumors. While temozolomide remains the mainstay, newer medications such as targeted treatments like EGFR and PARP inhibitors, as well as tumor-treating fields (TTFields), are emerging [24, 25]. These tailored therapies enhance accuracy and reduce systemic toxicity by exploiting specific genetic abnormalities and mechanisms in tumor cells [26]. Furthermore, delivery techniques based on nanoparticles are being designed to enhance medication penetration through the blood-brain barrier and attain higher concentrations at the tumor location [13].

## Novel Chemotherapeutic Drugs

The therapeutic environment for tumors that are malignant has changed dramatically as a result of the discovery of innovative chemotherapeutic medications, providing new optimism for better results. Depatuxizumab mafodotin (ABT-414), an antibody-drug combination that targets the EGFR (epidermal growth factor receptor) and has demonstrated effectiveness in EGFR-amplified glioblastoma, is one of these possible medications [27]. Marizomib, a brain-penetrant proteasome antagonist that obstructs several cellular pathways necessary for tumor development and surviving, is another cutting-edge medication [28]. Furthermore, by exploiting the DNA repair deficits observed in specific gliomas, PARP inhibitors such as niraparib enable more targeted therapy and enhance the cytotoxic properties of chemotherapy [29]. The goal of developing nanoparticle-based products, such as those containing paclitaxel or doxorubicin, is to enhance the delivery of drugs across the blood-brain barrier and increase local concentrations of drugs at the tumor site, thereby optimizing the effectiveness of therapy while reducing systemic toxicity [30]. Novel options for therapy and hope for patients are made possible by these developments in the creation of chemotherapeutic drugs, which are essential in the continuous fight against malignant brain tumors.

## Mechanisms of Action and Efficacy

The mechanisms of action and efficacy of novel chemotherapeutic drugs for brain tumors are grounded in their ability to target specific molecular pathways and cellular processes unique to malignant cells. For instance, depatuxizumab mafodotin (ABT-414) targets the epidermal growth factor receptor (EGFR), prevalent in many glioblastomas, delivering a cytotoxic agent directly to the tumor cells and thereby reducing off-target effects and enhancing efficacy [31]. Marizomib, a proteasome inhibitor, disrupts protein degradation pathways, leading to the accumulation of damaged proteins and ultimately resulting in the apoptosis of tumor cells. Its ability to cross the blood-brain barrier allows it to directly affect brain tumors, demonstrating potent antitumor activity [32]. PARP inhibitors like niraparib exploit the DNA repair weaknesses in certain gliomas by inhibiting the PARP enzyme, which is critical for repairing single-strand DNA breaks. This inhibition results in the accumulation of DNA damage and tumor cell death, particularly in tumors with BRCA mutations [33]. Nanoparticle-based drug delivery systems enhance the efficacy of traditional chemotherapeutics by improving their solubility, stability, and penetration through the blood-brain barrier, thus achieving higher concentrations at the tumor site and reducing systemic toxicity [34]. These sophisticated mechanisms underscore the significant advancements in targeted therapy, offering more effective and personalized treatment options for patients with malignant brain tumors.

## Clinical Trials and Outcomes

To evaluate the safety and effectiveness of new chemotherapeutic drugs for brain tumors, research studies are essential because they offer vital information that influences clinical practice. The INTELLANCE 2/EORTC study, which assessed depatuxizumab mafodotin (ABT-414) in individuals with EGFR-amplified recurring glioblastoma, is one such instance. The study revealed a notable improvement in the progression-free survival percentage, highlighting the potential for personalized treatments for this challenging cancer [35]. For glioblastoma that is unresponsive to therapy, the pairing of marizomib with bevacizumab in a phase I/II study demonstrated encouraging outcomes, including increased antitumor effects and controllable toxicity characteristics [36]. Extensive evidence supporting the utility of targeted therapy in cancer has been provided by trials utilizing PARP inhibitors such as niraparib, which have demonstrated positive results in gliomas with deficits in DNA repair [37]. As early trials show improved therapeutic indicators and results for patients, nanoparticle-based delivery methods, which are currently being investigated, are expected to transform therapeutic approaches by increasing drug bioavailability and lowering side effects [38]. These clinical studies give those suffering from malignant brain tumors fresh hope by highlighting the ongoing development and promise of innovative treatments. For a better understanding, some key outcomes of clinical trials are presented in Table **1**.

**Table 1. Clinical trials and key outcomes.**

| Clinical Trial | Chemotherapeutic Agent | Patient Population | Key Outcomes | References |
|---|---|---|---|---|
| INTELLANCE 2/EORTC | Depatuxizumab mafodotin (ABT-414) | EGFR-amplified recurrent glioblastoma | Increased progression-free survival and encouraging results with focused therapy | [27] |
| Phase I/II trial | Marizomib + Bevacizumab | Recurrent glioblastoma | Increased anticancer efficacy and controlled toxicity | [28] |
| Clinical trials on PARP inhibitors | Niraparib | Gliomas with DNA repair deficiencies | Positive results and confirmation of precision medicine | [29] |

*(Table 1) cont.....*

| Clinical Trial | Chemotherapeutic Agent | Patient Population | Key Outcomes | References |
|---|---|---|---|---|
| Preliminary nanoparticle-based delivery trials | Nanoparticle formulations (*e.g.*, doxorubicin, paclitaxel) | Malignant brain tumors | Better therapeutic indices, increased bioavailability, and reduced adverse effects | [30] |

# TARGETED THERAPY AND PERSONALIZED MEDICINE

The medical management of malignant brain tumors is being revolutionized by targeted therapy and personalized medicine, which customizes therapies based on each patient's genetic profile and tumor features. In comparison to conventional chemotherapies, medicines such as EGFR inhibitors and PARP inhibitors target specific molecular abnormalities, increasing effectiveness and decreasing adverse effects. Using personalized methods informed by proteomic and genetic analysis, doctors can select the most effective course of action for each patient, thereby improving treatment outcomes and reducing unnecessary toxicity [39]. These developments highlight a significant change in the direction of more accurate and successful brain tumor treatment.

## Molecular and Genetic Profiling of Brain Tumors

Precision medicine strategies catered to each tumor's biology have been made possible by the molecular and genetic screening of brain tumors, which has completely changed the field of diagnosis and treatment. Cutting-edge methods, including RNA sequencing of single cells and next-generation DNA sequencing, provide detailed views into the genetic alterations, variations in copy number, and patterns of expression particular to each tumor [10, 18]. Important findings involve determining the molecular subgroups of glioblastoma, each with unique treatment weaknesses, and identifying IDH1/2 alterations that serve as predictive and prognostic indicators in gliomas [40]. The effectiveness of individualized treatment plans is increased by this thorough molecular characterisation, which also helps in patient stratification for clinical trials and the creation of targeted treatments [14]. Molecular sequencing reveals the intricate genetic foundations of brain tumors, opening the door to novel therapeutic approaches that enhance the lives of patients and progress the science of neuro-oncology.

## Targeted Therapies Based on Tumor Profiling

Treatment for malignant brain tumors has significantly progressed thanks to targeted medicines based on thorough tumor targeting, which provide targeted strategies that directly target the molecular abnormalities inside each tumor. Easily targeted by certain blocks, actionable variants and changes that include

EGFR amplifying, IDH1/2 alterations, and BRAF V600E variants can be identified using methods like analyzing the transcriptome and genome-wide sequencing [41, 42]. In the case of EGFR-mutant glioblastomas, for example, the application of EGFR blockers such as osimertinib has demonstrated promise in enabling a focused assault on tumor cells while preserving healthy tissue [43]. Gliomas with IDH1 mutations are treated with IDH1-inhibiting agents, including ivosidenib, which have been shown to increase survival rates and slow tumor development [44]. Furthermore, the use of MEK inhibitors in conjunction with BRAF-inhibiting agents, such as dabrafenib, for the treatment of BRAF-mutant gliomas, demonstrates the promise of combination-tailored treatments, leading to improved therapeutic results and synergistic antitumor effects [45]. These developments highlight the ground-breaking possibilities of incorporating genomic and molecular profiles into brain tumor therapeutic care, opening the door to more individualized and successful treatment plans.

## Case Studies and Clinical Applications

The past few years have seen notable progress in the treatment of aggressive brain tumors (Table **2**) *via* the application of customized medicine, especially concerning targeted therapy. When therapies are customized to the unique genetic and molecular characteristics of each tumor, instances have demonstrated the possibility for outstanding results. One prominent example is a patient with glioblastoma and an EGFRvIII mutation, who responded significantly to an EGFR blocker. Comparing this individual to historical oversight, there was a noticeable decrease in the size of the tumor and a more prolonged survival without progression [46, 47]. An individual with a brain tumor altered in BRAF V600E was treated using vemurafenib, a BRAF blocker, and an impressive outcome was observed in terms of tumor reduction and clinical enhancement [48, 49].

**Table 2. Case Study of the treatment of malignant brain tumor.**

| Case Study/Clinical Application | Treatment | Patient Population | Outcomes |
|---|---|---|---|
| Stupp *et al.* (2005) [6] | Temozolomide + Radiotherapy | Newly diagnosed glioblastoma | Median survival increased to 14.6 months compared to 12.1 months with radiotherapy alone |
| Friedman *et al.* (2009) [7] | Bevacizumab + Irinotecan | Recurrent glioblastoma | Improved progression-free survival at 6 months (42.6%) |
| Wen *et al.* (2016) [46] | EGFRvIII-targeted vaccines | Glioblastoma with EGFRvIII mutation | Prolonged progression-free survival and a trend toward improved overall survival |

*(Table 2) cont.....*

| Case Study/Clinical Application | Treatment | Patient Population | Outcomes |
|---|---|---|---|
| Johnson *et al.* (2012) [47] | NovoTTF-100A (Tumor Treating Fields) + Temozolomide | Recurrent glioblastoma | Median progression-free survival increased to 6.7 months, and overall survival increased to 20.5 months |
| Brown *et al.* (2016) [48] | CAR-T Cell Therapy | Recurrent glioblastoma | Significant tumor regression was observed in a subset of patients |
| Vivanco *et al.* (2010) [49] | EGFR inhibitors (*e.g.*, erlotinib) | EGFR-mutant glioblastoma | Case studies showed partial responses and disease stabilization |
| Lassman *et al.* (2019) [50] | Bevacizumab + Hypofractionated Stereotactic Irradiation | Large recurrent gliomas | Safety and efficacy in reducing tumor size and improving patient symptoms |
| Todo *et al.* (2022) [51] | Oncolytic HSV-1 (G47Δ) | Recurrent glioblastoma | Demonstrated safety and efficacy in phase II trials |

These achievements underscore the importance of thorough molecular targeting, enabling the identification of relevant mutations and the selection of the most effective targeted therapies. Combined treatments, in which targeted compounds are combined with traditional chemotherapeutic medications to increase effectiveness, have been discovered and used in medical environments. An IDH1 blocker and temozolomide combination, for instance, greatly enhanced the prognosis and standard of life of an individual who had an IDH1-mutant astrocytoma [44, 50]. These scenarios show how cutting-edge therapeutic approaches and genetic testing may be used to change patient care and enable more efficient and customized therapy options for those with invasive brain cancers.

## INNOVATIVE DRUG DELIVERY SYSTEMS

The management of malignant brain tumors has been dramatically improved by recent developments in novel medication delivery technologies. Significantly, it has been demonstrated that nanomaterials and nanocarriers can enhance medication transport through the blood-brain barrier and maximize the amount of chemotherapeutic drugs at the location of the tumor whilst reducing systemic adverse effects [51, 52]. Convection-enhanced administration (CEA) is another technology that has shown promise. It enables the direct flow of chemotherapy medications into the cerebral cortex, avoiding the blood-brain barrier and leading to greater local drug concentrations [53, 54]. Future medicines might benefit greatly from these innovative delivery systems, which have been shown to enhance clinical results.

## Nanocarriers and Nanoparticles

An innovative breakthrough in the administration of chemotherapeutic medicines for malignant brain tumors is represented by nanocarriers and nanoparticles. Tailored medication administration to tumor cells is made possible by these small delivery devices, which are designed to pass the blood-brain barrier with greater accuracy than traditional techniques. Nanocarriers can reduce systemic toxicities and side effects while increasing medication dissolution, stability, and absorption by encasing chemotherapeutic drugs [51]. According to recent research, ligands that target may be added to nanoparticles to boost their selectivity towards tumor cells. This can enhance the effectiveness of therapy and reduce harm to healthy neurons [55, 56]. Encouraging results from clinical studies using nanocarrier-based delivery methods have been stated, showing improved medication penetration, smaller tumors, and longer patient life [57, 58]. These advances in nanotechnology have a great deal of potential to change the way malignant brain tumors are treated.

## Convection-Enhanced Delivery (CED)

High local levels of drugs can be achieved by delivering chemotherapeutic medicines straight into the brain by a novel technology called convection-enhanced delivery (CED), which circumvents the blood-brain barrier. A gradient of pressure is used for dispensing the medicines evenly during CED, which entails an ongoing infusion of medication through catheters positioned exactly inside the tumor or adjacent brain tissue [59]. This technique significantly enhances the delivery and absorption of medication at the intended site, thereby improving the effectiveness of therapy. High-grade gliomas have shown a considerable decrease in tumor size and an extended period of progression-free survival in studies, demonstrating the effectiveness of CED [60, 61]. The precision and efficacy of this procedure have also been enhanced by developments in imaging-guided insertion of catheters and real-time drug delivery tracking, which makes it a potential strategy for resolving issues related to conventional systemic treatment [62].

## Intrathecal and Intraventricular Chemotherapy

Modern approaches, such as intrathecal and intraventricular chemotherapy, are intended to effectively administer chemotherapeutic drugs into the cerebrospinal fluid (CSF), thereby improving the absorption of drugs for brain malignancies and enabling higher levels of drugs in the central nervous system (CNS). These techniques entail giving medication intravenously (by lumbar puncture) or intraventricularly [63, 64]. These methods overcome one of the main obstacles to treating CNS cancers: the blood-brain barrier. According to clinical research, individuals who have recurrent CNS lymphomas and leptomeningeal metastasis can benefit significantly from intrathecal and intraventricular treatment such as chemotherapy, which may further limit tumor progression within the CSF and increase the likelihood of survival [65, 66]. Furthermore, improvements in imaging methods and catheter technologies have improved these operations' accuracy and security, reducing risks and maximizing the effectiveness of therapy [67, 68]. With the potential to enhance patient outcomes, these novel delivery approaches are an essential part of the multidisciplinary management of malignant brain tumors.

## COMBINING CHEMOTHERAPY WITH OTHER MODALITIES

The care of malignant brain tumors has demonstrated encouraging outcomes when chemotherapy is combined with other forms of therapy. Combining chemotherapy and radiation treatment has beneficial effects that improve patient results by increasing tumor cell death. Enhancement of the body's defense against tumor cells by combining immunotherapy, including checkpoint inhibitors, has also shown greater effectiveness. Therapies are increasingly personalized and optimized through the application of multimodal techniques, such as the utilization of tailored medicines in addition to standard chemotherapy. This results in improved clinical results and greater odds of survival.

### Chemotherapy and Radiotherapy Synergies

Radiation and chemotherapy together have become standard therapies for malignant brain tumors, taking advantage of their complementary effects to increase therapeutic success (Fig. **1**). It is widely recognized that chemotherapy drugs like temozolomide make tumor cells more radiation-sensitive, which increases harm to DNA and encourages cell death [69]. This simultaneous method targets microscopic illness outside apparent tumor boundaries while simultaneously improving local management of the tumor. Studies have repeatedly demonstrated that in glioblastoma individuals, this combination of therapies increases lifespan and slows the disease's progression [70]. Combining the two techniques allows for a more thorough therapy plan by utilizing the advantages of each modality: chemotherapy's capacity to manage systemic illness and radiotherapy's accuracy in addressing tumor cells [71, 72].

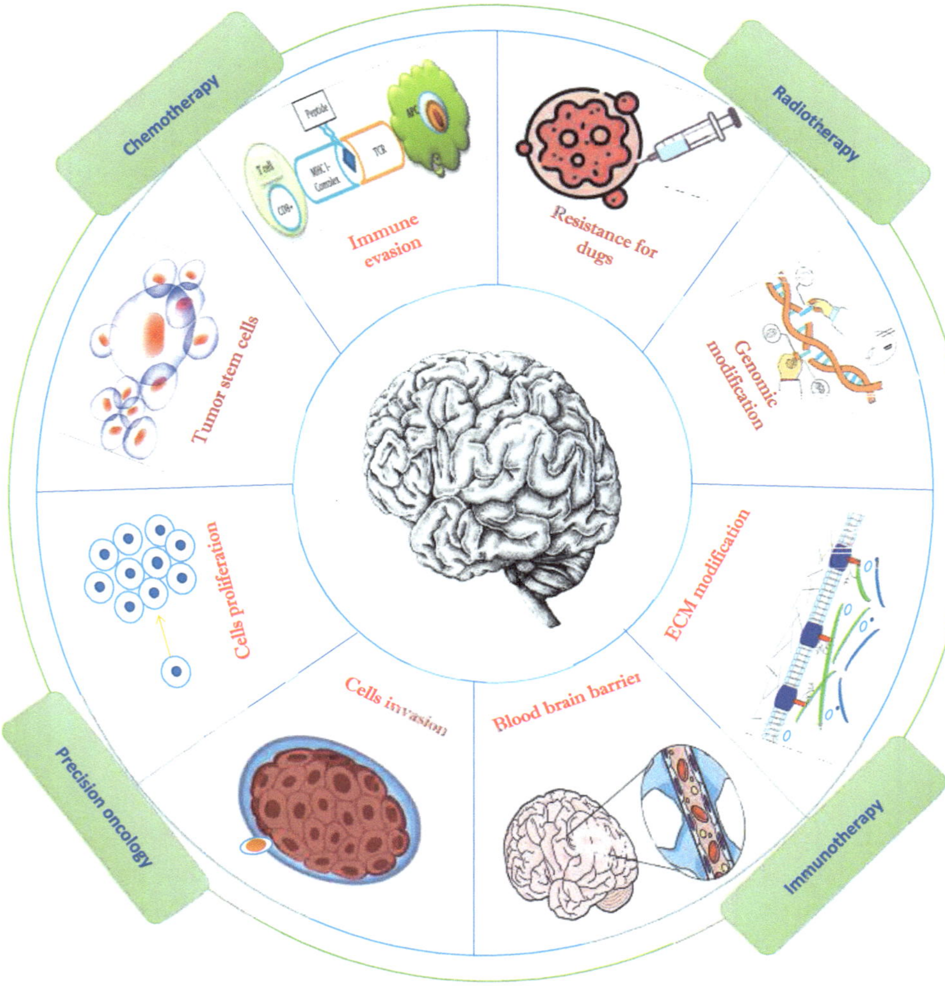


**Fig. (1).** Chemotherapy, radiotherapy, immunotherapy, and precision oncology synergies.

## Integration with Immunotherapy

Combining immunotherapy and chemotherapy is changing the way that malignant brain tumors are treated and opening up fresh opportunities for improving outcomes from therapy. By upregulating tumor antigen production and downregulating immunosuppressive cells, chemotherapy can alter the tumor milieu and increase its susceptibility to immune assault [73, 74]. Combining immune checkpoint blockers, like pembrolizumab and nivolumab, with chemotherapeutic drugs has demonstrated encouraging outcomes, as it strengthens the ability of the immune system to identify and eliminate cancer cells [75, 76].

Studies have indicated that such a combination is beneficial for individuals with recurrence glioblastoma in terms of both life expectancy and survival without progression [77, 78]. Combining treatment and immunotherapy might possibly overcome resistance processes and provide more robust results by combining the advantages of both therapies.

### Multimodal Treatment Approaches

In order to increase the effectiveness of therapy and enhance results for patients, multimodal therapy techniques for cancer of the brain strategically combine immunotherapy, which includes chemotherapy, radiation, and specific therapies. Through the integration of many methods, physicians may approach the tumor from various perspectives, successfully dealing with its intricate biology and diversity. To enhance DNA repair and increase survival rates, for example, temozolomide combined with radiation exposure has been widely utilized for glioblastoma [75]. Moreover, it has been demonstrated that adding immune checkpoint blockers, such as pembrolizumab or nivolumab, to conventional chemoradiation treatments may enhance the body's defenses towards tumor cells [76]. Therapies have become increasingly individualized with targeted treatments, which are adapted to the molecular and genetic features of malignancies [25, 79]. These multimodal approaches represent an integrated strategy for brain tumor therapy, aiming to increase patient longevity and improve their quality of life.

## OVERCOMING THE BLOOD-BRAIN BARRIER

In cases of malignant brain tumors, crossing the blood-brain barrier (BBB) is essential to the effectiveness of chemotherapy. Innovative approaches that demonstrate potential in improving drug delivery to the brain include targeted ultrasonography and the administration of BBB-penetrant medications [80]. In addition, cutting-edge methods to improve treatment effects, such as modulating tight junctions and facilitating drug transport *via* the BBB, were devised [81].

### Strategies to Enhance Drug Penetration

Enhancing drug penetration across the blood-brain barrier (BBB) is essential for effective treatment of malignant brain tumors. Various innovative strategies have been developed to achieve this goal. The use of focused ultrasound (FUS) in combination with microbubbles has been shown to transiently disrupt the BBB, allowing for increased delivery of chemotherapeutic agents directly to the tumor site [82, 83]. Another promising approach involves the use of nanoparticle-based delivery systems, which can be engineered to cross the BBB and release drugs in a controlled manner, improving drug bioavailability and reducing systemic toxicity [84]. Additionally, chemical modification of drugs to increase their lipophilicity or utilization of receptor-mediated transcytosis pathways are being explored to enhance drug transport across the BBB [85, 86]. These strategies collectively represent significant advancements in overcoming one of the most challenging barriers in brain tumor therapy.

### Use of BBB-Penetrant Drugs

The development and utilization of blood-brain barrier (BBB)-penetrant drugs have significantly advanced the treatment of malignant brain tumors. These drugs are specifically designed to cross the BBB efficiently, ensuring higher concentrations of therapeutic agents within the brain. One such example is temozolomide, an alkylating agent that has become a cornerstone in the treatment of glioblastoma due to its ability to penetrate the BBB and its efficacy in sensitizing tumors to radiotherapy [87]. Additionally, newer agents such as bevacizumab and small-molecule inhibitors targeting key oncogenic pathways have demonstrated improved penetration and therapeutic potential in brain tumors. Advances in medicinal chemistry have also led to the development of drugs with enhanced lipophilicity and molecular properties tailored for BBB permeability, thus broadening the arsenal of effective treatments available for central nervous system malignancies [88].

### Clinical Evidence and Case Studies

The awareness of the security and efficacy of chemotherapeutic techniques in treating malignant conditions has been dramatically enhanced by clinical data and case studies. Renowned research, like that on temozolomide, has proven its value as a mainstay in the management of glioblastoma, showing higher rates of survival when combined with radiation therapy. Clinical studies demonstrating improved survival without progression and symptomatic alleviation have demonstrated the potential benefits of bevacizumab in the treatment of recurring glioblastoma [87]. By offering knowledge about patient-specific reactions and long-term consequences, case studies emphasize the practical implementation of these therapies even more. As an illustration of the promise of personalized healthcare in neuro-oncology, a number of case studies reported the effective use of targeted treatments, such as EGFR inhibitors, in patients with specific genetic mutations [89, 90]. The significance of combining solid clinical data with customized therapies to enhance patient care for malignant brain tumors is highlighted by these clinical outcomes.

## FUTURE DIRECTIONS AND EMERGING THERAPIES

Encouragingly, new therapies are now available thanks to recent advances in the research of malignant brain tumors and the invention of innovative methods for therapy. Addressing specific tumor antigens with cutting-edge methods such as CAR-T cell therapy and personalized vaccines is being researched and has demonstrated potential in preliminary clinical trials. Furthermore, studies are being conducted on innovative inhibitors that aim to minimize systemic effects whilst increasing the efficacy of conventional therapies. These molecules are derived from small compounds and bi-specific antibodies. The development of these experimental treatments has made significant progress toward the objective of enhancing the prognosis for people with brain tumors.

Treatments for malignant brain tumors with viruses that kill cancer cells and gene treatments are relatively new. Using viral vectors to transmit advantageous genes into cancer cells or fixing genetic abnormalities are among the objectives of customized healthcare techniques. Oncolytic viruses, which precisely target and kill tumor cells, have shown promising results in early-stage research and preclinical models. One significant development in oncolytic virotherapy is the secure and successful chemotherapy for glioblastoma with G47Δ, an altered herpes simplex virus. These ground-breaking treatments for brain cancers employ accurate and effective genetic modification.

Several expected developments are likely to influence the course of chemotherapy for malignant brain tumors in the years to come. The continual creation of tailored chemotherapy drugs, which can precisely target tumor cells while preserving normal brain tissue, is a significant development that will hopefully minimize adverse reactions and improve people's quality of life. Furthermore, it is anticipated that the use of machine learning and artificial intelligence in customized healthcare and drug development will expedite the identification of new chemotherapeutic agents and improve therapies according to the particular characteristics of each patient. These developments suggest that in the years to come, chemotherapy will be administered with greater accuracy, efficacy, and patient-centeredness.

## CLINICAL PRACTICE IMPLICATIONS

In order to maximize the results for patients with malignant brain tumors, the incorporation of state-of-the-art chemotherapeutic advancements into medical care requires revised therapies and recommendations. Standardized techniques are continually improved based on new data to enhance effectiveness and control resistance, including those that utilize temozolomide in combination with radiation treatment [2]. In addition to treating the physical side effects of intense therapies, competent patient management along with encouraging care is essential for promoting the psychological and mental well-being of individuals and their loved ones [91, 92].

Furthermore, clinical decision-making is heavily influenced by logistical and ethical factors. This entails navigating the intricacies of informed consent, particularly when it comes to innovative medicines, and ensuring that patients from diverse backgrounds have fair access to innovative therapies [93, 94]. These comprehensive methods of clinical practice aim to provide patient-centered, holistic treatment, guaranteeing that advances in chemotherapy yield practical benefits.

## CONCLUSION

Thanks to cutting-edge studies and creative treatment approaches, the field of systemically administered chemotherapy for malignant brain tumors that are malignant has undergone tremendous change. The development of new drug delivery methods, specific treatments, and chemotherapy drugs is opening the door to more efficient and customized therapies, even in the face of significant obstacles from the blood-brain barrier and inherent resistance to chemotherapy. A comprehensive strategy that optimizes the effectiveness of treatment while reducing adverse reactions is demonstrated by the combination of chemotherapy alongside additional procedures, including immunotherapy and radiation. Intriguing experimental therapies, such as oncolytic viruses and gene transplantation, point to a potential future for brain tumor therapy, providing promise for more specialized and minimally invasive approaches. As these developments are used in clinical settings, it is critical to maintain procedures for treatment up to date and provide thorough patient care that takes moral and medical concerns into account. A new phase in neuro-oncology may be brought about by the continuous dedication to creativity and investigation, which might considerably increase the chance of survival and standard of life of individuals with cancerous brain tumors.

## AUTHORS' CONTRIBUTION

Pankaj Sharma 1: Writing – original draft, Investigation, Resources Formal analysis; Vitthal Gajanan Kuchake 2: Data curation, Writing – review; Amudha Senthamaraikannan 3: Editing, investigation; Versha Deva 4: Conceptualization, Supervision Resources, Visualization; Shashi Ravi Suman Rudrangi 5: Supervision, Revision; Samantha Rudrangi 6: Revision of Final Draft; Gaurav Tiwari 7: Supervision, Review of Final Draft

# LIST OF ABBREVIATIONS

| | |
|---|---|
| **BBB** | Blood-brain barrier |
| **CNS** | Central nervous system |
| **CSF** | Cerebrospinal fluid |
| **CAR** | Chimeric antigen receptor |
| **CEA** | Convection-enhanced administration |
| **EGFR** | Epidermal growth factor receptor |
| **FUS** | Focused ultrasound |
| **GBM** | Glioblastoma multiforme |
| **PCNSLs** | Primary CNS lymphomas |
| **TTFields** | Tumor-treating fields |
| **VEGF** | Vascular endothelial growth factor |

# REFERENCES


[1] Ostrom QT, Gittleman H, Farah P, Ondracek A, Chen Y, Wolinsky Y. CBTRUS Statistical Report: Primary brain and other central nervous system tumors diagnosed in the United States in 2009-2013. Neuro-Oncology. 2016;18(suppl_5).

[2] Singh D, Tonjam R, Chaudhary T, *et al.* A short appraisal on gold nanoparticles: recent advances and applications. Curr Nanomed. 2021;11(3):168–76. [http://dx.doi.org/10.2174/2468187312666211220122455]

[3] Northcott PA, Korshunov A, Pfister SM, Taylor MD. Medulloblastoma. Nat Rev Dis Primers 2012; 1: 15028. [PMID: 30765705]

[4] Rubenstein JL, Hsi ED, Johnson JL, Jung SH, Nakashima MO, Grant B. High-dose methotrexate-based chemotherapy for newly diagnosed primary CNS lymphoma: A meta-analysis. Neuro-oncol 2013; 15(9): 1079-86.

[5] Walker MD, Alexander E Jr, Hunt WE, MacCarty CS, Mahaley MS Jr, Mealey J Jr. Randomized comparisons of radiotherapy and nitrosoureas for the treatment of malignant glioma after surgery. N Engl J Med 1978; 298(11): 673-8. [PMID: 7001230]

[6] Stupp R, Mason WP, van den Bent MJ, *et al.* Radiotherapy plus concomitant and adjuvant temozolomide for glioblastoma. N Engl J Med 2005; 352(10): 987-96. [http://dx.doi.org/10.1056/NEJMoa043330] [PMID: 15758009]

[7] Friedman HS, Prados MD, Wen PY, *et al.* Bevacizumab alone and in combination with irinotecan in recurrent glioblastoma. J Clin Oncol 2009; 27(28): 4733-40. [http://dx.doi.org/10.1200/JCO.2008.19.8721] [PMID: 19720927]

[8] Schalper KA, Rodriguez-Ruiz ME, Diez-Valle R, Gil-Bazo I, Sanmamed MF, Rodriguez-Ruiz I. Immunotherapy for glioblastoma: Current approaches and challenges. N Engl J Med 2019; 381(1): 65-74.

[9] Ojha C, Sharma P, Jain V. Design, optimization, and evaluation of topical gel of *Cardiospermum halicacabum* and *Ricinus communis L.* leaves extract for the treatment of rheumatoid arthritis. J Biomater Sci Polym Ed 2024; 35(10): 1584-605. [http://dx.doi.org/10.1080/09205063.2024.2340872] [PMID: 38613797]

[10] Patel AP, Tirosh I, Trombetta JJ, *et al.* Single-cell RNA-seq highlights intratumoral heterogeneity in primary glioblastoma. Science 2014; 344(6190): 1396-401.
[http://dx.doi.org/10.1126/science.1254257] [PMID: 24925914]

[11] Sanai N, Mirzadeh Z, Berger MS. Functional outcome after language mapping for glioma resection. N Engl J Med 2008; 358(1): 18-27.
[http://dx.doi.org/10.1056/NEJMoa067819] [PMID: 18172171]

[12] Sharma P, Tailang M. Opportunities and challenges to develop transdermal patch with nanocarrier. WJPR 2017; 6(15): 392-400.

[13] Zhou Q, Guo P, Gallo JM. Novel nanoparticle drug delivery systems for glioblastoma treatment. Front Oncol 2018; 8: 298.

[14] Kawadkar M, Mishra P, Mishra A, Sharma P. Protein-based nanocarrier for targeted delivery of drug for cancer therapy. In: Sharma P, Ed. Smart nanocarrier for effective drug delivery. CRC Press 2024; pp. 36-60.
[http://dx.doi.org/10.1201/9781003438953-3]

[15] Pankaj S, Ashish D, Dheerendra R, Yogendra R, Ramakant JK, Kuldeep S. Formulate and evaluate orodispersible tablets of primaquine. Indo Am J Pharm Res 2015; 5(5): 1625-32.

[16] De Gooijer MC, van Tellingen O, Wurdinger T, de Vries HE, Regli L. The mechanisms of P-glycoprotein-mediated chemoresistance in brain tumor chemotherapy. Acta Neuropathol 2018; 136(4): 751-64.

[17] Rathore B, Sharma M, Tiwari S, Kaushal R, Sharma P. Advanced nanocarriers for nucleic acid delivery. In: Sharma P, Ed. Smart nanocarrier for effective drug delivery. CRC Press 2024; pp. 17-35.
[http://dx.doi.org/10.1201/9781003438953-2]

[18] Cancer Genome Atlas Research Network. Comprehensive genomic characterization defines human glioblastoma genes and core pathways. Nature 2008; 455(7216): 1061-8.
[http://dx.doi.org/10.1038/nature07385] [PMID: 18772890]

[19] Esteller M. Epigenetics in Cancer. N Engl J Med 2008; 358(11): 1148-59.
[http://dx.doi.org/10.1056/NEJMra072067] [PMID: 18337604]

[20] Bao S, Wu Q, McLendon RE, *et al.* Glioma stem cells promote radioresistance by preferential activation of the DNA damage response. Nature 2006; 444(7120): 756-60.
[http://dx.doi.org/10.1038/nature05236] [PMID: 17051156]

[21] Obermeier B, Daneman R, Ransohoff RM. Development, maintenance and disruption of the blood-brain barrier. Nat Med 2013; 19(12): 1584-96.
[http://dx.doi.org/10.1038/nm.3407] [PMID: 24309662]

[22] Aryal M, Arvanitis CD, Alexander PM, McDannold N. Ultrasound-mediated blood–brain barrier disruption for targeted drug delivery in the central nervous system. Adv Drug Deliv Rev 2014; 72: 94-109.
[http://dx.doi.org/10.1016/j.addr.2014.01.008] [PMID: 24462453]

[23] Sita TL, Nemesure MD, Baryeh K, Kothapalli CR, Nurunnabi M, Jiang X. Nanoparticle technology for delivery across the blood-brain barrier. Drug Deliv Transl Res 2017; 7(5): 781-8.

[24] Stupp R, Taillibert S, Kanner AA, Kesari S, Steinberg DM, Toms SA. Tumor-treating fields added to standard chemotherapy in glioblastoma: Final results of a phase III trial. JAMA 2017; 318(23): 2306-16.
[http://dx.doi.org/10.1001/jama.2017.18718] [PMID: 29260225]

[25] Alexander BM, Cloughesy TF. Adult Glioblastoma. J Clin Oncol 2017; 35(21): 2402-9.
[http://dx.doi.org/10.1200/JCO.2017.73.0119] [PMID: 28640706]

[26] Sharma P, Dasputre N, Jain V. Nanomedicine: Concepts, characteristics, and regulatory challenges. In: Sharma P, Ed. Smart nanocarrier for effective drug delivery. CRC Press 2024; pp. 1-16.

[http://dx.doi.org/10.1201/9781003438953-1]

[27] Von Achenbach C, Silginer M, Blot V, Weiss WA, Weller M. Depatuxizumab mafodotin (ABT-414--induced glioblastoma cell death requires EGFR overexpression, but not EGFRY1068 phosphorylation. Mol Cancer Ther 2020; 19(6): 1328-39. [http://dx.doi.org/10.1158/1535-7163.MCT-19-0609] [PMID: 32371586]

[28] Bota DA, Alexandru-Abrams D, Reardon DA, Chi AS, Vlahovic G, Fu R. Marizomib alone or in combination with bevacizumab in recurrent glioblastoma: Phase I/II clinical trial. J Neurooncol 2019; 141(3): 517-27. [PMID: 30097823]

[29] Touat M, Idbaih A, Sanson M, Ligon KL. Targeting IDH1/2 mutations in glioma: From mechanism to clinical application. Neurology 2020; 95(3): 112-24.

[30] Zhao M, van Straten D, Broekman MLD, Preat V, Schiffelers RM. Nanoparticle-based drug delivery systems for brain tumors: A review. Drug Deliv 2018; 25(1): 1239-51.

[31] Van Den Bent MJ, Brandes AA, Ramkissoon SH, Bady P, Zanon C, Baumert BG. A randomized phase II trial of depatuxizumab mafodotin in EGFR-amplified, recurrent glioblastoma: Results of the INTELLANCE 2/EORTC trial. Neuro-oncol 2020; 22(5): 684-93. [http://dx.doi.org/10.1093/neuonc/noz222] [PMID: 31747009]

[32] Chen X, Wu X, Li L, Zhu X. Development of proteasome inhibitors for cancer therapy. Int J Drug Discov Pharmacol. 2024; March 18:100004. [http://dx.doi.org/10.53941/ijddp.2024.100004]

[33] Sim HW, Galanis E, Khasraw M. PARP inhibitors in glioma: A review of therapeutic opportunities. Cancers (Basel) 2022; 14(4): 1003. [http://dx.doi.org/10.3390/cancers14041003] [PMID: 35205750]

[34] Li J, Cai S, Zhang Y, Zhang X, Zhu X. Nanoparticle drug delivery system for glioma and its efficacy improvement strategies: A comprehensive review. Int J Nanomed. 2020; April 17:2563-82. [http://dx.doi.org/10.2147/IJN.S243223]

[35] Lassman AB, Pugh SL, Wang TJC, *et al.* Depatuxizumab mafodotin in EGFR-amplified newly diagnosed glioblastoma: A phase III randomized clinical trial. Neuro-oncol 2023; 25(2): 339-50. [http://dx.doi.org/10.1093/neuonc/noac173] [PMID: 35849035]

[36] Akram W, Joshi R, Haider T, *et al.* Blockchain technology: A potential tool for the management of pharma supply chain. Res Social Adm Pharm 2024; 20(6): 156-64. [http://dx.doi.org/10.1016/j.sapharm.2024.02.014] [PMID: 38423927]

[37] Brown JS, O'Carrigan B, Jackson SP, Yap TA. Targeting DNA repair in cancer: Beyond PARP inhibitors. Cancer Discov 2017; 7(1): 20-37. [http://dx.doi.org/10.1158/2159-8290.CD-16-0860] [PMID: 28003236]

[38] Yang F, Xue J, Wang G, Diao Q. Nanoparticle-based drug delivery systems for the treatment of cardiovascular diseases. Front Pharmacol 2022; 13: 999404. [http://dx.doi.org/10.3389/fphar.2022.999404] [PMID: 36172197]

[39] Suva ML, Tirosh I. Single-cell RNA sequencing in cancer: Lessons learned and emerging challenges. Mol Cell 2020; 80(3): 452-60. [PMID: 33157015]

[40] Sharma P, Jain V, Tailang M. How artificial intelligence is transforming medicine: The future of pharmaceutical research. In: Tailang M, Ed. Industry 40 and healthcare: Impact of artificial intelligence. Springer Nature Singapore 2023; pp. 133-50. [http://dx.doi.org/10.1007/978-981-99-1949-9_7]

[41] Ceccarelli M, Barthel FP, Malta TM, *et al.* Molecular profiling reveals biologically discrete subsets and pathways of progression in diffuse glioma. Cell 2016; 164(3): 550-63. [http://dx.doi.org/10.1016/j.cell.2015.12.028] [PMID: 26824661]

[42] Sharma P. Nanostructured biomaterials in drug delivery: current trends and upcoming possibilities. Engineered biomaterials: synthesis and applications. Singapore: Springer Nature Singapore 2023; pp. 261-80.
[http://dx.doi.org/10.1007/978-981-99-6698-1_9]

[43] Reardon DA, Kim L, Baron NL, Moore K, Kaley TJ, Lim M. A phase 2 trial of osimertinib in EGFR-mutant glioblastoma. J Clin Oncol 2020; 38(15) (Suppl.): 2504.

[44] Mellinghoff IK, Penas-Prado M, Peters KB, Burris HA, Maher EA, Ahluwalia MS. Ivosidenib in isocitrate dehydrogenase 1–mutated advanced glioma. JCO Precis Oncol 2021; 5: 1-11.

[45] Kaley TJ, Touat M, Subbiah V, Hollebecque A, Rodon J, Lockhart AC. Phase II trial of combined BRAF and MEK inhibition in BRAF V600E mutant glioma. J Clin Oncol 2018; 36(15) (Suppl.): 2001.

[46] Wen PY, Reardon DA, Armstrong TS, Phuphanich S, Aiken RD, Fink KL. A randomized double-blind placebo-controlled phase 2 study of rindopepimut (CDX-110) in patients with relapsed EGFRvIII-expressing glioblastoma (ReACT). Lancet Oncol 2016; 17(3): 426-35.

[47] Sharma P. Design, optimization, and characterization of herbal gel for facial application. Asian J Pharm 2023; 17(3).

[48] Brown CE, Alizadeh D, Starr R, *et al.* Regression of glioblastoma after chimeric antigen receptor T-cell therapy. N Engl J Med 2016; 375(26): 2561-9.
[http://dx.doi.org/10.1056/NEJMoa1610497] [PMID: 28029927]

[49] Vivanco I, Mellinghoff IK. EGFR inhibitors in glioblastoma: lessons from preclinical studies and clinical trials. Neuro-oncol 2010; 12(4): 299-310.

[50] Lassman AB, Pugh SL, Wang TJC, Aldape KD, Gan HK, Preusser M. Updated results of NRG Oncology/RTOG 0913, a phase II study of concurrent bevacizumab and hypofractionated stereotactic irradiation (HFSRT) in large recurrent gliomas. Neuro-Oncol. 2019;21(Suppl_6).

[51] Todo T, Ito H, Ino Y, Ohtsu H, Ota Y, Shibahara J. Results of phase II clinical trial of oncolytic HSV-1, G47Δ, in glioblastoma. J Clin Oncol 2022; 40(10): 1000-9.

[52] Alexis F, Rhee JW, Richie JP. Nanoparticles for targeted delivery of chemotherapeutics: novel strategies for glioblastoma multiforme treatment. J Control Release 2019; 310: 131-43.

[53] Bobo RH, Laske DW. Convection-enhanced delivery of macromolecular therapeutics: translation from the laboratory to clinical practice. J Neurosurg 2016; 127(2): 295-302.

[54] Sampson JH, Brady ML. Convection-enhanced delivery for the treatment of malignant gliomas: a review of clinical outcomes. Neuro-oncol 2018; 20(2): 211-23.

[55] Sharma P, Jain V. An overview of current advances and pharmaceutical uses of 3D and 4D printing. Explor Med 2023; 4: 560-75.
[http://dx.doi.org/10.37349/emed.2023.00161]

[56] Cheng Y, Morshed RA, Auffinger B, Tobias AL, Lesniak MS. Multifunctional nanoparticles for brain tumor imaging and therapy. Adv Drug Deliv Rev 2014; 66: 42-57.
[http://dx.doi.org/10.1016/j.addr.2013.09.006] [PMID: 24060923]

[57] Wainer IW, Somberg JC. Clinical applications of nanoparticle drug delivery systems for the treatment of glioblastoma. Clin Pharmacokinet 2017; 56(1): 27-35.

[58] Shi J, Kantoff PW, Wooster R, Farokhzad OC. Cancer nanomedicine: progress, challenges and opportunities. Nat Rev Cancer 2017; 17(1): 20-37.
[http://dx.doi.org/10.1038/nrc.2016.108] [PMID: 27834398]

[59] Asthagiri AR, Walbridge S, Heiss JD, Lonser RR. Convection-enhanced delivery of targeted toxins for the treatment of malignant gliomas. Neurosurg Focus 2017; 20(4).

[60] Sharma P, Jain V, Tailang M. Advancement of nanocarrier-based Sharma P, Jain V, Tailang M. Advancement of nanocarrier-based engineering for specific drug delivery for cancer therapy. Targeted

cancer therapy in biomedical engineering. Singapore: Springer Nature Singapore 2023; pp. 465-86. [http://dx.doi.org/10.1007/978-981-19-9786-0_13]

[61] Sharma P, Jain V, Jain A, Tailang M. Role of swarm intelligence for health monitoring and related actions. Internet of Things and Fog Computing-enabled solutions for real-life challenges. CRC Press 2022; pp. 95-118.
[http://dx.doi.org/10.1201/9781003230236-6]

[62] Morrison PF, Chen MY. Direct delivery of therapeutics to brain tumors by convection-enhanced delivery: the translational potential of real-time monitoring. Neurotherapeutics 2018; 15(3): 379-88.

[63] Glantz MJ, LaFollette S. Intrathecal chemotherapy: current applications and future directions. J Neurooncol 2016; 128(3): 255-64.

[64] Groves MD, Glantz MJ. Intraventricular chemotherapy for the treatment of neoplastic meningitis. J Clin Oncol 2018; 16(5): 1561-9.

[65] Chamberlain MC, Johnston SK. Leptomeningeal metastasis: a complication of systemic cancer. Oncologist 2017; 15(8): 1103-13.

[66] Rubenstein JL, Combs D. Primary CNS lymphoma: advances in diagnosis and treatment. J Clin Oncol 2018; 27(12): 1617-24.

[67] Packer RJ, Krailo M. Phase III study of intrathecal chemotherapy with methotrexate in children with newly diagnosed primary CNS lymphoma. J Clin Oncol 2017; 25(3): 432-7.

[68] Sharma P, Tailang M. Primaquine-loaded transdermal patch for treating malaria: design, development, and characterization. Future J Pharm Sci 2022; 8(1): 43.
[http://dx.doi.org/10.1186/s43094-022-00433-5]

[69] Gilbert MR, Wang M, Aldape KD, *et al.* Dose-dense temozolomide for newly diagnosed glioblastoma: a randomized phase III clinical trial. J Clin Oncol 2013; 31(32): 4085-91.
[http://dx.doi.org/10.1200/JCO.2013.49.6968] [PMID: 24101040]

[70] Sharma P, Jain V, Tailang M. Swarm intelligence and machine learning algorithms for cancer diagnosis. Swarm intelligence and machine learning. CRC Press 2022; pp. 34-50.
[http://dx.doi.org/10.1201/9781003240037-3]

[71] Mallick S, Benson R, Hakim A, Rath GK. Management of glioblastoma after recurrence: A changing paradigm. J Egypt Natl Canc Inst 2016; 28(4): 199-210.
[http://dx.doi.org/10.1016/j.jnci.2016.07.001] [PMID: 27476474]

[72] Baumann M, Krause M. Radiochemotherapy in malignant gliomas: current standards and future perspectives. Strahlenther Onkol 2016; 192(8): 535-43.

[73] Kumar PB, Kadiri SK, Khobragade DS, *et al.* Synthesis, characterization and biological investigations of some new Oxadiazoles: *In-vitro* and *In-Silico* approach. Results in Chemistry. 2024;7:101241.
[http://dx.doi.org/10.1016/j.rechem.2023.101241]

[74] Zitvogel L, Apetoh L, Ghiringhelli F, André F, Tesniere A, Kroemer G. The anticancer immune response: indispensable for therapeutic success? J Clin Invest 2008; 118(6): 1991-2001.
[http://dx.doi.org/10.1172/JCI35180] [PMID: 18523649]

[75] Reardon DA, Wucherpfennig KW. Immunotherapy for glioblastoma: on the road to clinical success? J Immunother Cancer 2016; 4(1): 1-11.

[76] Sharma P. Applications of statistical tools for optimization and development of smart drug delivery system. Smart drug delivery. London: IntechOpen 2022; p. 183.
[http://dx.doi.org/10.5772/intechopen.99632]

[77] Omuro A, Vlahovic G, Lim M. Nivolumab with or without ipilimumab in patients with recurrent glioblastoma: results from CheckMate 143. J Clin Oncol 2018; 36(15) (Suppl.): 2018.
[PMID: 29106665]

[78] Schalper KA, O'Rourke DM. Pembrolizumab for recurrent glioblastoma: results of a multicenter phase II trial (KEYNOTE-028). J Clin Oncol 2019; 37(15) (Suppl.): 2019.

[79] Batra R, Jain V, Sharma P. Adenosine: a partially discovered medicinal agent. Future J Pharm Sci 2021; 7(1): 214.
[http://dx.doi.org/10.1186/s43094-021-00353-w] [PMID: 34697594]

[80] Sarin H, Kanevsky AS. Effective trans-blood-brain barrier delivery of chemotherapeutics by localized ultrasound-induced hyperthermia. J Transl Med 2008; 6: 27.

[81] Sahoo SK, Labhasetwar V. Enhanced antiproliferative activity of transferrin-conjugated paclitaxel-loaded nanoparticles is mediated *via* sustained intracellular drug retention. Mol Pharm 2005; 2(5): 373-83.
[http://dx.doi.org/10.1021/mp050032z] [PMID: 16196490]

[82] Sharma P, Jain V, Tailang M. Selection and role of polymers for designing of a drug carrier. 2022.
[http://dx.doi.org/10.5772/intechopen.103125]

[83] Timbie KF, Mead BP, Price RJ. Drug and gene delivery across the blood–brain barrier with focused ultrasound. J Control Release 2015; 219: 61-75.
[http://dx.doi.org/10.1016/j.jconrel.2015.08.059] [PMID: 26362698]

[84] Sharma P. Modification of human behavior due to coronavirus outbreak: a brief study on current scenario. Asian J Pharm 2021; 15(3).

[85] Sharma P, Jain V, Jain S. Fabrication, optimization and evaluation of chronotropic drug delivery system of captopril. Int J Pharm Sci Res 2021; 12: 2203-10.

[86] Sharma P, Tailang M. Design, optimization, and evaluation of hydrogel of primaquine loaded nanoemulsion for malaria therapy. Future J Pharm Sci 2020; 6(1): 26.
[http://dx.doi.org/10.1186/s43094-020-00035-z]

[87] Sharma P, Tailang M. Design, optimization and evaluation of buccal drug delivery system of propranolol for hypertension treatment. Int J Pharm Sci Res 2020; 11(1): 301-11.

[88] Kreisl TN, Kim L, Moore K, *et al.* Phase II trial of single-agent bevacizumab followed by bevacizumab plus irinotecan at tumor progression in recurrent glioblastoma. J Clin Oncol 2009; 27(5): 740-5.
[http://dx.doi.org/10.1200/JCO.2008.16.3055] [PMID: 19114704]

[89] Sharma P, Tailang M. pH dependent release potential of natural polymers in sustained release of ornidazole from colon targeted delivery system. Int J Pharm Sci Res 2019; 9: 1-0.

[90] Sharma P, Tailang M. *In-vivo* study of orodispersible tablet of primaquine. Int J Pharm Sci Res 2018; 9(8): 3506-10.

[91] Khan F, Amatya B, Ng L. Multidisciplinary rehabilitation in the treatment of glioblastoma multiforme: a randomized controlled trial. J Neurooncol 2014; 116(3): 529-36.

[92] Pace A, Di Lorenzo C, Guariglia L. Home care for brain tumor patients. Neurooncol Pract 2017; 4(4): 199-208.
[PMID: 31385989]

[93] Sharma P, Bhargava S, Parashar D, Mangal A. Formulation and evaluation of nanoparticles containing cyclophosphamide. World J Pharm Res 2017; 7(1): 785-98.

[94] Koffman J, Higginson IJ. Ethical issues in palliative care for patients with advanced cancer. J Med Ethics 2002; 28(6): 370-4.

# CHAPTER 6

# Dendrimer Technology: Current Advancements and Future Opportunities in Brain Tumour Detection and Management

**Anindita Behera[1,*], Sandesh Lodha[2], Nishigandha Sa[3], Sudhanshu Sekhar Rout[3], Gajanan Kalyankar[2], Nikita Panchal[2]** and **Manoj Kumar Sabnani[4]**

[1] *Amity Institute of Pharmacy, Amity University, Kolkata, West Bengal, India*

[2] *Maliba Pharmacy College, Uka Tarsadia University, Tarsadi, Dist-Surat, Gujarat, India*

[3] *School of Pharmaceutical Sciences, Siksha' O' Anusandhan Deemed to be University, Bhubaneswar, Odisha, India*

[4] *Aakha Biologics, Frisco Square Blvd, Frisco, USA*

**Abstract:** A brain tumour is an uncontrolled cell proliferation, forming a mass of tissue composed of cells that grow and divide abnormally, seemingly beyond the control of the body's normal regulatory processes. Approximately 70% of primary malignant brain tumors diagnosed each year originate from glial cells. The physiological blood-brain barrier (BBB) impairs drug distribution to the tumour microenvironment and complicates the treatment of malignant brain tumours. Most of the conventional chemotherapeutics lack specificity and lead to serious systemic toxicity. However, nanocarriers have shown efficient therapeutic efficacy in delivering medications to the brain tumour cells. The targeting of nanocarriers to the tumour sites can be achieved by active or passive targeting. The dimensions and the physicochemical properties of the nanocarriers significantly affect brain permeability. Among various nanotools, branched PAMAM, PPL, and PPI dendrimers possess great efficacy in transporting chemotherapeutic agents across the BBB for treating brain tumours. This chapter discusses the various generations of dendrimers, their synthesis techniques, and the passive and active targeting strategies used to deliver chemotherapeutics to the tumour sites. The chapter also includes dendrimers as diagnostics and contrast agents in brain tumour diagnosis. Dendrimers have been established as remarkable in diagnosing and treating brain tumours, as they can transport the therapeutically active agents across the BBB to the cancer cells after systemic administration. Different dendrimers like PAMAM, PLL, PPI, carbosilanes, and phosphorus-based are used to develop novel therapeutics having prolonged and controlled drug release, immunotherapy, and anticancer activity. This chapter can provide remarkable guidance to scientists working on brain-targeting delivery systems.

* **Corresponding author Anindita Behera:** Amity Institute of Pharmacy, Amity University, Kolkata, West Bengal, India; E-mail: abehera@kol.amity.edu

**Prashant Tiwari, Pankaj Kumar Singh & Sunil Kumar Kadiri (Eds.)**

**Keywords:** Brain tumour, Dendrimers, Imaging, Passive and active targeting, Therapeutics.

## INTRODUCTION

Dendrimers are branched macromolecules with bulky end-functional groups and a compressed molecular assembly. Each dendrimer has a central core molecule from which numerous branches extend, forming a well-defined and hierarchical architecture. The branches comprise recurring units or monomers chemically connected in a controlled fashion [1].

The chemical configuration and generation number of core, branches, and functional groups on the surface determine the dendrimer's shape, size, and reactivity. The word dendrimer was derived from a Greek term called "dendron", which means "branching of a tree". Dendrimers are branched, spherical, and symmetrical structures with defined shape and specificity, varying sizes from 1 to 15 nm. Dendrimers are primarily used to enhance the specific properties of a compound [2, 3] (Fig. **1**).

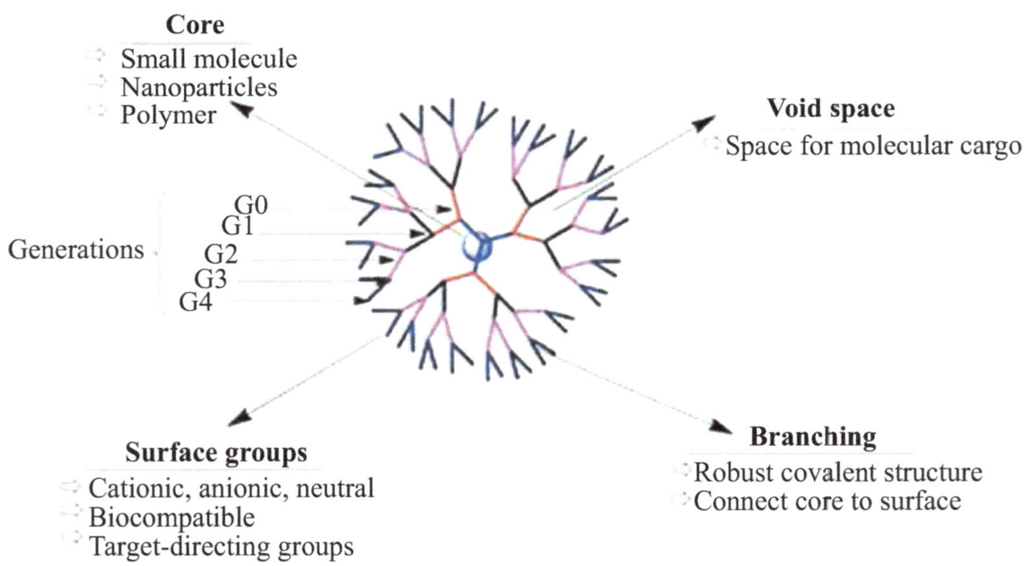


**Fig. (1).** Structure of Dendrimer (The dendrimer consists of three components: a core, interior layers with branching, and exterior surface functional groups).

Dendrimers are a fascinating class of molecules with unique properties that make them highly versatile for various applications. The key properties of dendrimers are as follows [2, 4]:

- **Well-defined Structure:** The dendrimer molecules of a particular generation are nearly identical in size and shape. This precise control over their architecture enables the tailoring of their properties with precision. Dendrimers can be synthesized as monodispersed materials.
- **Size and Generation:** The size of dendrimers is precisely controlled by the number of branching units, referred to as generation. Each generation features a doubling of the surface area and functionality compared to the previous one.
- **Functionality:** The outermost layer of a dendrimer, known as the periphery or surface groups, can be chemically modified with various functional groups. This allows researchers to design dendrimers for specific purposes, such as attaching drugs for targeted drug delivery or incorporating reactive groups for catalysis.
- **Host-Guest Inclusion:** The dense, branched structure of dendrimers creates cavities within their interior. These cavities can be hosts for encapsulating guest molecules, including drugs, imaging agents, or catalysts.
- **Biocompatibility:** Some dendrimers exhibit low cytotoxicity (minimal toxicity to living cells), making them potentially suitable for biomedical applications.

Dendrimers, due to their unique properties, are utilized in various fields. Their well-defined structure allows precise control, and their size can be tailored in generations. The external surface can be customized with specific functionalities for desired tasks, like attaching drugs. They can be tuned for water or oil solubility and hold guest molecules inside. Some are biocompatible, making them promising for medical applications.

Despite their advantages, dendrimers have some drawbacks. Certain types, particularly those of higher generation and cationic dendrimers, can be toxic to cells. Their complex structure makes them expensive to synthesize, and their behaviour in biological systems is still under investigation. This necessitates further research to optimize their biocompatibility and ensure safe use in various applications [5].

Dendrimers excel in nanotechnology due to their precisely tunable structure. Their hollow core and modifiable surface enable them to function as tiny delivery vehicles. They can carry drugs inside, protecting them until they reach their target. Surface groups can be designed to target specific cells, improving drug delivery and reducing side effects. This makes them promising for medicine, but they also have potential in sensors and catalysts. They are also being explored for gene therapy and diagnostics (Table **1**). However, research is ongoing to optimize their biocompatibility [6].

**Table 1. Types of dendrimers in biomedical applications.**

| Type of Dendrimers | Example | Core | Branches | Limitations | Applications | References |
|---|---|---|---|---|---|---|
| PAMAM (Poly(amidoamine)) Dendrimers | Starburst® dendrimers | ethylenediamine or ammonia | Amidoamine units | Cause organ and tissue toxicities on prolonged administration | Drug delivery, gene transfection, imaging | [25] |
| PPI (Poly(propylene imine)) Dendrimers/ DAB-AM dendrimers | DAB-AM-64 (5th generation PPI dendrimer) | Diaminobutane | Propylene imine units | Low water solubility, toxicity, haemolytic effect | Catalysis, molecular recognition, drug delivery | [8] |
| Phosphorus-containing dendrimers | Phosphorus dendrimers with aldehyde surface groups | $P(S)Cl_3$ | Phosphorus-containing units | Toxicity | Catalysis, materials science, nanomedicine | [11] |
| Carbosilane/ Silicon-based dendrimers with carbon-silicon bonds dendrimers | G4 carbosilane dendrimer with Si-H termini | Tetra allylsilane | $Si\text{-}CH_2\text{-}CH_2\text{-}CH_2\text{-}Si$ units | Toxicity | Catalysis, chemical sensors, drug delivery | [21] |
| Peptide Dendrimers | Multiple Antigen Peptide (MAP) systems | Lysine | Various amino acid sequences | Low water solubility and high nonspecific toxicity | Vaccine development, protein mimics, drug delivery | [56] |
| Polyester Dendrimers | Bis-MPA polyester dendrimers | Pentaerythritol or trimethylolpropane | 2,2-Bis(hydroxymethyl)propionic acid (bis-MPA) units | | Drug delivery, tissue engineering, biodegradable materials | [57] |
| Poly(benzyl ether) Dendrimers/ Fréchet-type dendrimers | G4 poly(benzyl ether) dendrimer | 1,3,5-trihydroxybenzene | 3,5-Dihydroxy benzyl alcohol units | Small yield, toxicity | Light harvesting, catalysis, drug delivery | [9] |
| Glycodendrimers | PAMAM-based glycodendrimer with maltose surface groups | Ethylenediamine (typical PAMAM core) | Amidoamine units | Stability problems | Lectin binding, drug delivery, anti-viral agents | [58] |
| Metallodendrimers | Ruthenium-centered phosphorus dendrimers | Ruthenium complex | Phosphorus-containing units | | Catalysis, photodynamic therapy, molecular electronics | [59] |
| Triazine Dendrimers | Melamine-based dendrimers | cyanuric chloride | 1,3,5-Triazine units | Low solubility, intrinsic toxicity | Drug delivery, light-harvesting, catalysis | [60] |
| Hybrid Dendrimers | PAMAM-PPI hybrid dendrimers | PAMAM structure | PPI structure | | Tailored drug delivery, combined properties of both dendrimer types | [61] |

# DENDRIMERS: AN EFFECTIVE NANOTOOL

## Concise Overview of the Structure of Dendrimers

The dendrimers consist of three components: a core with a central initiator molecule, interior branching units (generations), and an exterior layer containing terminal functional groups. This section provides a brief overview of the structural features and classifications of various dendrimers.

### *Core: Central Initiator Molecule*

The core of a dendrimer is the central initiator molecule from which the dendritic branches grow. It is crucial in determining the dendrimer's overall structure, properties, and functionality. Different core molecules can lead to dendrimers with varying characteristics and applications. The chemicals used for the core and its applications are explained below.

- **Ethylenediamine (EDA) Core**: One of the most common cores used in PAMAM dendrimers is EDA-core PAMAM dendrimers [7]. Its structure is based on a four-branched initiator and has applications in drug delivery, gene transfection, and other fields.
- **Ammonia Core:** This core is used in PAMAM and PPI dendrimers, such as ammonia-core PPI dendrimers [8]. Its structure is a three-branched initiator with applications in catalysis and molecular recognition.
- **Benzene Core**: The common dendrimer in this class is poly (benzyl ether) dendrimer with an example of 1,3,5-trisubstituted benzene core [9]. The structure is a three-branched aromatic initiator with applications in light harvesting, energy transfer, and other related fields.
- **Pentaerythritol Core:** It is used in polyester dendrimers, such as pentaerythritol-core polyester dendrimers [10]. The structure is a four-branched aliphatic initiator with applications in biodegradable materials, drug delivery, *etc.*
- **Cyclotriphosphazene Core:** This is common in phosphorus-containing dendrimers, such as cyclotriphosphazene-core dendrimers [11]. The structure is a six-branched inorganic initiator and is used in flame retardants, catalysis, *etc.*
- **Metal-Centered Core:** Dendrimers can be constructed around metal centres, resulting in unique properties [12]. A ruthenium-centred dendrimer is an example that has an octahedral structure of metal complex as an initiator with applications as photosensitizers, catalysis, *etc.*
- **Porphyrin Core:** Porphyrin-cored dendrimers have interesting photophysical properties, such as tetraphenylporphyrin-cored dendrimers [13]. The structure is

a four-branched planar aromatic initiator with applications in light-harvesting and photodynamic therapy, among others.

- **Carbohydrate Core:** The glucose-cored dendrimers are multibranched sugar initiators and have the potential for biological applications in drug delivery, biological recognition, *etc* [14].
- **Fullerene Core:** These dendrimers combine the properties of fullerenes with dendritic structures [15]. An example is C60-cored dendrimers, which feature spherical carbon cage initiator structures. Their applications include photovoltaics, electron transfer studies, *etc.*
- **DNA Core:** DNA can be a core for constructing a hybrid dendritic structure with a nucleic acid initiator [16]. The applications include gene delivery, biosensing, *etc.*

### *Effect of Core Multiplicity*

The number of reactive sites on the core (core multiplicity) affects dendrimer properties. The following is a comparison of 3-arm *vs.* 4-arm core PAMAM dendrimers [17]:

- 3-Arm core: More flexible, lower intrinsic viscosity.
- 4-Arm core: More rigid, higher intrinsic viscosity.

These demonstrate the diversity of core molecules used in dendrimer synthesis. The core choice can significantly influence the dendrimer's physical properties, reactivity, and potential applications. By selecting appropriate core molecules, researchers can tailor dendrimers for specific purposes across various fields, including materials science, biomedicine, and nanotechnology.

### *Interior Layers or Branching Units (Generations) of Dendrimers*

The interior layers of a dendrimer consist of repeating branching units that extend outward from the core. These layers are often called generations, with each new branching unit representing a new generation. The nature of these branching units significantly influences the properties and behaviour of the dendrimer. Generations and their examples are discussed below.

**A. PAMAM (Poly(amidoamine)) Dendrimers:** Polyamidoamine dendrimers contain an ethylenediamine group as a core with amide groups on the surface, which gives a spherical structure, and each terminal possesses amine functional groups at the surface [7].

Examples of PAMAM dendrimer generations are as follows:

- G0: 4 surface functional groups
- G1: 8 surface functional groups
- G2: 16 surface functional groups
- G3: 32 surface functional groups
- G4: 64 surface functional groups

**B. PPI (Poly(propylene imine)) Dendrimers:** Butylene diamine is the core, and repeated propylene imine is the branching unit of poly (propylene imine) dendrimers. This is achieved by the sequential Michael's addition reaction of acrylonitrile with a primary amino group and then hydrogenating the nitrile (C≡N) groups to form primary amino ($NH_2$) groups [18, 19].

Examples of PPI dendrimer generations are as follows:

- G1: 4 surface functional groups
- G2: 8 surface functional groups
- G3: 16 surface functional groups
- G4: 32 surface functional groups
- G5: 64 surface functional groups

**C. Poly(benzyl ether) Dendrimers:** These dendrimers closely resemble polyester dendrimers, having aromatic core bonded to peripheral 3,5-dihydroxybenzyl alcohol units by ester linkage [9].

Examples of Fréchet-type dendrimer generations are as follows:

- G1: 6 surface functional groups
- G2: 12 surface functional groups
- G3: 24 surface functional groups
- G4: 48 surface functional groups

**D. Phosphorus-Containing Dendrimers:** In this type of dendrimer, phosphorus is present at each branch point, and the surface functional groups containing 4-hydroxybenzaldehyde provide a hydrophobic backbone and hydrophilic shell to the dendrimer structure [20].

Examples of phosphorus dendrimer generations are as follows:

- G1: 6 surface functional groups
- G2: 12 surface functional groups
- G3: 24 surface functional groups
- G4: 48 surface functional groups

**E. Carbosilane Dendrimers:** Carbosilane dendrimers contain carbon and silicon molecules as building blocks. The carbosilane dendrimers having hydrophobic scaffolds possess higher thermal stability. The nucleophilic molecules easily reach the electrophilic silicon ($Si^+$). Hence, silicon is used in the preparation of carbosilane dendrimers. The C-Si bond in the dendrimers provides low polarity and high energy, which makes them more hydrophobic [21].

Examples of carbosilane dendrimer generations are as follows:

- G1: 4 surface functional groups
- G2: 8 surface functional groups
- G3: 16 surface functional groups
- G4: 32 surface functional groups

***Key Insights into the Impact of Generations***

***A. Generation-Dependent Properties***: The properties of dendrimers often change with increasing generation numbers [17].

Example: PAMAM dendrimers

- G0-G3: Open, flexible structures
- G4-G7: Transition to more globular shapes
- G8-G10: Dense, spheroidal structures

***B. Asymmetric Branching:*** Some dendrimers incorporate asymmetric branching units.

Example: Janus dendrimers [22]

- One side: Hydrophilic branching units
- Other side: Hydrophobic branching units

***C. Biodegradable Branching Units:*** Dendrimers with biodegradable branching units are used for biomedical applications [23].

Example: Polyester dendrimers

- Branching unit: 2,2-bis(hydroxymethyl)propionic acid
- Biodegradable ester linkages between generations

***D. Chiral Branching Unit:*** Dendrimers with chiral branching units are used for applications in asymmetric catalysis [24].

Example: BINOL-based chiral dendrimers

- Branching unit: 1,1'-Bi-2-naphthol (BINOL) derivatives
- Applications in enantioselective catalysis

***E. Generation-Dependent Reactivity:*** The reactivity of interior branching units can change with generation [25].

Example: PAMAM dendrimers

- Lower generations: Higher reactivity of interior amines
- Higher generations: Decreased reactivity due to steric hindrance

These examples illustrate the diversity and complexity of branching units in dendrimer interiors. The choice of branching unit and the number of generations have a significant influence on the dendrimer's physical properties, reactivity, and potential applications. By carefully designing the interior layers, researchers can create dendrimers with tailored characteristics for specific uses in fields, such as drug delivery, catalysis, and materials science.

***Exterior: Terminal Functional Groups***

The exterior of a dendrimer consists of terminal functional groups that form the outermost layer of the macromolecule. These groups determine the dendrimer's surface properties, reactivity, and potential applications. The nature of these terminal groups can be tailored to achieve specific characteristics or functionalities.

***Amine-Terminated Dendrimers***

The common amine-terminated dendrimers are PAMAM and PPI dendrimers. PAMAM dendrimers with primary amine surface groups have applications in gene delivery, drug delivery, and other fields. These dendrimers are cationic at physiological pH and have high water solubility [26].

***Hydroxyl-Terminated Dendrimers***

These are often used for biocompatible applications, such as drug solubilization and coatings. The example includes a hydroxyl-terminated PAMAM dendrimer with properties like neutral charge and high water solubility [27].

***Carboxyl-Terminated Dendrimers***

They are useful for pH-responsive applications, controlled drug release, and metal ion chelation, among other purposes. Carboxyl-terminated PAMAM dendrimers are anionic at physiological pH and show pH-responsive behaviour [28].

***Alkyl-Terminated Dendrimers***

These dendrimers exhibit hydrophobic character, enabling applications, such as catalysis in organic solvents and molecular recognition. An example is alkyl-terminated PPI dendrimers [29]. These dendrimers are hydrophobic in nature and soluble in organic solvents.

***PEGylated Dendrimers***

These dendrimers enhance biocompatibility and circulation time in the body. The applications include long-circulating drug carriers with protein-resistant surfaces. Examples include PEG-grafted PAMAM dendrimers with increased water solubility and reduced toxicity [30].

These examples demonstrate the variety of terminal functional groups that can be incorporated into dendrimer structures. The choice of terminal groups allows precise control over the dendrimer's surface properties, reactivity, and functionality. By carefully selecting and modifying these exterior groups, researchers can design dendrimers for specific applications in drug delivery, catalysis, sensing, and materials science.

## Synthesis of Dendrimers

The main approaches to dendrimer synthesis are divergent and convergent methods. However, the other two methods, the click-chemistry approach and accelerated synthesis, are also explained in this chapter.

### *Divergent Synthesis*

In divergent synthesis, dendrimers are built from the core outwards. The process begins with a multifunctional core molecule and proceeds with the stepwise addition of branching units to create successive generations.

Steps involved:

The following steps are involved:

a. Start with a core molecule.
b. React the core with monomers to create the first-generation.
c. Activate the end groups of the first-generation.
d. React with more monomers to create the next generation.
e. Repeat steps c and d to reach the desired generation number.

Example: Poly(amidoamine) (PAMAM) dendrimers.

PAMAM dendrimers were first synthesized using the divergent approach [7]. The synthesis starts with ethylenediamine as the core, followed by Michael addition with methyl acrylate and amination with ethylenediamine.

***Advantages of Divergent Synthesis***

- Can produce large quantities of dendrimers.
- Suitable for high-generation dendrimers.

***Disadvantages of Divergent Synthesis***

- Potential for structural defects in higher generations due to incomplete reactions.
- Purification can be challenging.

***Convergent Synthesis***

In convergent synthesis, dendrimers are built from the periphery inwards. The process begins with synthesizing dendritic wedges (dendrons), which are then attached to a multifunctional core.

Steps involved:

The following steps are involved:

a. Synthesize dendrons starting from the periphery
b. Build up the dendrons to the desired size
c. Attach the dendrons to a multifunctional core
   Example: Fréchet-type polyether dendrimers.

These dendrimers were first synthesized by Jean Fréchet and colleagues using the convergent approach [9]. The synthesis involves building benzyl ether dendrons and attaching them to a multifunctional core.

Advantages of convergent synthesis:

- Better control over the final structure.

- Easier purification of intermediates.
- Allows for the synthesis of asymmetric dendrimers.

Disadvantages of convergent synthesis:

- Limited to lower generations due to steric hindrance.
- Generally lower yields compared to divergent synthesis.

### *Click-Chemistry Approach*

In recent years, click-chemistry has been applied to synthesize dendrimers, offering a more efficient and versatile method. These dendrimers are synthesized using copper-catalyzed cycloaddition azide-alkyne (CuAAC) [31].

Example: Triazole dendrimers.

### *Accelerated Synthesis*

This approach combines aspects of both divergent and convergent methods to expedite the synthesis process. The method utilizes a hypercore to accelerate the synthesis of PPI dendrimers [32].

Example: Rapid synthesis of polypropylene imine (PPI) dendrimers.

These are the primary methods for dendrimer synthesis. Each method has its advantages and is chosen based on the specific dendrimer structure desired, the scale of production, and the intended application.

## Properties of Dendrimers

### *Physical Properties*

A. Size and Shape

Dendrimers can increase in size and change in shape with generation number. For example, PAMAM dendrimers change from open structures to densely packed spheroids [17]. Dendrimers' size and shape evolution with increasing generation number is a fundamental characteristic that influences many of their properties and applications. As the generation number increases, dendrimers become larger and transition from open, flexible structures to more compact, spherical shapes. A few examples are explained below:

PAMAM (Poly(amidoamine)) Dendrimers

PAMAM dendrimers are among the most well-studied and provide a clear example of size and shape evolution [33].

- Generation 0-2: Open, highly asymmetric structures.
- Generation 3-6: Transition to more spheroidal structures.
- Generation 7-10: Dense, spherical structures.

The diameter increases from approximately 1.5 nm for G1 to 14.5 nm for G10.

PPI (Poly(propylene imine)) Dendrimers

PPI dendrimers show a similar trend but with slightly different size ranges [34].

- G1: 1.4 nm diameter.
- G3: 2.8 nm diameter.
- G5: 5.4 nm diameter.

Carbosilane Dendrimers

Carbosilane dendrimers exhibit a more gradual transition in shape [35].

- G1-G3: Extended, open structures.
- G4-G6: Transition to more globular shapes.
- G7+: Compact, dense structures.

Phosphorus-Containing Dendrimers

These dendrimers exhibit a unique "onion-like" layered structure as the number of generations increases [36].

- G1-G4: Relatively open structures.
- G5-G8: Development of distinct concentric layers.
- G9+: Highly compact, multi-layered structures.

Peptide Dendrimers

Peptide dendrimers often exhibit more irregular shapes due to their amino acid building blocks [37].

- G1-G2: Extended, asymmetric structures.
- G3-G4: Transition to more compact forms.

- G5+: Globular structures, but often less symmetric than PAMAM or PPI dendrimers.

B. Solubility

They may have high solubility in various solvents due to multiple surface groups. The example includes PPI dendrimers with alkyl surface groups soluble in nonpolar solvents [38]. The solubility of dendrimers is primarily determined by their surface groups, which can be tailored during synthesis to achieve desired solubility properties. This characteristic allows dendrimers to be soluble in various solvents, making them versatile for multiple applications. A few examples are explained below:

PAMAM dendrimers with different terminal groups demonstrate varying solubility [39].

- Amine-terminated: Highly water-soluble.
- Hydroxyl-terminated: Soluble in polar solvents like water and methanol.
- Carboxyl-terminated: pH-dependent solubility in water.

Example: G4 PAMAM dendrimers.

- Amine-terminated: >150 mg/mL in water.
- Hydroxyl-terminated: >100 mg/mL in water and methanol.

These examples demonstrate the versatility of dendrimer solubility, which can be precisely tuned through surface group modifications. This property makes dendrimers valuable in various fields, including drug delivery, catalysis, and materials science.

C. Viscosity

The intrinsic viscosity of dendrimers is a unique physical property that sets them apart from linear polymers. This characteristic is attributed to their highly branched, compact structure, which results in a smaller hydrodynamic volume than linear polymers of similar molecular weight. PAMAM dendrimers show significantly lower intrinsic viscosity compared to linear polymers.

Example: Comparison of PAMAM dendrimers with linear poly(ethylene glycol) (PEG) [40].

- PAMAM G5 (MW ~28,800): Intrinsic viscosity = 4.1 mL/g.

• Linear PEG (MW ~29,000): Intrinsic viscosity = 19.0 mL/g.

Generation-Dependent Intrinsic Viscosity:

The intrinsic viscosity of dendrimers typically increases with generation number but at a much slower rate compared to the increase in molecular weight [25].

Example: PAMAM dendrimers.

• G3 (MW ~6,900): Intrinsic viscosity = 1.8 mL/g.
• G5 (MW ~28,800): Intrinsic viscosity = 4.1 mL/g.
• G7 (MW ~116,000): Intrinsic viscosity = 7.3 mL/g.

These examples and studies collectively demonstrate that the lower intrinsic viscosity of dendrimers compared to linear polymers of similar molecular weight is a consistent feature across different types of dendrimers. This property arises from their compact, highly branched structure. It has important implications for their behaviour in solution and potential applications in drug delivery, rheology modifiers, and nanocomposites.

D. Thermal Properties

The glass transition temperature (Tg) is a crucial physical property of polymers, including dendrimers. For dendrimers, the Tg often increases with the generation number due to increased molecular weight, decreased chain mobility, and an increased density of branch points [41, 42]. Poly(benzyl ether) dendrimers show a clear trend of increasing Tg with the following generation:

• Generation 1: Tg = 34°C.
• Generation 2: Tg = 43°C.
• Generation 3: Tg = 52°C.
• Generation 4: Tg = 60°C.
• Generation 5: Tg = 66°C.

These examples demonstrate that increasing glass transition temperature with increasing generation is common across different dendrimers. The increase in magnitude varies depending on the chemical composition and structure of dendrimers. This property has important implications for dendrimers' thermal and mechanical behaviour and their potential applications in various fields.

E. Encapsulation Ability

The encapsulation of guest molecules in the dendrimers is one of the most attractive and useful properties. This ability stems from their unique architecture, which creates internal cavities or voids that can host various molecules. This property has significant implications for catalysis, drug delivery, and environmental remediation applications. PAMAM dendrimers are widely studied for their encapsulation abilities [43].

Example: Encapsulation of nonsteroidal anti-inflammatory drugs (NSAIDs) as G4 PAMAM dendrimer.

- Ibuprofen: ~28 molecules per dendrimer.
- Ketoprofen: ~20 molecules per dendrimer.

Generation-Dependent Encapsulation:

The encapsulation capacity often increases with dendrimer generation due to larger internal volumes [44].

Example: PAMAM dendrimers encapsulating 5-fluorouracil (5-FU).

- G4: ~20 5-FU molecules per dendrimer.
- G5: ~25 5-FU molecules per dendrimer.

These examples demonstrate the versatile encapsulation abilities of dendrimers. The ability to encapsulate guest molecules within interior voids is influenced by factors, such as dendrimer generation, chemical structure, pH, and the properties of the guest molecules. This characteristic makes dendrimers promising candidates for various applications, particularly in drug delivery, catalysis, and nanotechnology.

F. Light Scattering

Light scattering is a powerful technique for characterizing dendrimers' size, shape, and molecular weight. As dendrimer generation increases, their size and density also increase, leading to changes in their light-scattering behaviour. PAMAM dendrimers show a clear trend of increased light scattering with generation.

Example: Radius of gyration (Rg) and molecular weight (Mw) from light scattering [45].

- G3: Rg = 1.58 nm, Mw = 6,900 g/mol.
- G5: Rg = 2.41 nm, Mw = 28,800 g/mol.
- G7: Rg = 3.28 nm, Mw = 116,000 g/mol.
- G9: Rg = 4.16 nm, Mw = 467,000 g/mol.

These studies demonstrate that the increase in light scattering with generation number is consistent across different types of dendrimers. This property arises from their increasing size and density as the generation number increases. Light scattering techniques provide valuable insights into dendrimers' size, shape, and interactions, making them crucial tools for characterizing these unique macromolecules and understanding their behaviour in solution.

***Biological Properties of Dendrimers***

Due to their unique properties, dendrimers have attracted significant attention in biological and biomedical applications. Their shape, size, surface chemistry, and generation number largely influence their biological properties.

***A. Cellular Uptake and Internalization:*** Dendrimers can be internalized by cells through various mechanisms [46].

Example: Cellular uptake of PAMAM dendrimer.

- Lower generations (G0-G3): Primarily through passive diffusion.
- Higher generations (G4-G8): Endocytosis-mediated uptake.

***B. Cytotoxicity:*** Dendrimer cytotoxicity generally increases with generation number and cationic surface charge [47].

Example: Cytotoxicity of amine-terminated PAMAM dendrimers.

- G4 and below: Generally low cytotoxicity.
- G5 and above: Increased cytotoxicity.

***C. Blood-Brain Barrier (BBB) Penetration:*** Some dendrimers can cross the BBB, making them potential carriers for CNS drug delivery [48].

Example: G4 PAMAM dendrimer conjugated with transferrin.

- Enhanced BBB penetration compared to non-conjugated dendrimers.
- Potential for targeted drug delivery to the brain.

***D. Immunogenicity:*** Dendrimers generally have low immunogenicity, but this can be modulated [49].

Example: PEGylated PAMAM dendrimers.

- Reduced immunogenicity compared to non-PEGylated dendrimers.
- Longer circulation time *in vivo.*

***E. Biodistribution and Pharmacokinetics:*** Dendrimer biodistribution and pharmacokinetics depend on size, surface charge, and functionality [50].

Example: G6 PAMAM dendrimer with different surface groups.

- Amine-terminated: Rapid liver accumulation.
- Acetylated: Prolonged blood circulation.

***F. Haemolytic Activity:*** Dendrimers (cationic) can interact with RBC membranes, potentially causing haemolysis [51].

Example: Comparison of PAMAM and PPI dendrimers.

- PAMAM dendrimers: Lower haemolytic activity.
- PPI dendrimers: Higher haemolytic activity.

***G. Antimicrobial Activity:*** Some dendrimers exhibit inherent antimicrobial properties:

Example: Poly(lysine) dendrimers [52].

- Broad-spectrum antimicrobial activity against both Gram-negative and positive bacteria.

***H. Gene Transfection Efficiency:*** Cationic dendrimers can complex with DNA and facilitate gene delivery [53].

Example: G7 PAMAM dendrimer for gene delivery.

- Higher transfection efficiency compared to lower generations.
- Comparable efficiency to commercial transfection agents.

***I. Protein Interactions:*** Dendrimers can interact with proteins, potentially altering their structure and function [54].

Example: Interaction of human serum albumin (HSA) with G4 PAMAM dendrimer.

- Formation of stable dendrimer-HSA complexes.
- Potential implications for drug delivery and toxicity.

***J. Biocompatibility and Biodegradability:*** Some dendrimers are designed to be biocompatible and biodegradable [23].

Example: Polyester dendrimers.

- Biodegradable through hydrolysis of ester bonds.
- Reduced long-term toxicity compared to non-degradable dendrimers.

***K. Immunomodulation:*** Certain dendrimers can modulate immune responses [55].

Example: Phosphorus-containing dendrimers.

- Activation of human monocytes.
- Potential applications in vaccine development and immunotherapy.

These examples illustrate the diverse biological properties of dendrimers. Their interactions with biological systems are complex and dependent on various structural factors. Understanding these properties is crucial for developing dendrimer-based biomedical applications, including diagnostic imaging, drug delivery, and gene therapy. Ongoing research continues to explore and optimize the biological behaviour of dendrimers for enhanced safety and efficacy in medical applications.

## TARGETING STRATEGIES OF DENDRIMERS FOR BRAIN TUMOUR

Cancer cell-targeting drugs must be designed to be cell-specific, highly bioavailable, and minimize off-target toxicity. Two strategies adopted for targeted drug delivery are passive and active targeting [62]. During cancer progression, the vasculature surrounding the cancer cells is highly leaky, allowing anticancer drugs to enter the cancer cells freely and remain for an extended period, thereby exhibiting their anticancer activity. The drugs exhibit therapeutic activity owing to the enhanced permeation and retention (EPR) effect of passive targeting (Fig. **2**). However, the limitation of passive targeting is its lack of cell specificity; hence, active targeting solves the problem of nonspecific drug delivery associated with passive targeting. In active targeting, drugs targeted for cancer cells are bound to ligands specific to the proteins or overexpressed receptors on these cells. Each type of cancer has specific receptors that are overexpressed, and the ligand-conjugated drug delivery system delivers the drug to only cancer cells by binding to these receptors. The drug is then internalized into the cancer cells through receptor-mediated endocytosis (Fig. **2**). Hence, active targeting is more advantageous than passive targeting with respect to cell specificity, high bioavailability, and less toxicity to normal cells [63, 64]. Some studies reporting the passive and active targeting of anticancer drugs for brain tumours are listed in Table **2**.

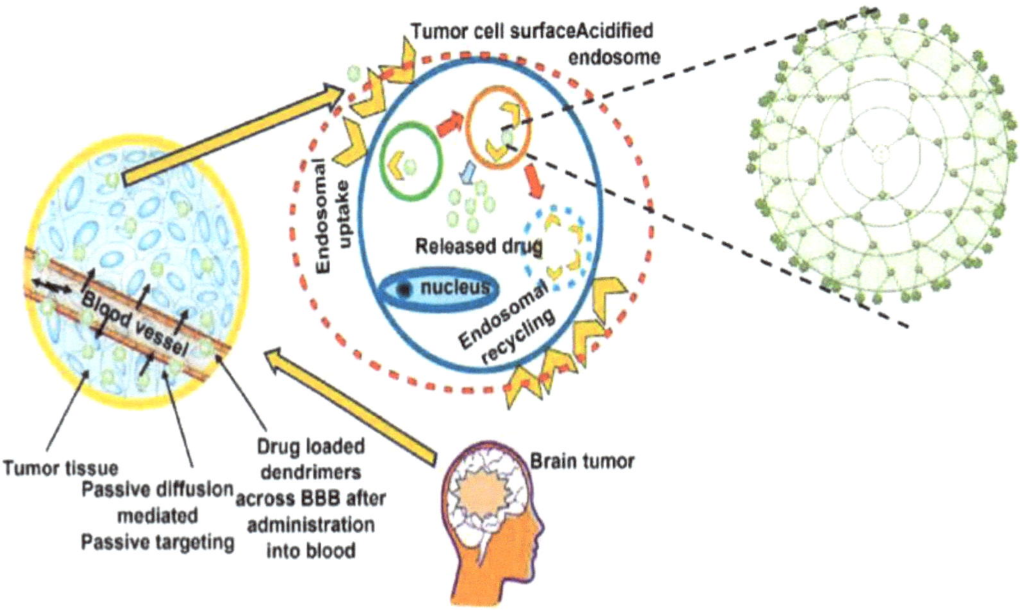


**Fig. (2).** Mechanism of active and passive Targeting of anticancer drugs to the brain (Taken from the open resource, Kaurav *et al.*, 2023) [65].

## Passive Targeting

The largest problem in cancer detection and treatment is developing drug or gene delivery techniques that specifically target tumour cells without harming healthy cells. Effective anticancer drug delivery to tumour cells can be achieved by two strategies, *i.e.*, passive and active targeting [66]. Passively or actively, nanocarriers carry medications or genes to cancer cells or tissues [67]. Iyer *et al.* demonstrated that passive targeting optimizes medicine delivery to specific locations by considering the size of dendrimers, circulation time, metabolic conditions (hypoxia, inflammation), and tumour biology (vascular structure, leakiness). EPR effect aids passive targeting [68]. In 1986, Matsumura and Maeda reported that the EPR effect is due to reduced lymphatic leakage and improved tumour vascular permeability. Poor blood artery architecture and high vascular permeability factors increase permeability in most solid cancers. Macromolecules over 40 kDa leak from tumour capillaries and accumulate in tumour tissue due to the EPR effect. The EPR effect does not transport drugs in normal tissues [69, 70].

EPR effect delivery is 20–30% higher in tumour cells than in normal tissues; however, selectivity is low. The EPR effect depends on intrinsic tumour biology, including lymphangiogenesis and angiogenesis, perivascular tumour growth, stromal response density, and intratumour pressure. All these parameters and nanocarrier physicochemical properties impact the efficiency of drug delivery [71].

**Table 2. Passive and active targeted dendrimer systems for brain tumour treatment.**

| **Passive Targeting** | | | |
|---|---|---|---|
| Types of dendrimers | Name of the drug | Study outcome | References |
| PAMAM | Doxorubicin | Increased delivery of drugs by dendrimer complexes to brain tumours, leading to enhanced therapeutic activity. | [92] |
| | Doxorubicin | iRGD-conjugated G4 PAMAM dendrimer showed increased accumulation of doxorubicin by enhanced permeability in the tumour and decreased tumour size. | [104] |
| | Docetaxel | Docetaxel hybrid dendrimer showed cytotoxicity in glioblastoma cell lines (U87MGMG) but was not toxic to normal cells. | [93] |
| PLL | TRAIL and herpes simplex virus thymidine kinase gene (HSV-TK) | Combination therapy of PPL-PEI dendrimers provided cost-effective treatment of glioblastoma. | [93] |
| | Minocycline | Neuroinflammation is reduced at lower doses. | [94] |
| Peptide dendrimers | $N^1$-butyl and $N^1$-aminopentane- tryptophan dendrimers | These amphiphilic peptide dendrimers possessed high antioxidant activity in glioblastoma cells, and retention of astrocyte viability was 85–95%. | [95] |
| Phosphorus- Triazine dendrimers | Gold (III) or copper (II) complexes and branched hydrophobic fragments | Showcased greater cytotoxicity on glioblastoma stem cells (JHH520, BTSC233, SF188, and NCH644) as compared to standard temozolomide. | [96] |

(Table 2) cont.....

| Active targeting | | | |
|---|---|---|---|
| PAMAM | Paclitaxel | Developed pH-sensitive folic acid-polymethacrylate-PAMAM dendrimers, which released the drug in the acidic tumour microenvironment *via* active targeting. | [97] |
| | Paclitaxel | Enhanced penetration of paclitaxel by 12-fold through brain endothelial cells. | [98] |
| | Apoptin | Apoptin loaded with peptide PAMAM dendrimer (arginine, histidine, and phenylalanine) induced apoptosis in primary glioma cells. | [99] |
| | Pentapeptide | PAMAM designed with CREKA internalized deeply into the brain tumour tissues with increased retention. | [100] |
| | Peripheral-type benzodiazepine receptor | Translocator protein (TSPO) - targeted G(4)-PAMAM- fluorescein isothiocyanate dendrimer targeting the mitochondria of the glioma cells and internalized into the cancer cells by the endocytosis pathway. | [101] |
| | RNA (siRNA) and Adriamycin | Dendrimers administered adriamycin and siRNA to resistant cancer cells. | [102] |
| | Doxorubicin | Poly (2-methacryloyloxyethyl phosphorylcholine) conjugated G3-PAMAM dendrimers showcased enhanced tumour targeting in the U-87 model with lesser toxicity. | [103] |
| | | Folic acid-borneol conjugate-modified G-5 PAMAM dendrimers augmented the doxorubicin accumulation in the C6 glioma model with extended survival time. | [105] |
| | Cy5-NHS ester | Saccharides (glucose, galactose, and mannose) conjugated PAMAM dendrimers significantly enhanced TMA (tumour-associated macrophages) and microglia targeting *via* enhancing penetration and cellular internalization into the brain. | [106] |
| Carbosilane | SiRNA | Crossed the blood-brain barrier without cytotoxicity. | [107] |
| | SiRNA (HIV-1 Nef silencing) | Silencing of genes in human astrocytes without cytotoxicity. | [108] |

(Table 2) cont.....

| | | | |
|---|---|---|---|
| PLL | Angiopep-2 and HSV-TK | The dendrimers had good biodegradability, high transfection efficiency, anti-glioma effect, and enhanced survival possibilities in the human glioblastoma animal model. | [109] |
| Poly (propyleneimine) | Hepcidin and β-Galactosidase | A gene delivery system that utilized a polypropylenimine dendrimer-bearing transferrin. | [110] |
| Peptide dendrimers | Histidine and nitro-arginine or proline functionalized ornithine dendrimers | Dendrimers showcased targeted toxicity on glioblastoma cell lines (U87 and T98G). | [111] |
| Polyether-copolyester dendrimers | Methotrexate-conjugated glucosylated and non-glucosylated dendrimers | Glycosylated methotrexate polyether-polyester dendrimers showed good permeability in the blood-brain barrier and were endocytosed effectively in U343 MGa and U87 MG cells. | [112] |
| Amphiphilic glyco dendrimers | Terminal functional groups are mannose and glucose | These dendrimers exhibited effective cellular uptake in astrocytes, microglia, and tumour cells. | [113] |

The most commonly utilized dendrimers for cancer and gliomas include PPI, PLL, PAMAM, polyesters (PGLSA-OH), amino bis(methylene phosphonic acid), and poly(2,2-bis(hydroxymethyl)propionic acid scaffold dendrimers (b-MPA) [72]. Brain tumours and gliomas are best treated with PPI, PAMAM, and PLL dendrimers [73, 74].

The internal core of dendrimers can encapsulate anticancer drugs, enabling controlled drug release to tumours. Another less common method is the use of micellar formulations of dendrimers for drug entrapment. Small hydrophobic anticancer drugs are typically encapsulated within dendrimers, enhancing their toxicity against cancer cells by increasing their solubility. Dendrimers of generations 4 to 6 (G4 to G6) are particularly effective [75]. Encapsulating the drug in PEGylated PAMAM dendrimers (G3 and G4) improves its solubility and reduces its off-target side effects. The size and PEG chain length of the dendrimers affect the drug encapsulation ability; with PEG 2000 and PAMAM dendrimers (G4), 6.5 doxorubicin molecules were encapsulated per molecule [76]. Encapsulating 5-fluorouracil in PEGylated PAMAM (G4) increased its water solubility and drug loading 12-fold over the non-PEGylated dendrimer [77].

In mice, encapsulated cisplatin in PAMAM dendrimers (20-25 wt% platinum) enhanced drug accumulation in B16F10 melanomas, increased permeability and retention, and reduced toxicity to healthy tissues by 3- to 15-fold compared to free cisplatin in solution [78]. Polymer-drug conjugates and drugs encapsulated in polymeric nanoparticles, dendrimers, and micelles exhibit passive targeting or EPR effects due to their selectivity for solid tumour accumulation [79 - 81, 222].

The molecular weight of polymers (N-(2-hydroxypropyl) methacrylamide (HPMA)) is significant for passive targeting *via* the EPR effect [82]. Etrych *et al.* grafted semitelechelic HPMA copolymer onto the PAMAM dendrimer core to develop biodegradable star-shaped polymeric doxorubicin (DOX) to passively target solid tumours [83]. Dendronized nanoparticles of 10–100 nanometers can be passively targeted by the EPR effect [84]. Targeted nanocarrier retention and localization depend on size range [85]. Heavy plasma proteins, biomolecules, and dendrimers can reduce plasma clearance and increase the therapeutic half-life in the blood. These conjugation approaches deliver medications to tumours constantly and precisely. It has been reported that 3.5G PAMAM dendrimer-enclosed cisplatin accumulated 50 times higher at the tumour site *via* the EPR effect than free cisplatin [86].

2,2-Bis(hydroxymethyl) propionic acid-containing aliphatic polyester dendrimers have shown promising anticancer activities. *In-vivo* and *in-vitro* studies have demonstrated the biocompatibility of water-soluble polyester dendrimers [87]. Slow dendrimer accumulation in vital organs allowed for long half-lives of EPR-mediated drug administration, enabling it to target malignancies [88]. Due to the increased secretion of vascular permeability mediators, tumour cells induce the widening of blood vessels [69, 89 - 91].

## Active Targeting

Active targeting improves therapeutic efficacy by increasing selectivity and delivering the cargo at overexpressed receptor sites on cancer cells. To increase the selectivity, the ligands are linked to drug-dendrimer conjugates. Aptamers, antibodies, small compounds, peptides, and DNA fragments target cancer cell surface membrane proteins or receptors. By targeting tissues with specific receptors, nanocarriers with these ligands reduce the adverse effects of conventional chemotherapy medicines [114]. Chemotherapeutic agents can be covalently bonded to the peripheral functional groups of dendrimers. Due to their multifunctional structure, dendrimers can attach multiple drug moieties and target specific ligands to enhance targeted drug delivery at tumour cell sites. The degradable chemical bonds between the dendrimer and the drug control the release of the drug [115]. Brain tumour capillary receptor overexpression guides ligand-anchored nanocarrier-based drug delivery to tumour tissues [116].

Sharma *et al.* used click chemistry to investigate three dendrimers designed with sugar moieties (glucose, mannose, or galactose) as targeting ligands. With systemic treatment, conjugating these sugars to dendrimers greatly improved tumour selectivity and accumulation kinetics. Sugar moieties also allowed receptor-specific interactions. The interaction of D-GLU with glucose transporters on tumour-associated macrophages (TAMs) promoted TAM-specific tumour accumulation. Targeted tumour microenvironment delivery was possible with D-GAL's galectin interactions with cancer cell surfaces [106].

Malignant gliomas, especially glioblastomas, are aggressive and difficult to treat due to poor drug targeting and BBB penetration. The current therapy has a 12 to 14-month survival rate. Improved medicine delivery systems use polymers and dendrimers. PAMAM dendrimers are promising yet toxic; however, PEGylation reduces their toxicity. This adjustment makes PAMAM a safer and more effective glioma drug carrier. PEGylated PAMAM, notably RGDyC-mPE--PAMAM/ATO, increases arsenic trioxide (ATO) targeting and circulation. Our customized delivery approach outperformed unencapsulated ATO *in-vivo* and *in-vitro* glioma models. RGDyC-mPEG-PAMAM/ATO may treat gliomas by bridging the BBB and delivering drugs to brain tumours [117].

PAMAM and PPI dendrimers release medication pH-dependently *via* terminal amine functional groups. Deprotonation of amine-terminated dendrimers' terminal tertiary amino groups in alkaline conditions (high pH) leads to reverse folding or collapse. Dendrimers entrap drug molecules during this phase. Protonating inner tertiary amino groups repel charges at acidic pH [118]. Dendrimers expand and release therapeutic compounds longer due to charge repulsion [119]. Such dendrimers attack tumour cells. For long-term tumor and glioma treatment, drugs are released in response to the acidic tumor microenvironment [74].

In another study, Patel *et al.* synthesized and characterized conjugated dendrimers, using them as vehicles for paclitaxel (PTX) delivery. *In vitro* and *in vivo* results showed significantly higher PTX transport with ligand-conjugated dendrimers than with free PTX or PTX in plain PPI. The ligands likely block the P-gp efflux system, thereby enhancing therapeutic effects and reducing distribution to excretory organs, which minimizes side effects. This suggests that sialic acid, glucosamine, and Con A are effective ligands for enhancing brain delivery of anticancer drugs and improving therapeutic outcomes [120].

Sharma *et al.* developed a new dendrimer conjugated to the translocator protein (TSPO) ligand 5,7-dimethylpyrazolo [1,5-a]pyrimidin-3ylacetamide (DPA). This dendrimer-DPA conjugate (D-DPA) significantly increased mitochondrial targeting *in-vitro* and enhanced antitumour immune signalling compared to the free TSPO ligand PK11195. After systemic administration, *in-vivo* studies exhibited that D-DPA targeted mitochondria within tumour-associated macrophages (TAMs). These findings suggest that dendrimers can be engineered for TAM-specific and mitochondria-specific drug delivery in glioblastoma, potentially improving treatment efficacy [121].

Polyamidoamine dendrimers and Tat peptides were coupled to bacterial magnetic nanoparticles (BMPs) to treat brain tumours with tailored gene delivery. Tat-BMPs-PAMAM was complexed with psiRNA-epidermal growth factor receptor to downregulate the epidermal growth factor receptor. PsiRNA-epidermal growth factor receptor delivered by Tat-BMPs-PAMAM significantly suppressed epidermal growth factor receptor expression and decreased U251 glioblastoma cell growth and invasion *in-vitro*. In mice, Tat-BMPs-PAMAM-psi-NA-epidermal growth factor receptor slowed tumour growth relative to controls. Apoptosis increased, and oncoprotein expression decreased. These findings support brain tumour-focused gene therapy with Tat-BMPs-PAMAM [122].

Bai *et al.* tested arginine-modified polyamidoamine dendrimer (PAMAM-R) for its ability to carry the human interferon beta (IFN-β) gene in a xenograft brain tumour model using U87MG cells injected subcutaneously into nude mice. They measured apoptosis *in-vitro* using DAPI staining, TUNEL assay, and caspase-3 activity assay after PAMAM-R/IFN-β gene polyplex transfection. Polyplex injection into xenograft brain tumours dramatically reduced tumour size in PAMAM-R/pORF-IFN-β-treated mice compared to controls. Both histology staining and epidermal growth factor receptor immunohistochemistry revealed tumour growth inhibition, while RT-PCR and TUNEL assay confirmed IFN-β expression and apoptosis induction. According to these findings, PAMAM-R/pORF-IFN-β is a promising treatment lead for glioblastoma multiforme [123].

Dendrimer-based gene and drug delivery to brain and cancer cells appears promising. Targeted intravenous delivery has improved delivery and safety, and technological advances have created 'next-generation' dendrimers with well-defined structures and specialized chemical groups for safe and effective distribution. Dendrimers with environmental-sensitive connections can limit medication release. Optimizing targeted techniques improves nucleic acid and medication delivery and treatment efficacy. Encapsulating or conjugating bioactive molecules with adjustable surfaces allows dendrimers to alter solubility, protection, release, and targeted delivery. Due to their multifunctional architecture, dendrimers are capable of targeting, imaging, diagnosing, and treating diseases. Future research may yield effective dendrimer-based nanomedicines for the treatment of cancer and neurological diseases [115].

## APPLICATION OF DENDRIMERS IN THE DETECTION OF BRAIN TUMOURS BY IMAGING

Cancer treatment typically involves a combination of conventional chemotherapy, surgery, and radiotherapy [124]. The limitations of these therapeutic strategies include a lack of selectivity, nonspecific targeting of tumor cells, and high cytotoxicity to normal cells. Hence, nanotechnology has been used to develop new drug delivery systems for detecting and treating cancer in recent years. These novel drug delivery systems offer promising possibilities for detecting and targeting drug delivery to cancer cells, thereby reducing off-target cytotoxicity. Dendrimers are becoming emerging drug delivery systems as novel strategies for diagnosis and therapy (theranostics) [125, 126]. Due to their loading capacity, high functionality, accurate molecular weight, and controlled chemical composition, dendrimer-based anticancer therapies present significant advantages over conventional dosage formulations. These advantages make dendrimers a consideration for advanced contrast agents in diagnosis [127], nanocarriers for drugs [125, 128], and theranostics [126]. The dendrimers even exhibit inherent antineoplastic and anti-metastatic properties without the need for therapeutic agents. In a study, the poly(acylthiourea) G4 dendrimer (PATU-PEG) demonstrated enhanced antineoplastic activity compared to Doxil (liposomal doxorubicin). The inert dendrimer system inhibited cell seeding, leading to a significant reduction in cell migration. This study unlocked new avenues for designing dendrimers having antineoplastic activity with high efficacy and least toxicity in which diagnostic agents can be incorporated for novel molecular imaging technologies [129].

When brain-targeted delivery systems are designed specifically for chemotherapy or diagnosis, studying anatomical and tumour microenvironments helps to develop a novel drug delivery system. The strategies for drug delivery, namely active or passive targeting, are selected based on the specific delivery requirements [63, 64, 130]. However, precise tumour margin discrimination for some brain tumours cannot be easily assessed by peritumoral oedema, making it difficult to quantify the actual tumour volume [131]. Therefore, a contrast agent may help estimate tumour margins from the central cross-sectional area [132, 133]. Early detection can eliminate the lethality associated with late-stage disease diagnosis, which is critical for effective treatment. The versatility and polyvalence property of dendrimers makes them highly valuable in diagnosis. The most common imaging procedures used for diagnostic purposes are computed tomography (CT), magnetic resonance imaging (MRI), and positron emission tomography (PET) with contrast agents (CAs). Additionally, CAs can be utilized in theranostic approaches *via* molecular imaging methods, such as photothermal therapy (PTT), photodynamic therapy (PDT), or neutron capture therapy (NCT)

[134]. Being noninvasive and providing high-clarity three-dimensional images without radiation damage, MRI is particularly useful for detecting solid tumours. The first synthesized and commercially available PAMAM dendrimers are the most extensively studied CAs for brain tumours.

Contrast agents (CAs) are necessary in certain body tissues to enhance signal intensity, improving image quality and method sensitivity. Gadolinium (Gd) chelates are the most commonly studied CAs for cancer diagnosis. However, properties like poor solubility, low penetration rate, and high toxicity limit the application of gadolinium-based CAs. $Gd^{3+}$ is a highly stable paramagnetic metal ion with seven unpaired electrons and a large, symmetrical magnetic moment that generates high imaging contrast. However, free $Gd^{3+}$ has high toxicity because it exhibits similarities with Ca2+ in terms of size, bonding, and coordination chemistry [135]. $Gd^{3+}$ ions can interfere with or replace $Ca^{2+}$ and other metal ions in metabolic processes, leading to neurological problems, neurotoxicity, and accumulation in bones and the liver [136, 137]. To neutralize the toxicity, various chelating agents are sequestered with $Gd^{3+}$. Many chelating agents have been developed, including diethyltriaminepentaacetic acid (DTPA) and 1,4,7,10-tetraazacyclododecane-1,4,7,10-tetraacetic acid (DOTA) and their derivatives [138]. Nanoparticles (NPs) and dendrimers have emerged as promising alternatives to address these issues. Noriega-Luna *et al.* studied Gd-conjugated chelates to a dendrimer, which showed an improved circulation time and specificity compared to free gadolinium chelates [139]. Wiener *et al.* created a novel category of MRI contrast agents with improved magnetic resonance characteristics by attaching a PAMAM G6 dendrimer to Gd(III)-DTPA through a thiourea bond, employing DTPA as a chelating agent. This development produced exceptional MRI images of blood vasculature and prolonged blood circulation times [127]. Misselwitz *et al.* designed PLL dendrimers as CAs for MRI through coupling with Gd-DTPA [140].

For clinical identification of abnormalities in the brain tissues and other organs, T1 relaxation-based CAs are used [141]. Magnevist® (Gd-DTPA, gadopentetate dimeglumine) produces positive contrast in T1-weighted images [142]. The advantage of Magnevist is its ability to easily penetrate the interstitium of both normal and tumor tissues from the bloodstream after 1–2 passes through the vasculature. In glioblastoma (GBM), angiogenesis forms convoluted and irregular leaky vasculature [143 - 145]. This defective vasculature allows NPs to accumulate and be retained in the tumour interstitium following systemic administration due to the enhanced permeability and retention (EPR) effect [146]. Hence, imaging agents in the nano dimension remain in the healthy tissues. In contrast, the leaky vasculature at the site of proliferating tumour cells helps escape contrasting agents in malignant tissues. Hence, numerous relaxation CAs are designed for leaky vascular systems by conjugating high molecular weight macromolecules, such as polylysine, albumin, or dendrimers, with a simple chelate like DTPA or DOTA to prepare large NPs, such as iron oxide particles. These larger particles tend to be taken up by the spleen and liver.

The type and stage of the tumour affect the size of pores and porosity of tumour vasculature. Similarly, the types of tumors, size, position, grades, and stages influence the efficacy of nano-drug delivery systems. Moreover, the size, core, and surface external groups affect the *in-vivo* efficacy of dendrimer contrast agents based on Gd-DTPA conjugates [147, 148]. The distribution of NPs across the blood-brain tumour barrier (BBTB) is relative to the properties of NPs to physiological pore sizes. Pharmacokinetic studies on glioma tumour-bearing rats were conducted across the blood-brain tumour barrier using different Gn-G--DTPA (n=1 to 8) nanoparticulate CAs based on PAMAM dendrimers [146, 149, 150]. The study discovered that gadolinium-chelated dendrimer NPs with core sizes less than 12 nm could penetrate the BBTB, whereas the larger nanoparticles had obstructed penetration [151].

Metal nanoparticles (NPs) entrapped or stabilized in dendrimers have been synthesized for improved tumour CT imaging. Dendrimer-entrapped gold nanoparticles (Au-DENPs) can be used for improved CT imaging applications. Gold nanoparticles (Au NPs)-based dendrimers show better X-ray attenuation properties and increased circulation time than conventional CAs [152, 153]. Au DENPs or dendrimer-stabilized Au NPs offer excellent colloidal stability and easier surface functionalization capabilities than free Au NPs [154]. Two elements, namely iodine and gold, have been incorporated into a single dendrimer platform to enhance the CT imaging sensitivity of tumors. Through EDC (1-ethyl-3-(3-dimethylamino propyl) carbodiimide hydrochloride) coupling reaction, an iodinated CT contrast agent (diatrizoic acid) was conjugated covalently to form dendrimer-stabilized gold nanoparticles (Au-DENPs) [155], or Au DENPs can be post-covalently linked with diatrizoic acid [156]. The dual radio-dense dendrimer-based gold NPs exhibited enhanced attenuated X-ray properties compared to Au DENPs, resulting in increased CT imaging sensitivity.

Increasing the concentration of Au NPs within the interior of dendrimers is another strategy to enhance the sensitivity of CT imaging, rather than loading dual radioactive elements into a single dendrimer. This can be achieved by stepwise reduction reactions. Amine-terminated generation 5 poly(amidoamine) dendrimers (G5.$NH_2$) serve as templates for preparing Au DENPs through a stepwise reduction of gold salt followed by complexation. Then, the terminal amines are acetylated [157]. After stepwise reduction and acetylation, the Au NPs can be tightly incorporated into the core of the dendrimer, resulting in improved stability. These nano complexes exhibit a two to three-fold increase in imaging sensitivity. PEGylation of the exterior of dendrimers is another technique of loading AuNPs into the core of the dendrimer due to the rise in the geometry of the dendrimer after PEGylation, which increases the sensitivity of CT imaging [157 - 159]. Bimodal or multimodal dendrimer-based contrast agents can improve imaging precision and detection accuracy [160].

## MRI Contrast Agents Based on Dendrimer Nanotechnology

Magnetic resonance imaging (MRI) provides outstanding clarity for visualizing soft tissue and the function of solid tumors. Therefore, developing nanotools with dual-mode CT/MRI capabilities could be a viable approach for achieving precise tumor diagnosis. Furthermore, dual-mode SPECT (single-photon emission computed tomography)/CT is a powerful molecular imaging technology because it combines functional information from SPECT with diagnostic anatomical information from CT simultaneously, significantly enhancing diagnostic accuracy [160]. Sometimes, MRI is conjugated with fluorescence imaging techniques in surgeries to identify the microscopic pathological tissues [142, 145]. This employs a synergistic approach by complexing a fluorescent dye and a T1-relaxation agent, or a PARAmagnetic chemical exchange saturation transfer agent, to the same dendrimer-based carrier. Karki *et al.* utilised this approach to design clinically applicable MRI CAs with Gd-DOTA and a 5$^{th}$ generation (G5) dendrimer with Dylight 680. A nanosized dual-modality imaging CA was synthesized by conjugating Eu-DOTA-Gly4, PARACEST MRI CA, and Dylight 680 (fluorescent agent) to a G5-PAMAM dendrimer [145]. Imaging using both *in-vivo* and *ex-vivo* techniques accurately identified the location of the brain tumour, demonstrating the benefit of exogenous CA in brain tumour detection. This helped identify the tumour location precisely due to the persistence of the MRI CAs throughout the glioma [142, 145].

The contrast agents commonly used today, such as Gd-DOTA and Gd-DTPA, are small, but have limitations, including non-specificity, rapid renal clearance, and low contrast efficiency. Nano-based macromolecular CAs, with their nano dimensions, can inhibit rapid renal clearance, prolonging their circulation half-life. Dendrimers are particularly noteworthy among the various nanosystems developed for MRI due to their well-defined structure and multivalent properties within nano dimensions [161 - 163]. Several CAs can be loaded within the interior or at a dendrimer's terminals, which enhances imaging capabilities. Additionally, by varying the dendrimer generations, the relaxivity of dendrimer-based CAs can be modified compared to small molecular agents.

Wiener *et al.*, in 1994, first studied dendrimer-conjugated MRI agents [127], in which the DTPA CA was linked to the terminal functional group, *i.e.*, the amine of 2[nd] and 6[th] generation PAMAM dendrimers, followed by complexing with Gadolinium(III) chloride to deliver the dendrimer contrast agent. These dendrimer-based agents showcased greater *in-vivo* imaging than the Gd-DTPA.

A perfect dendrimer-based CA for glioblastoma imaging should have some key characteristics, mainly biocompatibility, the capacity to transport and safeguard the imaging functionality from degradation and rapid clearance, and good permeability across the BBB and reach the brain. The CAs should perform specific targeting and delivery of the agent to the desired tumour cells, which should be excreted or degraded after imaging.

Additional studies using PAMAM and PPI dendrimers have reported that, with an increase in molecular weight and generation number of dendrimers, the molecular and ionic relaxivities also increase [164]. However, the water exchange rate at the metal centre does not significantly change when the $Gd^{3+}$-chelate is conjugated to a large dendrimer [165]. Additionally, protonated dendrimers possess a more rigid and open structure, increasing relaxivity [166]. A pH-responsive agent based on a 5[th]-generation PAMAM dendrimer incorporated a DOTA-derived chelating ligand, DOTA-4AmP, on its surface. The relaxivity of this agent changed more than two times upon changing the pH from 9 to 6 [167].

**Application Strategies for Dendrimer Contrast Agents for Diagnosis of Brain Tumour**

Brain tumours are characterised by leaky vasculature and a broken blood-brain barrier (BBB), which permits the passage of nanosized dendrimer agents across the BBB and to reach tumour sites through the leaking vasculature. Similarly, lymphatic drainage is often impaired in tumour tissues. Hence, nanoformulations reaching these cells are not easily cleared out and accumulate, causing enhanced permeability and retention (EPR) effects within the tumour. This phenomenon, known as passive targeting [70, 168], results in the accumulation of high-concentration nanosized agents within tumour lesions, thereby increasing imaging sensitivity, resolution, and quality. On the other hand, active targeting provides more effective and efficient delivery to the brain. Due to the tight junctions of BBB, paracellular diffusion is impossible for the passage of macromolecular dendrimer agents. In contrast, adsorptive-mediated transport has been used to deliver dendrimers across the BBB, which is a nonspecific process [169, 170]. Receptor-mediated endocytosis is based on ligating specially designed nanoformulations with ligands that interact with overexpressed receptors on the BBB endothelium. The overexpression of receptors on brain capillary endothelial cells (BCECs) has made receptor-mediated endocytosis a potential delivery route. This method utilizes ligands like aptamers, peptides, transferrin, folic acid, insulin-like growth factors, low-density lipoprotein receptor-related protein (LRP), and leptin [171].

### *Passive Targeting*

As brain cancer develops, it disrupts the integrity of the blood-brain barrier (BBB) and establishes the blood-brain-tumour barrier (BBTB) [172]. The BBTB often has larger pore sizes, increasing the permeability of nanosized macromolecules through the BBB. In a study, PAMAM dendrimers of generations 5 to 8 were labelled with Gd-DTPA at the periphery to evaluate the pore size of the BBTB. The transport of these dendrimers was studied in brain tumours using dynamic contrast-enhanced magnetic resonance imaging [146, 173]. The size of the G7 dendrimers was 11 ± 1 nm, and that of the G8 dendrimers was 13 ± 1 nm. The G5 to G7 functionalized dendrimers, having sizes less than 12 nm, easily passed through the BBTB pores in RG-2 glioma xenograft-bearing mice. However, the larger G8 dendrimers could not cross the BBTB. Similarly, dendrimers with longer circulation half-lives can accumulate more significantly within brain tumour cells. This study reported that the physiological pore size in the BBTB of the microvasculature in malignant solid tumours is approximately 12 nm. In another study, nanoparticles of different sizes were investigated regarding their permeability through solid tumors, and the optimal size for tumor penetration was found to be around 12 nm [174]. Hence, nanoparticles with dimensions less than 12 nm and a prolonged blood circulation time are suitable for effective delivery into malignant brain cancer cells [175].

Bertin *et al.* designed a dendritic $Mn^{2+}$-enhanced contrast agent for targeting the brain [176]. The glutamate and voltage-gated calcium channels in NMDA receptors are responsible for extracellular $Mn^{2+}$ uptake by neurons. This uptake is coupled with the physiological and functional activities of the cells. $Mn^{2+}$-based CAs are promising diagnostics for investigating the activation and architecture of neurons. The dendritic $Mn^{2+}$-conjugate has high water solubility and exhibits no *in vitro* neuronal toxicity (at a 1 mM concentration) or *in vivo* cytotoxicity (at 20 mM) in a rat model. This safety profile is due to the slow release of the paramagnetic ions.

### *Active Targeting*

Extensive work has been done to create active targeting strategies to target brain cancer cells and get beyond the BBB. As mentioned earlier, endogenous and exogenous chemicals can pass the blood-brain barrier through receptor-mediated transcytosis. Transferrin, low-density lipoprotein receptor-related protein (LRP), neurotoxins, peptides, nucleic acids, and antibodies are among the ligands investigated for their potential to target brain tumours through receptor-mediated transcytosis. By interacting with the receptors for these ligands on the surface of brain capillary endothelial cells, therapeutic medicines can be transported across the blood-brain barrier and directly to brain cancer cells [177].

Chlorotoxin (CTX), a 36-amino acid peptide, attaches particularly to brain glioma cells [178], thus penetrating gliomas through tissue barriers. Membrane-bound matrix metalloproteinase-2 isoforms (MMP-2) [179], which are overexpressed in gliomas, are the specific target of its selective binding. Clinical trials are presently being conducted on an iodine-131-labelled CTX to treat brain cancer [180]. A dendrigraft poly-L-lysine dendrimer (DGLs) outfitted with Gd-DTPA as the imaging CAs for MRI and CTX as the glioma tumour targeting ligand was described by Huang *et al.* [181]. This drug demonstrated better cellular absorption in C6 gliomas but not normal cells. Furthermore, the imaging contrast of mice treated with this drug was much better than that of the commercial control and its unmodified equivalent.

Proteases, lipoproteins, extracellular matrix proteins, and protease inhibitor complexes are among the ligands transported across the blood-brain barrier mediated by LRP [182]. Brain capillary endothelial cells (BCECs) and several forms of glioblastoma exhibit significant overexpression of LRP receptors [183]. Li *et al.* synthesized a dual MRI/fluorescent imaging agent by conjugating Angiopep-2 to a dendrimer of generation 5, Gd-DTPA, and the fluorescent dyes rhodamine and Cy5.5 [184]. This combination exhibited excellent sensitivity and enhanced permeability through the blood-brain barrier, with both passive and active targeting capabilities, enabling the imaging of glioblastomas. Li *et al.* coupled the cyclic RGD peptide to the dendrimer CAs to further enhance cancer targeting and improve image quality [185]. The αVβ3 integrin is strongly overexpressed in activated endothelial cells of the tumour neovasculature. The RGD tripeptide is the ligand for αVβ3 integrin [186]. The anticancer drugs for glioblastoma cancer models were successfully delivered by dendrimer nanosystems coated with the RGD targeting peptide [187, 188]. It was anticipated that the dual-targeting imaging agent created by Li *et al.* would bind to the cyclic RGD peptide and then target the αVβ3 integrin on the tumour vasculature [189]. The interaction of the BBB-permeable Angiopep-2 peptide with the LRP receptors on the vascular endothelial cells was facilitated by an elevated local concentration within the tumour vasculature, which promoted the transport through the BBB *via* LRP receptor-mediated endocytosis. According to *in-vivo* imaging investigations, the boundary of orthotopic U87MG human glioblastoma xenografts was precisely and sensitively detected.

A dendrimer nanosystem was used in a comparative investigation of liver and brain cancer models to target the overexpressed transferrin (Tf) receptor (TfR) in both cancer cells [190]. TfR is significantly overexpressed in hepatocytes and the BBB. The peptide T7 in the comparative investigation was decorated with MRI CAs based on PAMAM dendrimers that carried a Gd-DTPA with a high-affinity TfR ligand [191]. More efficiently than brain tumour cells, liver cancer was the target of this T7-conjugated dendrimer contrast agent [190]. These findings demonstrated that successfully delivering the brain requires overcoming the blood-brain barrier.

### ***Stimulus-Responsive Targeting***

Using physiologically responsive contrast agents, MRI can provide information at the molecular or cellular level in addition to morphological and functional imaging. The pH, biological metal ion concentration, enzymatic activity, and other particular local environmental changes can all affect the stimulus-responsive or smart CAs. They may report these changes by modifying the MRI signal [192 - 194]. One important criterion for evaluating the effectiveness of a stimulus-responsive contrast agent is the shift in relaxivity of these agents both before and following stimulation [195]. Here, we describe dendrimer agents that respond to stimuli and are sensitive to common *in vivo* microenvironmental factors, such as pH, metal ions, redox potential, and enzymes. These stimuli-responsive molecules may provide more precise and targeted functional MRIs.

### ***pH-Responsive***

The microenvironment surrounding cancer cells and inflammatory sites has a lower extracellular pH, and pH-responsive drug delivery systems can be utilized to diagnose disease or monitor treatment efficacy [221]. Hence, measuring cancer tissue pH and using noninvasive techniques to deliver drug cargo or diagnostic CAs properly is very important. Gd-DOTA-4AmP$^{5-}$, a pH-sensitive MRI agent, was thoroughly investigated by Zhang *et al.* [196]. This agent's pH-sensing mechanism was based on the protonation of the four phosphonate groups, which, over the pH range of 9.5–6.0, improved relaxivity by 1.5 times. When protonated, phosphonate groups, which have pKa values between 6.5 and 8.0, enable the catalytic exchange of protons from the $Gd^{3+}$-bound water molecule. Through conjugation to a PAMAM dendrimer, Zhang *et al.* showed enhanced relaxivity and responsiveness of Gd-DOTA-4AmP$^{5-}$. Over the pH range of 9.5–6.0, this pH-responsive dendrimer-based CA increased relaxivity by a factor of 2.2.

Furthermore, a dendrimer-based CA with Gd-DOTA-functionalized poly-L-ornithine was created by Aime *et al.* At low pH, the CA formed by this produced random assemblies; at higher pH, however, it developed organized helical structures [197]. These findings demonstrated the promise of efficient, pH-responsive magnetic resonance agents. Nonetheless, most physiological or pathological processes have relatively little pH change. It might not alter relaxivity sufficiently, which could be a drawback for CAs that react to pH.

## *Metal Ion-Responsive*

Redox homeostasis, signalling pathways, and biological metabolism depend on metal ions. Several MRI CAs have been created to react to biologically significant metal ions, such as Ca2+, Zn2+, Fe2+, and Cu+/2+ [198]. The limited sensitivity of MRI allows it to identify ions at concentrations higher than the micromolar range. Since $Ca^{2+}$ is important in neural signalling and is present in large concentrations, various research groups have created $Ca^{2+}$-responsive MRI CAs. The first $Ca^{2+}$-responsive CA was described by Meade *et al.* [199]. It comprised two Gd-DO3A chelates connected by a modified linker called BAPTA (1,2-bis(o-aminophenyl)-ethane-N, N, N′, N′-tetraacetic acid) [199]. The relaxivity of BAPTA carboxylate arms connected to chelated $Gd^{3+}$ was 3.26 $mM^{-1}s^{-1}$ in the absence of $Ca^{2+}$. In the meantime, the relaxivity rose to 5.76 $mM^{-1}s^{-1}$ in the presence of $Ca^{2+}$. The BAPTA linker's preference for $Ca^{2+}$ over $Mg^{2+}$, along with its suitable binding affinity for intracellular $Ca^{2+}$ detection, constitutes another noteworthy characteristic. A dendrimer contrast agent that is $Ca^{2+}$-sensitive was recently described by Gundez *et al.* [200]. MRI research conducted in the rat cerebral cortex revealed that these CAs have better diffusion characteristics than their monomeric counterparts.

## *Enzyme-Responsive*

Enzymatically regulated CAs can locate enzymes, measure their activity, and identify the physiological characteristics associated with each enzyme. Meade *et al.* created the first enzyme-sensitive CAs to illustrate the idea behind β-galactosidase's enzymatic activation of MRI CAs [201, 202]. Based on a fifth-generation PAMAM dendrimer that carried multiple copies of the Gd-DOTA payload, Cy5, the peptide ACPP, and the fluorescent dye, Chen *et al.* created MRI CAs. This peptide comprised the gelatinase-cleavable peptide segment and the cell-penetrating peptide Arg9 [203]. Matrix metalloproteinases (MMPs) implicated in neurovascular dysfunction following stroke include gelatinases (MMP2/MMP9). MMP2/MMP9 *in vivo* detection provides information on bleeding, BBB disruption, and neuronal death or injury. The dendrimer CAs, combined with fluorescent imaging and MRI agents, can monitor the brain's MMP2/MMP9 cellular level activity. As a result, this non-invasive, gelatinase-specific drug helped identify the brain tumor's clinical circumstances *in vivo* and confirmed potential therapeutic approaches. Table **3** lists some dendrimer-based contrast agents used for diagnosing brain tumors.

**Table 3. List of dendrimer-based diagnostic or contrast agents to diagnose brain tumour.**

| Types of Dendrimer | Imaging Technique | Contrast Agent | Conjugating Moiety | Targeting Cells | References |
|---|---|---|---|---|---|
| G5 PAMAM | MRI | DTPA-Gd | GdDTPA-PAMAM-PEG-T7 | The contrast agent with peptide T7 targets the cancer cells, which overexpress transferrin (Tfr) receptors. | [181] |
| | | SPIONs | 5NHAc-RGDFe$_3$O$_4$ NPs | Targeting tumour cells overexpressing αvβ3 integrin by RGD peptide. | [204] |
| | SPECT | $^{131}$I radionuclide | $^{131}$I-labeled dendrimers designed with LyP-1 peptide | Radionuclide therapy, dendrimeric nanodevice, antimetastasis therapy for cancer. | [205] |
| | | | G5-NH$_2$- PAMAM-PE- -CTX- HPAO-I$^{131}$ | Tumour cells with overexpressed MMP-2. | [206] |
| G3 Poly-- -lysine dendrigraft | MRI | DTPA-Gd | SP-PEG-DGL-DTPA-DACHPt | Substance P (SP), an endogenous neuropeptide, binds to the neurokinin-1 (NK-1) receptor. | [207] |
| | | | Gd-DTPA-D3-PEG-CTX | Chlorotoxin for MMP-2 receptors. | [208] |
| PAMAM | Ion imaging | Sodium dye | PAMAM dendrimers conjugated with sodium dye | Imaging with sodium-sensitive nanoprobe. | [209] |

*(Table 3) cont.....*

| Types of Dendrimer | Imaging Technique | Contrast Agent | Conjugating Moiety | Targeting Cells | References |
|---|---|---|---|---|---|
| G2 PAMAM | CT and MRI | Macrocyclic Mn (II) | RGD-Au–Mn DENP | Targeting tumour cells overexpressing αvβ3 integrin by RGD peptide. | [210] |
| Polyglycerol dendrimers | single-molecule optical imaging | Boron-dipyrromethene | Boron-dipyrromethene conjugated polyglycerol dendrimers | Fluorescent probe. | [211] |

## LIMITATIONS IN THE USE OF DENDRIMERS

Despite effective brain-targeting and imaging nanotools, dendrimers' limitations impede their therapeutic applicability. Numerous reports of cytotoxicity in mammalian cell lines have been published. The chemistry and surface functionality of dendrimers, as well as their production, concentrations, experimental cell types, and cell medium composition, all influence their cytotoxicity. Compared to PEGylated dendrimers and those bearing -OH groups, pyrrolidone, and anionic (such as carboxyl) functions of the same generation and concentration, amino-terminated dendrimers generally cause greater cytotoxicity. Catalytic dendrimers with surface-exposed main amino clusters are more harmful because they cause instability and perturbation of the plasma membrane. Due to these characteristics, important metabolites, such as ATP, NADPH, and other vital metabolic enzymes, may be lost intracellularly. A comparison of the cytotoxicity of PEG-terminated PPI dendrimers with PPI dendrimers with peripheral neutral acetamide groups was published by Stasko *et al.* Compared to PPI dendrimers, these dendrimers showed less cytotoxicity and very little time-dependent alterations in the HUVEC cell lines' plasma membrane [212]. The cytotoxicity of cationic dendrimers, such as melamine-based dendrimers with amino, carboxyl, sulfonate, guanidine, phosphonate, or PEGylated functionalities, was investigated by Chen *et al.* [213]. Similarly, up to 5 mg/mL of PAMAM and diaminobutane dendrimers with surface carboxyl functionality did not show cytotoxicity in CCRF, B16F10, and HepG2 cells [214].

Blood cells and different blood plasma proteins interact with the dendrimers. These interactions affect the pharmacokinetics and biodistribution of dendrimers in the body by modifying the structure and function of blood proteins, as well as the survivability and performance of blood cells. In one study, γ-globulins interacted with G5 PAMAM dendrimers with surface functional groups, such as carboxyl, amino, and hydroxyl, which altered immunoglobulins' secondary structure and conformation [215]. The stoichiometric complexation of G4

PAMAM dendrimers at (4-5):1 and (4-6):1 with bovine serum albumin and human serum albumin, respectively, was described in a study on the interaction of PAMAM dendrimers with blood proteins, such as serum albumin and fibrinogen [216, 217]. Interaction between G3, G4, G5, and G5-OH PAMAM dendrimers with fibrinogen resulted in changes to the structure and conformation. The high concentration of cationic dendrimers impaired fibrinogen polymerization, inhibiting the clotting factor and fibrinogen activities in whole-blood coagulation [218].

The dendrimers are also reported to exhibit hemolytic activities. Their concentration and generation number affect the hemolysis. Due to the combined effects of membrane-perturbing characteristics and electrostatic interactions, cationic dendrimers are susceptible to hemolysis, leading to alterations in membrane curvature and pore formation [219]. Moreover, the hemolytic activity of PAMAM and PPI dendrimers with diaminobutane and diaminoethane cores is not generation-dependent [220].

## CHALLENGES AND FUTURE PROSPECTIVES

Dendrimers are not considered suitable for clinical or hospital applications due to various challenges associated with transitioning from the laboratory to the clinic. Biocompatibility is a significant issue since dendrimers' distinct chemical and structural characteristics can trigger immunological and toxic reactions. A more trustworthy understanding of the safety and effectiveness of dendrimers for clinical use would be made possible by developing and adopting standardized evaluation techniques and criteria, which would also help assure consistency and comparability in assessments.

Another restriction on the clinical uses of dendrimers is the effectiveness of medication delivery. Multiple drug loading sites are provided by their branching structure, and the loading and release processes may be restricted, which could reduce the effectiveness of delivery. Therefore, advancements in dendrimer design and production techniques are required to increase drug loading and controlled release efficiency. The structure of the dendrimer can be customized to meet the specific needs of each therapeutic application, hence improving the efficacy and delivery of the loaded medicines [221].

Another difficulty is the stability of dendrimers *in vivo*. Dendrimers may experience degradation, metabolism, or excretion within a biological system, affecting their persistence and potential therapeutic benefits. Appropriate dendrimer materials and chemical modification techniques can enhance the stability and resistance to degradation of dendrimers *in vivo*. Furthermore, in some applications, selecting the appropriate dendrimer topologies and branching densities can help strike a balance between strength and the requirements of drug delivery. If these characteristics are optimised, dendrimers can be used more consistently and effectively as drug-delivery devices in clinical settings.

The therapeutic application of dendrimers faces significant challenges in terms of bulk production and standardization. Dendrimer synthesis typically involves intricate chemical reactions, a small manufacturing volume, and the absence of established techniques. Therefore, more effective synthesis techniques and procedures need to be developed to produce dendrimers on a large scale and standardize them. This could entail creating uniform operating processes, implementing quality control testing techniques, enforcing strict quality control and assurance policies, and carrying out ongoing monitoring and assessment. The clinical applicability of dendrimers can be advanced by improving their viability through enhanced production quality and consistency.

Dendrimer safety evaluation is another significant challenge for clinical applications. *In vitro* and *in vivo* studies are necessary for this purpose, and they must assess cytotoxicity, immunological responses, biodegradability, and other relevant factors. Furthermore, long-term surveillance and clinical observational studies are crucial for evaluating the safety of dendrimers. Results from safety assessments that are dependable and trustworthy can be achieved by establishing uniform assessment standards and procedures as well as by promoting cooperation and compliance [222].

Lastly, market regulation is another important issue affecting the clinical use of dendrimers. It is essential to consider and comply with relevant legal and regulatory requirements, clinical trial protocols, registration and approval procedures, and market access criteria. The effects of market regulation on the therapeutic use of dendrimers can be mitigated by streamlining approval processes, offering assistance and training, and promoting cooperation and compliance. By optimizing these procedures, dendrimers can be brought to market more quickly, thereby hastening their acceptance in medical contexts.

## CONCLUSION

According to several studies on animal studies, dendrimers are a viable option for systemic drug delivery in treating brain tumours. Dendrimers and nanotechnology have become the go-to treatments for brain tumours, providing a strong substitute for traditional chemotherapy. The spherical architecture and multivalent periphery of dendrimers enable their simultaneous use for diagnosis and treatment, offering reliable targeting of brain tumors with sustained drug release. Utilizing the multiple functions of BBB for treating and monitoring brain tumors has gained increasing attention from researchers in recent years, potentially leading to significant improvements in patient survival. With advancements in oncology research, dendrimer-based nanomedicines are gaining popularity in clinical settings due to their excellent targeting efficiency, efficacy, biocompatibility, and affordability. Nevertheless, additional, comprehensive research is necessary to provide precise data regarding the targeting, effectiveness, safety, and potential mortality issues of dendrimers, thereby certifying them as appropriate and feasible nanoarchitectures for brain cancer imaging and therapy.

Dendrimers appear to have a bright future from a global perspective, which is encouraging for successful clinical translation in brain tumours and other cancers. The timely administration of treatment is contingent upon prompt diagnosis; this objective can be realized through individualized treatment planning for brain tumors. This strategy uses medications with predictable adverse effects while enhancing tumour diagnostics. It has been demonstrated that improvements in medicine boost patients' contentment with their life expectancy. In the coming years, dendrimers are expected to be manufactured and tested in clinical trials, offering a promising approach to enhancing cancer therapy outcomes.

## AUTHORS' CONTRIBUTIONS

Anindita Behera and Sandesh Lodha contributed to conceptualization, supervision, and writing, including review and final editing. Nishigandha Sa, Sudhanshu Sekhar Rout, Gajanan Kalyankar, Nikita Panchal, and Manoj Kumar Sabnani were responsible for writing the original draft and data curation.

## REFERENCES

[1] Tomalia DA, Baker H, Dewald J, *et al.* A new class of polymers: starburst-dendritic macromolecules. Polym J 1985; 17(1): 117-32. [http://dx.doi.org/10.1295/polymj.17.117]

[2] Abbasi E, Aval SF, Akbarzadeh A, *et al.* Dendrimers: synthesis, applications, and properties. Nanoscale Res Lett 2014; 9(1): 247. [http://dx.doi.org/10.1186/1556-276X-9-247] [PMID: 24994950]

[3] Zahin N, Anwar R, Tewari D, *et al.* Nanoparticles and its biomedical applications in health and diseases: special focus on drug delivery. Environ Sci Pollut Res Int 2020; 27(16): 19151-68. [http://dx.doi.org/10.1007/s11356-019-05211-0] [PMID: 31079299]

[4] Pérez-Ferreiro M, M Abelairas A, Criado A, *et al.* Dendrimers: exploring their wide structural variety and applications. Polymers (Basel) 2023; 15(22): 4369. [http://dx.doi.org/10.3390/polym15224369] [PMID: 38006093]

[5] Chis AA, Dobrea C, Morgovan C, *et al.* Applications and limitations of dendrimers in biomedicine. Molecules 2020; 25(17): 3982. [http://dx.doi.org/10.3390/molecules25173982] [PMID: 32882920]

[6] Crintea A, Motofelea AC, Şovrea AS, *et al.* Dendrimers: advancements and potential applications in cancer diagnosis and treatment—an overview. Pharmaceutics 2023; 15(5): 1406. [http://dx.doi.org/10.3390/pharmaceutics15051406] [PMID: 37242648]

[7] Tomalia DA, Baker H, Dewald J, Hall M, Kallos G, Martin S. A new class of polymers: starburst-dendritic macromolecules. 1985;17(1):117-32.
[http://dx.doi.org/10.1295/polymj.17.117]

[8] De Brabander-van den Berg EMM, Meijer EW. Poly (propylene imine) dendrimers: large-scale synthesis by hetereogeneously catalyzed hydrogenations. Angew Chem Int Ed Engl 1993; 32(9): 1308-11.
[http://dx.doi.org/10.1002/anie.199313081]

[9] Hawker CJ, Frechet JMJ. Preparation of polymers with controlled molecular architecture. A new convergent approach to dendritic macromolecules. J Am Chem Soc 1990; 112(21): 7638-47.
[http://dx.doi.org/10.1021/ja00177a027]

[10] Ihre H, Padilla De Jesús OL, Fréchet JMJ. Fast and convenient divergent synthesis of aliphatic ester dendrimers by anhydride coupling. J Am Chem Soc 2001; 123(25): 5908-17.
[http://dx.doi.org/10.1021/ja010524e] [PMID: 11414823]

[11] Majoral JP, Caminade AM. Dendrimers containing heteroatoms (si, p, B, ge, or bi). Chem Rev 1999; 99(3): 845-80.
[http://dx.doi.org/10.1021/cr970414j] [PMID: 11749433]

[12] Gorman C. Metallodendrimers: structural diversity and functional behavior. Adv Mater 1998; 10(4): 295-309.
[http://dx.doi.org/10.1002/(SICI)1521-4095(199803)10:4<295::AID-ADMA295>3.0.CO;2-N]

[13] Dichtel WR, Serin JM, Edder C, *et al.* Singlet oxygen generation *via* two-photon excited FRET. J Am Chem Soc 2004; 126(17): 5380-1.
[http://dx.doi.org/10.1021/ja031647x] [PMID: 15113208]

[14] Turnbull WB, Stoddart JF. Design and synthesis of glycodendrimers. J Biotechnol 2002; 90(3-4): 231-55.
[PMID: 12071227]

[15] Nierengarten JF, Guttiérez-Nava M, Zhang S, *et al.* Fullerene-containing macromolecules for materials science applications. Carbon 2004; 42(5-6): 1077-83.
[http://dx.doi.org/10.1016/j.carbon.2003.12.003]

[16] Nilsen TW, Grayzel J, Prensky W. Dendritic nucleic acid structures. J Theor Biol 1997; 187(2): 273-84.
[http://dx.doi.org/10.1006/jtbi.1997.0446] [PMID: 9237897]

[17] Maiti PK, Çağın T, Wang G, Goddard WA. Structure of PAMAM Dendrimers: Generations 1 through 11. Macromolecules 2004; 37(16): 6236-54.
[http://dx.doi.org/10.1021/ma035629b]

[18] Buhleier E, Wehner W, Vögtle F. Cascade"-and" nonskid-chain-like" syntheses of molecular cavity topologies. Synthesis 1978; 1978(2): 155-8.
[http://dx.doi.org/10.1055/s-1978-24702]

[19] Sherje AP, Jadhav M, Dravyakar BR, Kadam D. Dendrimers: A versatile nanocarrier for drug delivery and targeting. Int J Pharm 2018; 548(1): 707-20.
[http://dx.doi.org/10.1016/j.ijpharm.2018.07.030] [PMID: 30012508]

[20] Launay N, Caminade AM, Majoral JP. Synthesis of bowl-shaped dendrimers from generation 1 to generation 8. J Organomet Chem 1997; 529(1-2): 51-8.
[http://dx.doi.org/10.1016/S0022-328X(96)06293-6]

[21] Van der Made AW, van Leeuwen PWNM. Silane dendrimers. J Chem Soc Chem Commun 1992; (19): 1400-1.
[http://dx.doi.org/10.1039/c39920001400]

[22] Percec V, Wilson DA, Leowanawat P, *et al.* Self-assembly of Janus dendrimers into uniform

dendrimersomes and other complex architectures. Science 2010; 328(5981): 1009-14. [http://dx.doi.org/10.1126/science.1185547] [PMID: 20489021]

[23] Singh D, Tonjam R, Chaudhary T, *et al.* A short appraisal on gold nanoparticles: recent advances and applications. Curr Nanomed. 2021;11(3):168–76. [http://dx.doi.org/10.2174/2468187312666211220122455]

[24] Oosterom GE, Reek JNH, Kamer PCJ, van Leeuwen PWNM. Transition metal catalysis using functionalized dendrimers. Angew Chem Int Ed 2001; 40(10): 1828-49. [http://dx.doi.org/10.1002/1521-3773(20010518)40:10<1828::AID-ANIE1828>3.0.CO;2-Y] [PMID: 11385652]

[25] Tomalia DA, Naylor AM, Goddard WA III. Starburst dendrimers: molecular-level control of size, shape, surface chemistry, topology, and flexibility from atoms to macroscopic matter. Angew Chem Int Ed Engl 1990; 29(2): 138-75. [http://dx.doi.org/10.1002/anie.199001381]

[26] Kukowska-Latallo JF, Bielinska AU, Johnson J, Spindler R, Tomalia DA, Baker JR Jr. Efficient transfer of genetic material into mammalian cells using Starburst polyamidoamine dendrimers. Proc Natl Acad Sci USA 1996; 93(10): 4897-902. [http://dx.doi.org/10.1073/pnas.93.10.4897] [PMID: 8643500]

[27] Kojima C, Kono K, Maruyama K, Takagishi TJBc. Synthesis of polyamidoamine dendrimers having poly (ethylene glycol) grafts and their ability to encapsulate anticancer drugs. 2000;11(6):910-7.

[28] Kono K, Kojima C, Hayashi N, *et al.* Preparation and cytotoxic activity of poly(ethylene glycol)-modified poly(amidoamine) dendrimers bearing adriamycin. Biomaterials 2008; 29(11): 1664-75. [http://dx.doi.org/10.1016/j.biomaterials.2007.12.017] [PMID: 18194811]

[29] Pan Y, Ford WT. Dendrimers with both hydrophilic and hydrophobic chains at every end. Macromolecules 1999; 32(16): 5468-70. [http://dx.doi.org/10.1021/ma990675q]

[30] Liu M, Kono K, Fréchet JMJJoCR. Water-soluble dendritic unimolecular micelles:: Their potential as drug delivery agents. 2000;65(1-2):121-31.

[31] Wu P, Feldman AK, Nugent AK, Hawker CJ, Scheel A, Voit B. Efficiency and fidelity in a click-chemistry route to triazole dendrimers by the copper (I)-catalyzed ligation of azides and alkynes. 2004;116(30):4018-22.

[32] Launay N, Caminade AM, Lahana R, Majoral JP. A general synthetic strategy for neutral phosphorus-containing dendrimers. Angew Chem Int Ed Engl 1994; 33(15-16): 1589-92. [http://dx.doi.org/10.1002/anie.199415891]

[33] Tomalia DA. Dendrimers as multi-purpose nanodevices for oncology drug delivery and diagnostic imaging. Nanomedicine 2006; 2(4): 309. [http://dx.doi.org/10.1016/j.nano.2006.10.125]

[34] Crooks R, Zhao MJAM. Dendrimer-encapsulated pt nanoparticles. Synthesis, Characterization, and Applications to Catalysis 1999; 11(3): 217-20.

[35] Cuadrado I, Morán M, Casado CM, Alonso B, Losada JJCCR. Organometallic dendrimers with transition metals. 1999;193:395-445. [http://dx.doi.org/10.1016/S0010-8545(99)00036-3]

[36] Caminade A-M, Maraval V, Laurent R, Turrin C-O, Sutra P, Leclaire J. Phosphorus dendrimers: from synthesis to applications. 2003;6(8-10):791-801. [http://dx.doi.org/10.1016/j.crci.2003.04.009]

[37] Sadler K, Tam JP. Peptide dendrimers: applications and synthesis. J Biotechnol 2002; 90(3-4): 195-229. [PMID: 12071226]

[38] Bosman dA, Janssen H, Meijer EJCr. About dendrimers: structure, physical properties, and applications. 1999;99(7):1665-88.

[39] Esfand R, Tomalia DAJDdt. Poly (amidoamine)(PAMAM) dendrimers: from biomimicry to drug delivery and biomedical applications. 2001;6(8):427-36.

[40] Frechet JMJS. Functional polymers and dendrimers: reactivity, molecular architecture, and interfacial energy. 1994;263(5154):1710-5.
[http://dx.doi.org/10.1126/science.8134834]

[41] Wooley KL, Hawker CJ, Frechet JMJJotACS. Unsymmetrical three-dimensional macromolecules: preparation and characterization of strongly dipolar dendritic macromolecules. 1993;115(24):11496-505.

[42] Wooley K, Hawker C, Pochan J, Frechet JJM. Physical properties of dendritic macromolecules: a study of glass transition temperature. 1993;26(7):1514-9.
[http://dx.doi.org/10.1021/ma00059a006]

[43] Kolhe P, Misra E, Kannan RM, Kannan S, Lieh-Lai MJIjop. Drug complexation, *in vitro* release and cellular entry of dendrimers and hyperbranched polymers. 2003;259(1-2):143-60.

[44] Szota M, Reczyńska-Kolman K, Pamuła E, Michel O, Kulbacka J, Jachimska B. Poly (Amidoamine) dendrimers as nanocarriers for 5-fluorouracil: Effectiveness of complex formation and cytotoxicity studies. Int J Mol Sci 2021; 22(20): 11167.
[http://dx.doi.org/10.3390/ijms222011167] [PMID: 34681827]

[45] Topp A, Bauer BJ, Prosa TJ, Scherrenberg R, Amis EJJM. Size change of dendrimers in concentrated solution. 1999;32(26):8923-31.
[http://dx.doi.org/10.1021/ma990433q]

[46] Albertazzi L, Serresi M, Albanese A, Beltram F. Dendrimer internalization and intracellular trafficking in living cells. Mol Pharm 2010; 7(3): 680-8.
[http://dx.doi.org/10.1021/mp9002464] [PMID: 20394437]

[47] Mukherjee SP, Lyng FM, Garcia A, Davoren M, Byrne HJ. Mechanistic studies of *in vitro* cytotoxicity of poly(amidoamine) dendrimers in mammalian cells. Toxicol Appl Pharmacol 2010; 248(3): 259-68.
[http://dx.doi.org/10.1016/j.taap.2010.08.016] [PMID: 20736030]

[48] Huang RQ, Qu YH, Ke WL, Zhu JH, Pei YY, Jiang C. Efficient gene delivery targeted to the brain using a transferrin-conjugated polyethyleneglycol-modified polyamidoamine dendrimer. FASEB J 2007; 21(4): 1117-25.
[http://dx.doi.org/10.1096/fj.06-7380com] [PMID: 17218540]

[49] Kobayashi H, Kawamoto S, Saga T, *et al.* Novel liver macromolecular MR contrast agent with a polypropylenimine diaminobutyl dendrimer core: Comparison to the vascular MR contrast agent with the polyamidoamine dendrimer core. Magn Reson Med 2001; 46(4): 795-802.
[http://dx.doi.org/10.1002/mrm.1259] [PMID: 11590657]

[50] Singh D, Tiwari P, Nagdev S. Particulate vaccine dispersions emerge as a novel carrier for deep pulmonary immunization. Curr Nanomedicine. 2023;13(2):71–4.
[http://dx.doi.org/10.2174/2468187313666230714124009]

[51] Jain K, Kesharwani P, Gupta U, Jain NK. Dendrimer toxicity: Let's meet the challenge. Int J Pharm 2010; 394(1-2): 122-42.
[http://dx.doi.org/10.1016/j.ijpharm.2010.04.027] [PMID: 20433913]

[52] Chen CZ, Cooper SL. Interactions between dendrimer biocides and bacterial membranes. Biomaterials 2002; 23(16): 3359-68.
[http://dx.doi.org/10.1016/S0142-9612(02)00036-4] [PMID: 12099278]

[53] Pandita D, Madaan K, Kumar S, Poonia N, Lather V. Dendrimers in drug delivery and targeting: Drug-dendrimer interactions and toxicity issues. J Pharm Bioallied Sci 2014; 6(3): 139-50.

[http://dx.doi.org/10.4103/0975-7406.130965] [PMID: 25035633]

[54] Giri J, Diallo MS, Simpson AJ, *et al.* Interactions of poly(amidoamine) dendrimers with human serum albumin: binding constants and mechanisms. ACS Nano 2011; 5(5): 3456-68. [http://dx.doi.org/10.1021/nn1021007] [PMID: 21438566]

[55] Poupot M, Griffe L, Marchand P, *et al.* Design of phosphorylated dendritic architectures to promote human monocyte activation. FASEB J 2006; 20(13): 2339-51. [http://dx.doi.org/10.1096/fj.06-5742com] [PMID: 17077311]

[56] Tam JP. Synthetic peptide vaccine design: synthesis and properties of a high-density multiple antigenic peptide system. Proc Natl Acad Sci USA 1988; 85(15): 5409-13. [http://dx.doi.org/10.1073/pnas.85.15.5409] [PMID: 3399498]

[57] Stenström P, Manzanares D, Zhang Y, Ceña V, Malkoch M. Evaluation of amino-functional polyester dendrimers based on Bis-MPA as nonviral vectors for siRNA delivery. Molecules 2018; 23(8): 2028. [http://dx.doi.org/10.3390/molecules23082028] [PMID: 30110914]

[58] Woller EK, Cloninger MJ. The lectin-binding properties of six generations of mannose-functionalized dendrimers. Org Lett 2002; 4(1): 7-10. [http://dx.doi.org/10.1021/ol016568+] [PMID: 11772077]

[59] Astruc D, Ruiz J. The redox functions of metallodendrimers. J Inorg Organomet Polym Mater 2015; 25(1): 2-11. [http://dx.doi.org/10.1007/s10904-014-0091-3]

[60] Steffensen MB, Hollink E, Kuschel F, Bauer M, Simanek EE. Dendrimers based on [1,3,5]-triazines. J Polym Sci A Polym Chem 2006; 44(11): 3411-33. [http://dx.doi.org/10.1002/pola.21333] [PMID: 19953202]

[61] Caminati G, Turro NJ, Tomalia DA. Photophysical investigation of starburst dendrimers and their interactions with anionic and cationic surfactants. J Am Chem Soc 1990; 112(23): 8515-22. [http://dx.doi.org/10.1021/ja00179a041]

[62] Padhi S, Behera A. Nanotechnology based targeting strategies for the delivery of Camptothecin. Sustainable Agriculture Reviews 44: Pharmaceutical Technology for Natural Products Delivery Vol 2 Impact of Nanotechnology. 2020:243-72. [http://dx.doi.org/10.1007/978-3-030-41842-7_7]

[63] Behera A, Padhi S. Passive and active targeting strategies for the delivery of the camptothecin anticancer drug: a review. Environ Chem Lett 2020; 18(5): 1557-67. [http://dx.doi.org/10.1007/s10311-020-01022-9]

[64] Patra N, Kar D, Pal A, Behera A. Antibacterial, anticancer, anti-diabetic and catalytic activity of bio-conjugated metal nanoparticles. Advances in Natural Sciences: Nanoscience and Nanotechnology 2018; 9(3): 035001. [http://dx.doi.org/10.1088/2043-6254/aad12d]

[65] Kaurav M, Ruhi S, Al-Goshae HA, *et al.* Dendrimer: An update on recent developments and future opportunities for the brain tumors diagnosis and treatment. Front Pharmacol 2023; 14: 1159131. [http://dx.doi.org/10.3389/fphar.2023.1159131] [PMID: 37006997]

[66] Khizar S, Ahmad NM, Zine N, Jaffrezic-Renault N, Errachid-el-salhi A, Elaissari A. Errachid-el-salhi A, Elaissari A. Magnetic nanoparticles: from synthesis to theranostic applications. ACS Appl Nano Mater 2021; 4(5): 4284-306. [http://dx.doi.org/10.1021/acsanm.1c00852]

[67] Kumar R, Singh A, Sharma N, Jain S. Advances in nanocarrier-based tumor targeting: emerging strategies and clinical perspectives. Adv Drug Deliv Rev. 2024;205:114567. [http://dx.doi.org/10.1016/j.addr.2023.114567]

[68] Iyer AK, Duan Z, Amiji MM. Nanodelivery systems for nucleic acid therapeutics in drug resistant tumors. Mol Pharm 2014; 11(8): 2511-26.

[http://dx.doi.org/10.1021/mp500024p] [PMID: 24661041]

[69] Iyer AK, Khaled G, Fang J, Maeda H. Exploiting the enhanced permeability and retention effect for tumor targeting. Drug Discov Today 2006; 11(17-18): 812-8. [http://dx.doi.org/10.1016/j.drudis.2006.07.005] [PMID: 16935749]

[70] Matsumura Y, Maeda H. A new concept for macromolecular therapeutics in cancer chemotherapy: mechanism of tumoritropic accumulation of proteins and the antitumor agent smancs. Cancer Res 1986; 46(12 Pt 1): 6387-92. [PMID: 2946403]

[71] Attia MF, Anton N, Wallyn J, Omran Z, Vandamme TF. An overview of active and passive targeting strategies to improve the nanocarriers efficiency to tumour sites. J Pharm Pharmacol 2019; 71(8): 1185-98. [http://dx.doi.org/10.1111/jphp.13098] [PMID: 31049986]

[72] Mignani S, El Kazzouli S, Bousmina M, Majoral JP. Expand classical drug administration ways by emerging routes using dendrimer drug delivery systems: A concise overview. Adv Drug Deliv Rev 2013; 65(10): 1316-30. [http://dx.doi.org/10.1016/j.addr.2013.01.001] [PMID: 23415951]

[73] Kesharwani P, Jain K, Jain NK. Dendrimer as nanocarrier for drug delivery. Prog Polym Sci 2014; 39(2): 268-307. [http://dx.doi.org/10.1016/j.progpolymsci.2013.07.005]

[74] Kesharwani P, Tekade RK, Jain NK. Formulation development and *in vitro-in vivo* assessment of the fourth-generation PPI dendrimer as a cancer-targeting vector. Nanomedicine (Lond) 2014; 9(15): 2291-308. [http://dx.doi.org/10.2217/nnm.13.210] [PMID: 24593000]

[75] Wolinsky J, Grinstaff M. Therapeutic and diagnostic applications of dendrimers for cancer treatment. Adv Drug Deliv Rev 2008; 60(9): 1037-55. [http://dx.doi.org/10.1016/j.addr.2008.02.012] [PMID: 18448187]

[76] Kojima C, Kono K, Maruyama K, Takagishi T. Synthesis of polyamidoamine dendrimers having poly(ethylene glycol) grafts and their ability to encapsulate anticancer drugs. Bioconjug Chem 2000; 11(6): 910-7. [http://dx.doi.org/10.1021/bc0000583] [PMID: 11087341]

[77] Bhadra D, Bhadra S, Jain S, Jain NK. A PEGylated dendritic nanoparticulate carrier of fluorouracil. Int J Pharm 2003; 257(1-2): 111-24. [http://dx.doi.org/10.1016/S0378-5173(03)00132-7] [PMID: 12711167]

[78] Malik N, Evagorou EG, Duncan R. Dendrimer-platinate. Anticancer Drugs 1999; 10(8): 767-76. [http://dx.doi.org/10.1097/00001813-199909000-00010] [PMID: 10573209]

[79] Puri A, Loomis K, Smith B, Lee J-H, Yavlovich A, Heldman E. Lipid-based nanoparticles as pharmaceutical drug carriers: from concepts to clinic. Critical Reviews™ in Therapeutic Drug Carrier Systems. 2009;26(6). [http://dx.doi.org/10.1615/CritRevTherDrugCarrierSyst.v26.i6.10]

[80] Shcharbina N, Shcharbin D, Bryszewska M. Nanomaterials in stroke treatment: perspectives. Stroke 2013; 44(8): 2351-5. [http://dx.doi.org/10.1161/STROKEAHA.113.001298] [PMID: 23715957]

[81] Kesharwani P, Banerjee S, Padhye S, Sarkar FH, Iyer AK. Parenterally administrable nano-micelles of 3,4-difluorobenzylidene curcumin for treating pancreatic cancer. Colloids Surf B Biointerfaces 2015; 132: 138-45. [http://dx.doi.org/10.1016/j.colsurfb.2015.05.007] [PMID: 26037703]

[82] Etrych T, Chytil P, Mrkvan T, Šírová M, Říhová B, Ulbrich K. Conjugates of doxorubicin with graft HPMA copolymers for passive tumor targeting. J Control Release 2008; 132(3): 184-92.

[http://dx.doi.org/10.1016/j.jconrel.2008.04.017] [PMID: 18534705]

[83] Etrych T, Strohalm J, Chytil P, *et al.* Biodegradable star HPMA polymer conjugates of doxorubicin for passive tumor targeting. Eur J Pharm Sci 2011; 42(5): 527-39. [http://dx.doi.org/10.1016/j.ejps.2011.03.001] [PMID: 21392579]

[84] She W, Li N, Luo K, *et al.* Dendronized heparin–doxorubicin conjugate based nanoparticle as pH-responsive drug delivery system for cancer therapy. Biomaterials 2013; 34(9): 2252-64. [http://dx.doi.org/10.1016/j.biomaterials.2012.12.017] [PMID: 23298778]

[85] Kaminskas LM, Boyd BJ, Porter CJH. Dendrimer pharmacokinetics: the effect of size, structure and surface characteristics on ADME properties. Nanomedicine (Lond) 2011; 6(6): 1063-84. [http://dx.doi.org/10.2217/nnm.11.67] [PMID: 21955077]

[86] Duncan R, Izzo L. Dendrimer biocompatibility and toxicity. Adv Drug Deliv Rev 2005; 57(15): 2215-37. [http://dx.doi.org/10.1016/j.addr.2005.09.019] [PMID: 16297497]

[87] Padilla De Jesús OL, Ihre HR, Gagne L, Fréchet JMJ, Szoka FC Jr. Polyester dendritic systems for drug delivery applications: *in vitro* and *in vivo* evaluation. Bioconjug Chem 2002; 13(3): 453-61. [http://dx.doi.org/10.1021/bc010103m] [PMID: 12009933]

[88] Gillies E, Fréchet J. Dendrimers and dendritic polymers in drug delivery. Drug Discov Today 2005; 10(1): 35-43. [http://dx.doi.org/10.1016/S1359-6446(04)03276-3] [PMID: 15676297]

[89] Greish K, Fang J, Inutsuka T, Nagamitsu A, Maeda H. Macromolecular Therapeutics. Clin Pharmacokinet 2003; 42(13): 1089-105. [http://dx.doi.org/10.2165/00003088-200342130-00002] [PMID: 14531722]

[90] Maeda H, Wu J, Sawa T, Matsumura Y, Hori K. Tumor vascular permeability and the EPR effect in macromolecular therapeutics: a review. J Control Release 2000; 65(1-2): 271-84. [http://dx.doi.org/10.1016/S0168-3659(99)00248-5] [PMID: 10699287]

[91] Maeda H, Sawa T, Konno T. Mechanism of tumor-targeted delivery of macromolecular drugs, including the EPR effect in solid tumor and clinical overview of the prototype polymeric drug SMANCS. J Control Release 2001; 74(1-3): 47-61. [http://dx.doi.org/10.1016/S0168-3659(01)00309-1] [PMID: 11489482]

[92] Li Q, Xu W. Novel anticancer targets and drug discovery in post genomic age. Curr Med Chem Anticancer Agents 2005; 5(1): 53-63. [http://dx.doi.org/10.2174/1568011053352631] [PMID: 15720261]

[93] Swami R, Singh I, Kulhari H, Jeengar MK, Khan W, Sistla R. p-Hydroxy benzoic acid-conjugated dendrimer nanotherapeutics as potential carriers for targeted drug delivery to brain: an *in vitro* and *in vivo* evaluation. J Nanopart Res 2015; 17(6): 265. [http://dx.doi.org/10.1007/s11051-015-3063-9]

[94] Sharma R, Kim SY, Sharma A, *et al.* Activated microglia targeting dendrimer–minocycline conjugate as therapeutics for neuroinflammation. Bioconjug Chem 2017; 28(11): 2874-86. [http://dx.doi.org/10.1021/acs.bioconjchem.7b00569] [PMID: 29028353]

[95] Sowińska M, Szeliga M, Morawiak M, Zabłocka B, Urbanczyk-Lipkowska Z. Design, synthesis and activity of new N 1-alkyl tryptophan functionalized dendrimeric peptides against glioblastoma. Biomolecules 2022; 12(8): 1116. [http://dx.doi.org/10.3390/biom12081116] [PMID: 36009010]

[96] Apartsin EK, Knauer N, Kahlert UD, Caminade AM. Amphiphilic triazine-phosphorus metallodendrons possessing anti-cancer stem cell activity. Pharmaceutics 2022; 14(2): 393. [http://dx.doi.org/10.3390/pharmaceutics14020393] [PMID: 35214126]

[97] Shen M, Huang Y, Han L, *et al.* Multifunctional drug delivery system for targeting tumor and its acidic microenvironment. J Control Release 2012; 161(3): 884-92.

[http://dx.doi.org/10.1016/j.jconrel.2012.05.013] [PMID: 22587941]

[98] Gupta U, Dwivedi SKD, Bid HK, Konwar R, Jain NK. Ligand anchored dendrimers based nanoconstructs for effective targeting to cancer cells. Int J Pharm 2010; 393(1-2): 186-97. [http://dx.doi.org/10.1016/j.ijpharm.2010.04.002] [PMID: 20382210]

[99] Lockman PR, Oyewumi MO, Koziara JM, Roder KE, Mumper RJ, Allen DD. Brain uptake of thiamine-coated nanoparticles. J Control Release 2003; 93(3): 271-82. [http://dx.doi.org/10.1016/j.jconrel.2003.08.006] [PMID: 14644577]

[100] Zhao J, Zhang B, Shen S, *et al.* CREKA peptide-conjugated dendrimer nanoparticles for glioblastoma multiforme delivery. J Colloid Interface Sci 2015; 450: 396-403. [http://dx.doi.org/10.1016/j.jcis.2015.03.019] [PMID: 25863222]

[101] Denora N, Laquintana V, Lopalco A, *et al. In vitro* targeting and imaging the translocator protein TSPO 18-kDa through G(4)-PAMAM–FITC labeled dendrimer. J Control Release 2013; 172(3): 1111-25. [http://dx.doi.org/10.1016/j.jconrel.2013.09.024] [PMID: 24096015]

[102] Biswas S, Deshpande PP, Navarro G, Dodwadkar NS, Torchilin VP. Lipid modified triblock PAMAM-based nanocarriers for siRNA drug co-delivery. Biomaterials 2013; 34(4): 1289-301. [http://dx.doi.org/10.1016/j.biomaterials.2012.10.024] [PMID: 23137395]

[103] Ban J, Li S, Zhan Q, *et al.* PMPC modified PAMAM dendrimer enhances brain tumor-targeted drug delivery. Macromol Biosci 2021; 21(4): 2000392. [http://dx.doi.org/10.1002/mabi.202000392] [PMID: 33506646]

[104] Wang K, Zhang X, Liu Y, Liu C, Jiang B, Jiang Y. Tumor penetrability and anti-angiogenesis using iRGD-mediated delivery of doxorubicin-polymer conjugates. Biomaterials 2014; 35(30): 8735-47. [http://dx.doi.org/10.1016/j.biomaterials.2014.06.042] [PMID: 25023394]

[105] Xu X, Li J, Han S, *et al.* A novel doxorubicin loaded folic acid conjugated PAMAM modified with borneol, a nature dual-functional product of reducing PAMAM toxicity and boosting BBB penetration. Eur J Pharm Sci 2016; 88: 178-90. [http://dx.doi.org/10.1016/j.ejps.2016.02.015] [PMID: 26965003]

[106] Sharma R, Liaw K, Sharma A, *et al.* Glycosylation of PAMAM dendrimers significantly improves tumor macrophage targeting and specificity in glioblastoma. J Control Release 2021; 337: 179-92. [http://dx.doi.org/10.1016/j.jconrel.2021.07.018] [PMID: 34274384]

[107] Serramía MJ, Álvarez S, Fuentes-Paniagua E, *et al. In vivo* delivery of siRNA to the brain by carbosilane dendrimer. J Control Release 2015; 200: 60-70. [http://dx.doi.org/10.1016/j.jconrel.2014.12.042] [PMID: 25559178]

[108] Sepúlveda-Crespo D, Gómez R, De La Mata FJ, Jiménez JL, Muñoz-Fernández MÁ. Polyanionic carbosilane dendrimer-conjugated antiviral drugs as efficient microbicides: Recent trends and developments in HIV treatment/therapy. Nanomedicine 2015; 11(6): 1481-98. [http://dx.doi.org/10.1016/j.nano.2015.03.008] [PMID: 25835558]

[109] Ke W, Shao K, Huang R, *et al.* Gene delivery targeted to the brain using an Angiopep-conjugated polyethyleneglycol-modified polyamidoamine dendrimer. Biomaterials 2009; 30(36): 6976-85. [http://dx.doi.org/10.1016/j.biomaterials.2009.08.049] [PMID: 19765819]

[110] Demeule M, Currie JC, Bertrand Y, *et al.* Involvement of the low-density lipoprotein receptor-related protein in the transcytosis of the brain delivery vector Angiopep-2. J Neurochem 2008; 106(4): 1534-44. [http://dx.doi.org/10.1111/j.1471-4159.2008.05492.x] [PMID: 18489712]

[111] Cieślak M, Ryszawy D, Pudełek M, *et al.* Bioinspired bola-type peptide dendrimers inhibit proliferation and invasiveness of glioblastoma cells in a manner dependent on their structure and amphipathic properties. Pharmaceutics 2020; 12(11): 1106. [http://dx.doi.org/10.3390/pharmaceutics12111106] [PMID: 33217976]

[112] Dhanikula RS, Argaw A, Bouchard JF, Hildgen P. Methotrexate loaded polyether-copolyester dendrimers for the treatment of gliomas: enhanced efficacy and intratumoral transport capability. Mol Pharm 2008; 5(1): 105-16.
[http://dx.doi.org/10.1021/mp700086j] [PMID: 18171013]

[113] Zhang W, Dhumal D, Zhu X, *et al.* Bola-amphiphilic glycodendrimers: new carbohydrate-mimicking scaffolds to target carbohydrate-binding proteins. Chemistry 2022; 28(58): e202201400.
[http://dx.doi.org/10.1002/chem.202201400] [PMID: 35820051]

[114] Kumar LA, Pattnaik G, Satapathy BS, Swapna S, Mohanty D. Targeting to brain tumor: Nanocarrier-based drug delivery platforms, opportunities, and challenges. J Pharm Bioallied Sci 2021; 13(2): 172-7.
[http://dx.doi.org/10.4103/jpbs.JPBS_239_20] [PMID: 34349476]

[115] Somani S, Dufès C. Applications of dendrimers for brain delivery and cancer therapy. Nanomedicine (Lond) 2014; 9(15): 2403-14.
[http://dx.doi.org/10.2217/nnm.14.130] [PMID: 25413857]

[116] Juillerat-Jeanneret L. The targeted delivery of cancer drugs across the blood–brain barrier: chemical modifications of drugs or drug-nanoparticles? Drug Discov Today 2008; 13(23-24): 1099-106.
[http://dx.doi.org/10.1016/j.drudis.2008.09.005] [PMID: 18848640]

[117] Lu Y, Han S, Zheng H, *et al.* A novel RGDyC/PEG co-modified PAMAM dendrimer-loaded arsenic trioxide of glioma targeting delivery system. Int J Nanomedicine 2018; 13: 5937-52.
[http://dx.doi.org/10.2147/IJN.S175418] [PMID: 30323584]

[118] Kesharwani P. Relative study of cancer targeting potential of engineered dendrimer. LAP Lambert Academic Publishing 2015.

[119] Kesharwani P, Iyer AK. Recent advances in dendrimer-based nanovectors for tumor-targeted drug and gene delivery. Drug Discov Today 2015; 20(5): 536-47.
[http://dx.doi.org/10.1016/j.drudis.2014.12.012] [PMID: 25555748]

[120] Patel HK, Gajbhiye V, Kesharwani P, Jain NK. Ligand anchored poly(propyleneimine) dendrimers for brain targeting: Comparative *in vitro* and *in vivo* assessment. J Colloid Interface Sci 2016; 482: 142-50.
[http://dx.doi.org/10.1016/j.jcis.2016.07.047] [PMID: 27501037]

[121] Sharma A, Liaw K, Sharma R, *et al.* Targeting mitochondria in tumor-associated macrophages using a dendrimer-conjugated TSPO ligand that stimulates antitumor signaling in glioblastoma. Biomacromolecules 2020; 21(9): 3909-22.
[http://dx.doi.org/10.1021/acs.biomac.0c01033] [PMID: 32786523]

[122] Han L, Zhang A, Wang H, *et al.* Tat-BMPs-PAMAM conjugates enhance therapeutic effect of small interference RNA on U251 glioma cells *in vitro* and *in vivo*. Hum Gene Ther 2010; 21(4): 417-26.
[http://dx.doi.org/10.1089/hum.2009.087] [PMID: 19899955]

[123] Bai CZ, Choi S, Nam K, An S, Park JS. Arginine modified PAMAM dendrimer for interferon beta gene delivery to malignant glioma. Int J Pharm 2013; 445(1-2): 79-87.
[http://dx.doi.org/10.1016/j.ijpharm.2013.01.057] [PMID: 23384727]

[124] Pérez-Herrero E, Fernández-Medarde A. Advanced targeted therapies in cancer: Drug nanocarriers, the future of chemotherapy. Eur J Pharm Biopharm 2015; 93: 52-79.
[http://dx.doi.org/10.1016/j.ejpb.2015.03.018] [PMID: 25813885]

[125] Quintana A, Raczka E, Piehler L, *et al.* Design and function of a dendrimer-based therapeutic nanodevice targeted to tumor cells through the folate receptor. Pharm Res 2002; 19(9): 1310-6.
[http://dx.doi.org/10.1023/A:1020398624602] [PMID: 12403067]

[126] Alibolandi M, Hoseini F, Mohammadi M, *et al.* Curcumin-entrapped MUC-1 aptamer targeted dendrimer-gold hybrid nanostructure as a theranostic system for colon adenocarcinoma. Int J Pharm 2018; 549(1-2): 67-75.

[http://dx.doi.org/10.1016/j.ijpharm.2018.07.052] [PMID: 30048777]

[127] Wiener E, Brechbiel MW, Brothers H, *et al.* Dendrimer-based metal chelates: A new class of magnetic resonance imaging contrast agents. Magn Reson Med 1994; 31(1): 1-8. [http://dx.doi.org/10.1002/mrm.1910310102] [PMID: 8121264]

[128] Wu Y, Sefah K, Liu H, Wang R, Tan W. DNA aptamer–micelle as an efficient detection/delivery vehicle toward cancer cells. Proc Natl Acad Sci USA 2010; 107(1): 5-10. [http://dx.doi.org/10.1073/pnas.0909611107] [PMID: 20080797]

[129] Shao S, Zhou Q, Si J, *et al.* A non-cytotoxic dendrimer with innate and potent anticancer and anti-metastatic activities. Nat Biomed Eng 2017; 1(9): 745-57. [http://dx.doi.org/10.1038/s41551-017-0130-9] [PMID: 31015667]

[130] Padhi S, Behera A. Advanced drug delivery systems in the treatment of ovarian cancer Advanced Drug Delivery Systems in the Management of Cancer. Elsevier 2021; pp. 127-39. [http://dx.doi.org/10.1016/B978-0-323-85503-7.00020-1]

[131] Shamji MF, Fric-Shamji EC, Benoit BG. Brain tumors and epilepsy: pathophysiology of peritumoral changes. Neurosurg Rev 2009; 32(3): 275-85. [http://dx.doi.org/10.1007/s10143-009-0191-7] [PMID: 19205766]

[132] Lehmann P, Vallée JN, Saliou G, *et al.* Dynamic contrast-enhanced T2*-weighted MR imaging: A peritumoral brain oedema study. J Neuroradiol 2009; 36(2): 88-92. [http://dx.doi.org/10.1016/j.neurad.2008.10.001] [PMID: 19054561]

[133] Schoenegger K, Oberndorfer S, Wuschitz B, *et al.* Peritumoral edema on MRI at initial diagnosis: an independent prognostic factor for glioblastoma? Eur J Neurol 2009; 16(7): 874-8. [http://dx.doi.org/10.1111/j.1468-1331.2009.02613.x] [PMID: 19473360]

[134] Kunjachan S, Jayapaul J, Mertens ME, Storm G, Kiessling F, Lammers T. Theranostic systems and strategies for monitoring nanomedicine-mediated drug targeting. Curr Pharm Biotechnol 2012; 13(4): 609-22. [http://dx.doi.org/10.2174/138920112799436302] [PMID: 22214503]

[135] Evans CH. Biochemistry of the Lanthanides. Springer Science & Business Media 2013.

[136] Bellin MF. MR contrast agents, the old and the new. Eur J Radiol 2006; 60(3): 314-23. [http://dx.doi.org/10.1016/j.ejrad.2006.06.021] [PMID: 17005349]

[137] Ersoy H, Rybicki FJ. Biochemical safety profiles of gadolinium-based extracellular contrast agents and nephrogenic systemic fibrosis. J Magn Reson Imaging 2007; 26(5): 1190-7. [http://dx.doi.org/10.1002/jmri.21135] [PMID: 17969161]

[138] Chang CA. Magnetic resonance imaging contrast agents. Design and physicochemical properties of gadodiamide. Invest Radiol 1993; 28 (Suppl. 1): S21-7. [http://dx.doi.org/10.1097/00004424-199303001-00003] [PMID: 8486500]

[139] Noriega-Luna B, Godínez LA, Rodríguez FJ, *et al.* Applications of dendrimers in drug delivery agents, diagnosis, therapy, and detection. J Nanomater 2014; 2014(1): 507273. [http://dx.doi.org/10.1155/2014/507273]

[140] Misselwitz B, Schmitt-Willich H, Ebert W, Frenzel T, Weinmann HJ. Pharmacokinetics of Gadomer-17, a new dendritic magnetic resonance contrast agent. MAGMA 2001; 12(2-3): 128-34. [http://dx.doi.org/10.1007/BF02668094] [PMID: 11390268]

[141] Bhuiyan MPI, Aryal MP, Janic B, *et al.* Concentration-independent MRI of pH with a dendrimer-based pH-responsive nanoprobe. Contrast Media Mol Imaging 2015; 10(6): 481-6. [http://dx.doi.org/10.1002/cmmi.1651] [PMID: 26173742]

[142] Karki K, Ewing JR, Ali MM. Targeting glioma with a dual mode optical and paramagnetic nanoprobe across the blood-brain tumor barrier. J Nanomed Nanotechnol 2016; 7(4): 395. [http://dx.doi.org/10.4172/2157-7439.1000395] [PMID: 27695645]

[143] Veiseh O, Sun C, Fang C, *et al.* Specific targeting of brain tumors with an optical/magnetic resonance imaging nanoprobe across the blood-brain barrier. Cancer Res 2009; 69(15): 6200-7. [http://dx.doi.org/10.1158/0008-5472.CAN-09-1157] [PMID: 19638572]

[144] Jain R, Ellika SK, Scarpace L, *et al.* Quantitative estimation of permeability surface-area product in astroglial brain tumors using perfusion CT and correlation with histopathologic grade. AJNR Am J Neuroradiol 2008; 29(4): 694-700. [http://dx.doi.org/10.3174/ajnr.A0899] [PMID: 18202239]

[145] Ali MM, Bhuiyan MPI, Janic B, *et al.* A nano-sized PARACEST-fluorescence imaging contrast agent facilitates and validates *in vivo* CEST MRI detection of glioma. Nanomedicine (Lond) 2012; 7(12): 1827-37. [http://dx.doi.org/10.2217/nnm.12.92] [PMID: 22891866]

[146] Sarin H, Kanevsky AS, Wu H, *et al.* Effective transvascular delivery of nanoparticles across the blood-brain tumor barrier into malignant glioma cells. J Transl Med 2008; 6(1): 80. [http://dx.doi.org/10.1186/1479-5876-6-80] [PMID: 19094226]

[147] Kobayashi H, Brechbiel MW. Dendrimer-based macromolecular MRI contrast agents: characteristics and application. Mol Imaging 2003; 2(1): 1-10. [http://dx.doi.org/10.1162/15353500200303100] [PMID: 12926232]

[148] Sato N, Kobayashi H, Hiraga A, *et al.* Pharmacokinetics and enhancement patterns of macromolecular MR contrast agents with various sizes of polyamidoamine dendrimer cores. Magn Reson Med 2001; 46(6): 1169-73. [http://dx.doi.org/10.1002/mrm.1314] [PMID: 11746584]

[149] Sarin H, Kanevsky AS, Fung SH, *et al.* Metabolically stable bradykinin B2 receptor agonists enhance transvascular drug delivery into malignant brain tumors by increasing drug half-life. J Transl Med 2009; 7(1): 33. [http://dx.doi.org/10.1186/1479-5876-7-33] [PMID: 19439100]

[150] Sarin H. Physiologic upper limits of pore size of different blood capillary types and another perspective on the dual pore theory of microvascular permeability. J Angiogenes Res 2010; 2(1): 14. [http://dx.doi.org/10.1186/2040-2384-2-14] [PMID: 20701757]

[151] Sarin H. Recent progress towards development of effective systemic chemotherapy for the treatment of malignant brain tumors. J Transl Med 2009; 7(1): 77. [http://dx.doi.org/10.1186/1479-5876-7-77] [PMID: 19723323]

[152] Kim D, Park S, Lee JH, Jeong YY, Jon S. Antibiofouling polymer-coated gold nanoparticles as a contrast agent for *in vivo* X-ray computed tomography imaging. J Am Chem Soc 2007; 129(24): 7661-5. [http://dx.doi.org/10.1021/ja071471p] [PMID: 17530850]

[153] Guo R, Wang H, Peng C, *et al.* X-ray attenuation property of dendrimer-entrapped gold nanoparticles. J Phys Chem C 2010; 114(1): 50-6. [http://dx.doi.org/10.1021/jp9078986]

[154] Bronstein LM, Shifrina ZB. Dendrimers as encapsulating, stabilizing, or directing agents for inorganic nanoparticles. Chem Rev 2011; 111(9): 5301-44. [http://dx.doi.org/10.1021/cr2000724] [PMID: 21718045]

[155] Peng C, Li K, Cao X, *et al.* Facile formation of dendrimer-stabilized gold nanoparticles modified with diatrizoic acid for enhanced computed tomography imaging applications. Nanoscale 2012; 4(21): 6768-78. [http://dx.doi.org/10.1039/c2nr31687k] [PMID: 23010987]

[156] Guo R, Wang H, Peng C, *et al.* Enhanced X-ray attenuation property of dendrimer-entrapped gold nanoparticles complexed with diatrizoic acid. J Mater Chem 2011; 21(13): 5120-7. [http://dx.doi.org/10.1039/c0jm04094k]

[157] Xiao T, Wen S, Wang H, *et al.* Facile synthesis of acetylated dendrimer-entrapped gold nanoparticles with enhanced gold loading for CT imaging applications. J Mater Chem B Mater Biol Med 2013; 1(21): 2773-80.
[http://dx.doi.org/10.1039/c3tb20399a] [PMID: 32260984]

[158] Peng C, Zheng L, Chen Q, *et al.* PEGylated dendrimer-entrapped gold nanoparticles for *in vivo* blood pool and tumor imaging by computed tomography. Biomaterials 2012; 33(4): 1107-19.
[http://dx.doi.org/10.1016/j.biomaterials.2011.10.052] [PMID: 22061490]

[159] Wang H, Zheng L, Peng C, *et al.* Computed tomography imaging of cancer cells using acetylated dendrimer-entrapped gold nanoparticles. Biomaterials 2011; 32(11): 2979-88.
[http://dx.doi.org/10.1016/j.biomaterials.2011.01.001] [PMID: 21277019]

[160] Louie A. Multimodality imaging probes: design and challenges. Chem Rev 2010; 110(5): 3146-95.
[http://dx.doi.org/10.1021/cr9003538] [PMID: 20225900]

[161] Villaraza AJL, Bumb A, Brechbiel MW. Macromolecules, dendrimers, and nanomaterials in magnetic resonance imaging: the interplay between size, function, and pharmacokinetics. Chem Rev 2010; 110(5): 2921-59.
[http://dx.doi.org/10.1021/cr900232t] [PMID: 20067234]

[162] Langereis S, Dirksen A, Hackeng TM, van Genderen MHP, Meijer EW. Dendrimers and magnetic resonance imaging. New J Chem 2007; 31(7): 1152-60.
[http://dx.doi.org/10.1039/b616960k]

[163] Qiao Z, Shi X. Dendrimer-based molecular imaging contrast agents. Prog Polym Sci 2015; 44: 1-27.
[http://dx.doi.org/10.1016/j.progpolymsci.2014.08.002]

[164] Bryant LH Jr, Brechbiel MW, Wu C, Bulte JWM, Herynek V, Frank JA. Synthesis and relaxometry of high-generation (G = 5, 7, 9, and 10) PAMAM dendrimer-DOTA-gadolinium chelates. J Magn Reson Imaging 1999; 9(2): 348-52.
[http://dx.doi.org/10.1002/(SICI)1522-2586(199902)9:2<348::AID-JMRI30>3.0.CO;2-J] [PMID: 10077036]

[165] Tóth É, Pubanz D, Vauthey S, Helm L, Merbach AE. The role of water exchange in attaining maximum relaxivities for dendrimeric MRI contrast agents. Chemistry 1996; 2(12): 1607-15.
[http://dx.doi.org/10.1002/chem.19960021220]

[166] Laus S, Sour A, Ruloff R, Tóth É, Merbach AE. Rotational dynamics account for pH-dependent relaxivities of PAMAM dendrimeric, Gd-based potential MRI contrast agents. Chemistry 2005; 11(10): 3064-76.
[http://dx.doi.org/10.1002/chem.200401326] [PMID: 15776490]

[167] Ali MM, Woods M, Caravan P, *et al.* Synthesis and relaxometric studies of a dendrimer-based pH-responsive MRI contrast agent. Chemistry 2008; 14(24): 7250-8.
[http://dx.doi.org/10.1002/chem.200800402] [PMID: 18601236]

[168] Maeda H, Nakamura H, Fang J. The EPR effect for macromolecular drug delivery to solid tumors: Improvement of tumor uptake, lowering of systemic toxicity, and distinct tumor imaging *in vivo*. Adv Drug Deliv Rev 2013; 65(1): 71-9.
[http://dx.doi.org/10.1016/j.addr.2012.10.002] [PMID: 23088862]

[169] Gupta B, Levchenko TS, Torchilin VP. TAT peptide-modified liposomes provide enhanced gene delivery to intracranial human brain tumor xenografts in nude mice. Oncol Res 2006; 16(8): 351-9.
[http://dx.doi.org/10.3727/000000006783980946] [PMID: 17913043]

[170] Oller-Salvia B, Sánchez-Navarro M, Giralt E, Teixidó M. Blood–brain barrier shuttle peptides: an emerging paradigm for brain delivery. Chem Soc Rev 2016; 45(17): 4690-707.
[http://dx.doi.org/10.1039/C6CS00076B] [PMID: 27188322]

[171] Huang R, Ke W, Liu Y, Jiang C, Pei Y. The use of lactoferrin as a ligand for targeting the polyamidoamine-based gene delivery system to the brain. Biomaterials 2008; 29(2): 238-46.

[http://dx.doi.org/10.1016/j.biomaterials.2007.09.024] [PMID: 17935779]

[172] Nagy JA, Chang S-H, Dvorak AM, Dvorak HF. Why are tumour blood vessels abnormal and why is it important to know? Br J Cancer 2009; 100(6): 865-9.
[http://dx.doi.org/10.1038/sj.bjc.6604929] [PMID: 19240721]

[173] Sarin H, Kanevsky AS, Wu H, *et al.* Physiologic upper limit of pore size in the blood-tumor barrier of malignant solid tumors. J Transl Med 2009; 7(1): 51.
[http://dx.doi.org/10.1186/1479-5876-7-51] [PMID: 19549317]

[174] Chauhan VP, Stylianopoulos T, Martin JD, *et al.* Normalization of tumour blood vessels improves the delivery of nanomedicines in a size-dependent manner. Nat Nanotechnol 2012; 7(6): 383-8.
[http://dx.doi.org/10.1038/nnano.2012.45] [PMID: 22484912]

[175] Chauhan VP, Jain RK. Strategies for advancing cancer nanomedicine. Nat Mater 2013; 12(11): 958-62.
[http://dx.doi.org/10.1038/nmat3792] [PMID: 24150413]

[176] Bertin A, Steibel J, Michou-Gallani AI, Gallani JL, Felder-Flesch D. Development of a dendritic manganese-enhanced magnetic resonance imaging (MEMRI) contrast agent: synthesis, toxicity (*in vitro*) and relaxivity (*in vitro, in vivo*) studies. Bioconjug Chem 2009; 20(4): 760-7.
[http://dx.doi.org/10.1021/bc8004683] [PMID: 19368343]

[177] Nakamura Y, Mochida A, Choyke PL, Kobayashi H. Nanodrug delivery: is the enhanced permeability and retention effect sufficient for curing cancer? Bioconjug Chem 2016; 27(10): 2225-38.
[http://dx.doi.org/10.1021/acs.bioconjchem.6b00437] [PMID: 27547843]

[178] Lyons SA, O'Neal J, Sontheimer H. Chlorotoxin, a scorpion-derived peptide, specifically binds to gliomas and tumors of neuroectodermal origin. Glia 2002; 39(2): 162-73.
[http://dx.doi.org/10.1002/glia.10083] [PMID: 12112367]

[179] Veiseh M, Gabikian P, Bahrami SB, *et al.* Tumor paint: a chlorotoxin:Cy5.5 bioconjugate for intraoperative visualization of cancer foci. Cancer Res 2007; 67(14): 6882-8.
[http://dx.doi.org/10.1158/0008-5472.CAN-06-3948] [PMID: 17638899]

[180] Mamelak AN, Jacoby DB. Targeted delivery of antitumoral therapy to glioma and other malignancies with synthetic chlorotoxin (TM-601). Expert Opin Drug Deliv 2007; 4(2): 175-86.
[http://dx.doi.org/10.1517/17425247.4.2.175] [PMID: 17335414]

[181] Huang R, Han L, Li J, *et al.* Chlorotoxin-modified macromolecular contrast agent for MRI tumor diagnosis. Biomaterials 2011; 32(22): 5177-86.
[http://dx.doi.org/10.1016/j.biomaterials.2011.03.075] [PMID: 21531455]

[182] Shibata M, Yamada S, Kumar SR, *et al.* Clearance of Alzheimer's amyloid-β1-40 peptide from brain by LDL receptor–related protein-1 at the blood-brain barrier. J Clin Invest 2000; 106(12): 1489-99.
[http://dx.doi.org/10.1172/JCI10498] [PMID: 11120756]

[183] Maletínská L, Blakely EA, Bjornstad KA, Deen DF, Knoff LJ, Forte TM. Human glioblastoma cell lines: levels of low-density lipoprotein receptor and low-density lipoprotein receptor-related protein. Cancer Res 2000; 60(8): 2300-3.
[PMID: 10786698]

[184] Yan H, Wang J, Yi P, *et al.* Imaging brain tumor by dendrimer-based optical/paramagnetic nanoprobe across the blood-brain barrier. Chem Commun (Camb) 2011; 47(28): 8130-2.
[http://dx.doi.org/10.1039/c1cc12007g] [PMID: 21629879]

[185] Yan H, Wang L, Wang J, *et al.* Two-order targeted brain tumor imaging by using an optical/paramagnetic nanoprobe across the blood brain barrier. ACS Nano 2012; 6(1): 410-20.
[http://dx.doi.org/10.1021/nn203749v] [PMID: 22148835]

[186] Brooks PC, Clark RAF, Cheresh DA. Requirement of Vascular Integrin $\alpha_v \beta_3$ for Angiogenesis. Science 1994; 264(5158): 569-71.
[http://dx.doi.org/10.1126/science.7512751] [PMID: 7512751]

[187] Cui Q, Yang S, Ye P, *et al.* Downregulation of TLX induces TET3 expression and inhibits glioblastoma stem cell self-renewal and tumorigenesis. Nat Commun 2016; 7(1): 10637. [http://dx.doi.org/10.1038/ncomms10637] [PMID: 26838672]

[188] Liu X, Liu C, Chen C, *et al.* Targeted delivery of Dicer-substrate siRNAs using a dual targeting peptide decorated dendrimer delivery system. Nanomedicine 2014; 10(8): 1627-36. [http://dx.doi.org/10.1016/j.nano.2014.05.008] [PMID: 24965758]

[189] Schottelius M, Laufer B, Kessler H, Wester HJ. Ligands for Mapping $\alpha_v \beta_3$ -Integrin Expression *in Vivo*. Acc Chem Res 2009; 42(7): 969-80. [http://dx.doi.org/10.1021/ar800243b] [PMID: 19489579]

[190] Han L, Li J, Huang S, *et al.* Peptide-conjugated polyamidoamine dendrimer as a nanoscale tumor-targeted T1 magnetic resonance imaging contrast agent. Biomaterials 2011; 32(11): 2989-98. [http://dx.doi.org/10.1016/j.biomaterials.2011.01.005] [PMID: 21277017]

[191] Lee JH, Engler JA, Collawn JF, Moore BA. Receptor mediated uptake of peptides that bind the human transferrin receptor. Eur J Biochem 2001; 268(7): 2004-12. [http://dx.doi.org/10.1046/j.1432-1327.2001.02073.x] [PMID: 11277922]

[192] Angelovski G. Heading toward macromolecular and nanosized bioresponsive MRI probes for successful functional imaging. Acc Chem Res 2017; 50(9): 2215-24. [http://dx.doi.org/10.1021/acs.accounts.7b00203] [PMID: 28841293]

[193] Angelovski G. What we can really do with bioresponsive MRI contrast agents. Angew Chem Int Ed 2016; 55(25): 7038-46. [http://dx.doi.org/10.1002/anie.201510956] [PMID: 27112329]

[194] Heffern MC, Matosziuk LM, Meade TJ. Lanthanide probes for bioresponsive imaging. Chem Rev 2014; 114(8): 4496-539. [http://dx.doi.org/10.1021/cr400477t] [PMID: 24328202]

[195] Caravan P. Strategies for increasing the sensitivity of gadolinium based MRI contrast agents. Chem Soc Rev 2006; 35(6): 512-23. [http://dx.doi.org/10.1039/b510982p] [PMID: 16729145]

[196] Zhang S, Wu K, Sherry AD. A novel pH-sensitive MRI contrast agent. Angew Chem Int Ed 1999; 38(21): 3192-4. [http://dx.doi.org/10.1002/(SICI)1521-3773(19991102)38:21<3192::AID-ANIE3192>3.0.CO;2-#] [PMID: 10556899]

[197] Aime S, Fedeli F, Sanino A, Terreno E. A R2/R1 ratiometric procedure for a concentration-independent, pH-responsive, Gd(III)-based MRI agent. J Am Chem Soc 2006; 128(35): 11326-7. [http://dx.doi.org/10.1021/ja062387x] [PMID: 16939235]

[198] Que EL, Chang CJ. Responsive magnetic resonance imaging contrast agents as chemical sensors for metals in biology and medicine. Chem Soc Rev 2010; 39(1): 51-60. [http://dx.doi.org/10.1039/B914348N] [PMID: 20023836]

[199] Li W, Fraser SE, Meade TJ. A calcium-sensitive magnetic resonance imaging contrast agent. J Am Chem Soc 1999; 121(6): 1413-4. [http://dx.doi.org/10.1021/ja983702l]

[200] Gündüz S, Nitta N, Vibhute S, *et al.* Dendrimeric calcium-responsive MRI contrast agents with slow *in vivo* diffusion. Chem Commun (Camb) 2015; 51(14): 2782-5. [http://dx.doi.org/10.1039/C4CC07540D] [PMID: 25383973]

[201] Louie AY, Hüber MM, Ahrens ET, *et al.* *In vivo* visualization of gene expression using magnetic resonance imaging. Nat Biotechnol 2000; 18(3): 321-5. [http://dx.doi.org/10.1038/73780] [PMID: 10700150]

[202] Moats RA, Fraser SE, Meade TJ. “smart” magnetic resonance imaging agent that reports on specific

enzymatic activity. Angew Chem Int Ed Engl 1997; 36(7): 726-8. [http://dx.doi.org/10.1002/anie.199707261]

[203] Chen S, Cui J, Jiang T, *et al.* Gelatinase activity imaged by activatable cell-penetrating peptides in cell-based and *in vivo* models of stroke. J Cereb Blood Flow Metab 2017; 37(1): 188-200. [http://dx.doi.org/10.1177/0271678X15621573] [PMID: 26681768]

[204] Yang J, Luo Y, Xu Y, *et al.* Conjugation of iron oxide nanoparticles with RGD-modified dendrimers for targeted tumor MR imaging. ACS Appl Mater Interfaces 2015; 7(9): 5420-8. [http://dx.doi.org/10.1021/am508983n] [PMID: 25695661]

[205] Song N, Zhao L, Xu X, *et al.* LyP-1-modified multifunctional dendrimers for targeted antitumor and antimetastasis therapy. ACS Appl Mater Interfaces 2020; 12(11): 12395-406. [http://dx.doi.org/10.1021/acsami.9b18881] [PMID: 32077680]

[206] Rasouli R, Zaaeri F, Rajabi AB, Darbandi-Azar A, Faridi-Majidi R, Ardestani MS. 99mTc-anionic linear globular dendrimer-G2-phenylalanine conjugate: Novel brain tumor SPECT imaging. Biointerface Res Appl Chem 2021; 11(4): 11244-55.

[207] Sun T, Jiang X, Wang Q, *et al.* Substance P mediated DGLs complexing with DACHPt for targeting therapy of glioma. ACS Appl Mater Interfaces 2017; 9(40): 34603-17. [http://dx.doi.org/10.1021/acsami.7b05997] [PMID: 28925679]

[208] Zhao L, Zhu J, Cheng Y, *et al.* Chlorotoxin-conjugated multifunctional dendrimers labeled with radionuclide 131I for single photon emission computed tomography imaging and radiotherapy of gliomas. ACS Appl Mater Interfaces 2015; 7(35): 19798-808. [http://dx.doi.org/10.1021/acsami.5b05836] [PMID: 26291070]

[209] Lamy CM, Sallin O, Loussert C, Chatton JY. Sodium sensing in neurons with a dendrimer-based nanoprobe. ACS Nano 2012; 6(2): 1176-87. [http://dx.doi.org/10.1021/nn203822t] [PMID: 22288942]

[210] Xu X, Liu K, Wang Y, *et al.* A multifunctional low-generation dendrimer-based nanoprobe for the targeted dual mode MR/CT imaging of orthotopic brain gliomas. J Mater Chem B Mater Biol Med 2019; 7(23): 3639-43. [http://dx.doi.org/10.1039/C9TB00416E]

[211] Yang SK, Shi X, Park S, Ha T, Zimmerman SC. A dendritic single-molecule fluorescent probe that is monovalent, photostable and minimally blinking. Nat Chem 2013; 5(8): 692-7. [http://dx.doi.org/10.1038/nchem.1706] [PMID: 23881501]

[212] Stasko NA, Johnson CB, Schoenfisch MH, Johnson TA, Holmuhamedov EL. Cytotoxicity of polypropylenimine dendrimer conjugates on cultured endothelial cells. Biomacromolecules 2007; 8(12): 3853-9. [http://dx.doi.org/10.1021/bm7008203] [PMID: 18004811]

[213] Chen HT, Neerman MF, Parrish AR, Simanek EE. Cytotoxicity, hemolysis, and acute *in vivo* toxicity of dendrimers based on melamine, candidate vehicles for drug delivery. J Am Chem Soc 2004; 126(32): 10044-8. [http://dx.doi.org/10.1021/ja048548j] [PMID: 15303879]

[214] Malik N, Wiwattanapatapee R, Klopsch R, *et al.* Dendrimers. J Control Release 2000; 65(1-2): 133-48. [http://dx.doi.org/10.1016/S0168-3659(99)00246-1] [PMID: 10699277]

[215] Lin J, Hua W, Zhang Y, *et al.* Effect of poly(amidoamine) dendrimers on the structure and activity of immune molecules. Biochim Biophys Acta, Gen Subj 2015; 1850(2): 419-25. [http://dx.doi.org/10.1016/j.bbagen.2014.11.016] [PMID: 25463324]

[216] Shcharbin D, Janicka M, Wasiak M, *et al.* Serum albumins have five sites for binding of cationic dendrimers. Biochim Biophys Acta Proteins Proteomics 2007; 1774(7): 946-51. [http://dx.doi.org/10.1016/j.bbapap.2007.04.016]

[217] Klajnert B, Stanisławska L, Bryszewska M, Pałecz B. Interactions between PAMAM dendrimers and bovine serum albumin. Biochim Biophys Acta Proteins Proteomics 2003; 1648(1-2): 115-26. [http://dx.doi.org/10.1016/S1570-9639(03)00117-1]

[218] Fu Y, Hu R, Li C, Wang Q, Liu Z, Xue W. Effects of poly(amidoamine) dendrimers on the structure and function of key blood components. J Bioact Compat Polym 2014; 29(2): 165-79. [http://dx.doi.org/10.1177/0883911514521921]

[219] Asthana A, Chauhan AS, Diwan PV, Jain NK. Poly(amidoamine) (PAMAM) dendritic nanostructures for controlled sitespecific delivery of acidic anti-inflammatory active ingredient. AAPS PharmSciTech 2005; 6(3): E536-42. [http://dx.doi.org/10.1208/pt060367] [PMID: 16354015]

[220] Eliyahu H, Servel N, Domb AJ, Barenholz Y. Lipoplex-induced hemagglutination: potential involvement in intravenous gene delivery. Gene Ther 2002; 9(13): 850-8. [http://dx.doi.org/10.1038/sj.gt.3301705] [PMID: 12080379]

[221] Behera A, Padhi S. pH-Sensitive Polymeric Nanoparticles for Cancer Treatment. In:Polymeric nanoparticles for the treatment of solid tumors. 2022 Nov 15:401-25.

[222] Mishra S, Behera A. Applications of biodegradable polymeric biomaterials in biomedical science. In Advances in Green and Sustainable Nanomaterials 2023 Jul 17 (pp. 229-277). Apple Academic Press. [http://dx.doi.org/10.1201/9781003328322-10]

# CHAPTER 7

# Exploring the Role of Nutraceuticals in Brain Cancer Prevention and Treatment

**Vidhi V. Mandoth**[1], **Ravindranath B.S.**[2] and **Hema Kumar C.**[1,*]

[1] *Department of Biotechnology, Dayananda Sagar College of Engineering (Affiliated to Visveshwaraya Technological University, Belagavi) Kumar Swamy Layout, Bangalore-560111, Karnataka, India*

[2] *Manipal Institute of Technology, Manipal Academy of Higher Education (MAHE), Manipal, 576104, Karnataka, India*

**Abstract:** Brain cancer remains one of the most challenging malignancies to treat, with limited therapeutic options and poor prognosis. Brain cancer, particularly gliomas, presents significant challenges due to its aggressive nature and limited treatment options. In recent years, nutraceuticals—bioactive compounds derived from food sources—have gained attention for their potential in preventing and treating various forms of cancer, including brain tumours. The growing importance of nutraceuticals in the treatment and prevention of brain cancer is examined in this chapter. Important components of nutraceuticals, like vitamins, carotenoids, polyphenols, and omega-3 fatty acids, show promise in the fight against cancer by regulating cell signalling pathways involved in tumour growth and metastasis, acting as antioxidants, and exhibiting anti-inflammatory effects. The chapter further delves into the dietary sources rich in these nutraceuticals, such as curcumin, green tea, ashwagandha, mushrooms, berberin, gasotile, and silibinin. By understanding the molecular mechanisms through which these compounds act, we can better harness their potential in integrative strategies for brain cancer management, ultimately improving patient outcomes. Understanding the molecular mechanism of action of any compounds at the cellular level is very important, as signalling proteins or proteins involved in the cell cycle can serve as therapeutic targets and also be used as biomarkers. Cell signalling proteins, such as ERK, JNK, p53, AKT, NF-kappa B, and cellular proteins CDKs, DAPK, and caspases, along with their roles, are discussed in this chapter. It also highlights the molecular changes caused by the nutraceuticals in brain cancer cell lines.

**Keywords:** Berberine, Brain cancer, Curcumin, Chemotherapy, Cell signalling, Epicatechin, ERK, Gastrodin, Nutraceuticals, p53, Silibinin, JNK.

[*] **Corresponding author Hema Kumar C.:** Department of Biotechnology, Dayananda Sagar College of Engineering (Affiliated to Visveshwaraya Technological University, Belagavi) Kumar Swamy Layout, Bangalore-560111, Karnataka, India; Tel: 91-7019238054; E-mail: hemakum2014@gmail.com

**Prashant Tiwari, Pankaj Kumar Singh & Sunil Kumar Kadiri (Eds.)**

# INTRODUCTION

Brain cancer, or primary brain tumors, refers to abnormal growths of cells within the brain or its surrounding structures. These tumors can be classified into primary brain tumors, which originate in the brain, and secondary (or metastatic) brain tumors, which spread from other parts of the body. Brain tumours vary widely in terms of their cell type, location, and growth rate. They are categorized into different grades based on their aggressiveness, ranging from benign (low-grade) to malignant (high-grade). Common types of primary brain tumours include gliomas (such as glioblastomas), meningiomas, and medulloblastomas. Secondary brain tumours often originate from cancers of the lung, breast, or melanoma. Brain cancer is challenging to treat due to its complex location and the sensitive nature of the surrounding brain tissue. The incidence of brain cancer worldwide is 3.0 per 100,000 for women and 3.9 per 100,000 for men, according to 2020 reports from the International Agency for Research on Cancer (IARC) [1]. Over 97% of glioblastoma patients pass away within five years after being diagnosed, according to reports [2]. Bioactive substances obtained from dietary sources are called nutraceuticals, and they offer additional health benefits. By combining the words "nutrition" and "pharmaceutical," the term highlights their possible medicinal applications. For example, curcumin, resveratrol, green tea extract, and ashwagandha. It has been found that nutraceutical compounds can helpprevent chronic diseases and disorders, such as cardiovascular disease, diabetes, and cancer. Flavonoids, such as luteolin [3], fisetin [4], acetylapoaranotin [5], and astaxanthin [3], as well as lycopene from various sources including tomatoes, watermelons, grapes, and papaya [6], and GABA [7] are some examples of secondary metabolites exhibiting potential anticancer properties.

Bioactive components obtained from natural sources, known as anticancer nutraceuticals, have been shown to influence a number of biological processes that contribute to the development and growth of tumors, thereby potentially preventing cancer. These nutraceuticals can be found in foods, dietary supplements, and herbal preparations. Additionally, dietary antioxidants are known to be significant preventive factors against a number of cancers, such as liver cancer [8], colorectal cancer [9], and breast cancer [10]. However, the effect of dietary antioxidant vitamins against brain malignancies, especially gliomas, has received less academic attention [11].

Both the prevention and treatment of brain cancer can benefit significantly from the use of nutraceuticals. Recently, increasing attention has been directed toward nutraceuticals, natural substances extracted from plants, fruits, or cereals. These compounds are well-known for their antibacterial, anti-inflammatory, antioxidant, and antitumor properties [12]. Certain nutraceuticals may help lower the probability of brain tumors by preventing the early stages of the disease's development. This is known as their potential chemopreventive qualities. Alongside standard treatments like chemotherapy, radiation, and surgical procedures, nutraceuticals may enhance therapeutic efficacy, minimize side effects, and improve overall quality of life. Moreover, chemotherapy and nutraceuticals together enable lower dosages of synthetic medications, reducing their severe adverse effects [12].

Certain nutraceuticals may specifically target cancer cells or molecular pathways involved in brain tumour growth and progression, offering a complementary approach to traditional treatments. A variety of nutraceuticals have recently been considered due to their unique ability to prevent tumor growth, including gliomas, by blocking the angiogenic process, cancer metastases, and apoptosis. By altering various pathways, nutraceuticals can cause cancer cells to malfunction. In reality, these substances have the ability to slow the development of cancer cells, prevent their angiogenesis and proliferation, and cause them to undergo apoptosis. They are also crucial at different phases of the therapeutic process. Regulation of cellular signaling pathways, including PI3K/Akt/mTORC1, JAK/STAT, and GSK-3, as well as other mechanisms, such as cytokine receptors, inflammatory pathways, reactive oxygen species, and miRNAs, may be one of the main objectives of nutraceuticals [13].

A list of nutraceuticals obtained from various sources that possess anticancer properties is mentioned in Fig. (**1**).

A potential alternative strategy for preventing brain tumours is offered using nutraceuticals.

By harnessing the therapeutic potential of bioactive compounds found in foods and supplements, they may provide supportive benefits to conventional therapies, improve patient outcomes, and contribute to overall cancer management strategies.

**Fig. (1).** Nutraceuticals (curcumin, berberin, mushrooms, green tea, ashwagandha, and silibinin).

## CURCUMIN

### Curcumin as an Important Nutraceutical for Lifestyle Disorders, Including Brain Cancer

The rhizome of Curcuma longa L. (turmeric) yields the yellow pigment curcumin, which has been extensively used in Southeast Asian and Indian Ayurvedic medicine for its anti-inflammatory properties [14]. Current research suggests that curcumin may be beneficial in managing a wide range of human conditions, including diabetes, arthritis, cardiovascular diseases, Parkinson's disease, Alzheimer's disease, and various carcinomas, such as brain tumors [15]. Numerous molecular pathways regulated by curcumin have been the focus of intensive investigation. Its anticancer effects are supported by downstream modulation of enzymes, growth factors, receptors, kinases, inflammatory cytokines, and transcription factors [16]. Curcumin has been shown to selectively induce apoptosis in cancer cells, enhance the efficacy of radiation and chemotherapy while sparing healthy tissues, and suppress the malignant characteristics of Glioblastoma (GBM) stem cells [17 - 20]. Experimental evidence demonstrates that curcumin protects neurons from oxygen and glucose deprivation in an ischemia-reperfusion *in vitro* model using rat cortical neurons. Additionally, curcumin may reduce ATP and glutathione levels by inhibiting cellular glyoxalases, which could potentially impact cellular metabolism and contribute to its antitumor and anti-inflammatory properties [21].

In an *in vitro* study using the U87 human glioma cell line, curcumin (5–10 μM) was found to inhibit cell proliferation. The treated cells were arrested in the G2/M phase, associated with increased DUSP-2 expression and decreased phosphorylation of JNK and ERK [22, 23]. Notably, G2/M phase arrest may enhance cellular sensitivity to radiation. Curcumin's radiosensitizing effects were also observed in neuroblastoma (SK-N-MC) cells, where pretreatment with 100 nM curcumin led to increased caspase activation, reduced anti-apoptotic protein levels, and decreased NF-κB activity. Multiple studies have reported curcumin's efficacy in cancer therapy [15]. Curcumin induces G2/M cell cycle arrest and apoptosis in cancer cells primarily by upregulating Death-Associated Protein Kinase 1 (DAPK1), a key mediator of its pro-apoptotic effects. Elevated DAPK1 levels correlate with increased apoptosis and reduced phosphorylation of NF-κB and STAT3 [24]. Furthermore, curcumin enhances the expression of tumor suppressors, such as p53 and p21, and activates Egr-1 *via* ERK and JNK pathways, contributing to cell cycle arrest and inhibition of proliferation. It also promotes caspase activation and suppresses anti-apoptotic factors, thereby sensitizing cells to radiation and improving radiotherapy outcomes. Curcumin triggers apoptosis in various cancer cell lines by activating caspase 3/7, inducing DNA damage, and modulating apoptosis-related proteins including survivin, Bcl-2, and BAX. It also reduces mitochondrial membrane potential and inhibits NF-κB signaling.

In animal models, curcumin treatment significantly reduced tumor burden, with treated mice developing fewer tumors than the control mice. These findings suggest that curcumin exerts anticancer effects both by inducing apoptosis *in vitro* and by inhibiting tumor growth *in vivo* [15]. Additionally, curcumin stimulates autophagy, a critical process for cellular energy regulation and homeostasis. Laboratory studies show that curcumin-induced autophagy helps control the proliferation of glioma-initiating cells. Animal studies further revealed that curcumin prolongs lifespan and induces autophagy by activating the ERK1/2 pathway while inhibiting the AKT/mTOR/p70S6K signaling axis. Curcumin also causes G2/M cell cycle arrest in GBM cells, underscoring its therapeutic potential [15]. Curcumin disrupts key molecular pathways in glioblastoma and medulloblastoma. It inhibits NF-κB and AP-1, thereby reducing cancer cell viability and enhancing responsiveness to treatments such as cisplatin. Curcumin also targets glucose-6-phosphate translocase (G6PT), which may help overcome drug resistance. By suppressing STAT3, SHH, and GLI1 signaling, curcumin slows tumor progression and improves survival in animal models. Moreover, it inhibits Wnt/β-catenin signaling and promotes tubulin acetylation, contributing to its anticancer activity. Curcumin also inhibits glioma invasion by downregulating matrix metalloproteinases (MMP-1, MMP-3, MMP-9, and MMP-14) *via* AP-1 and MAP kinase pathways. It prevents the nuclear translocation of RelA/NF-κB, thereby suppressing urokinase-type Plasminogen Activator (uPA). Additionally, curcumin reduces the expression of HIF1A and VEGFA in pituitary adenoma cells, thereby inhibiting angiogenesis [15].

Curcumin (Fig. **2**) demonstrates significant therapeutic potential for Central Nervous System (CNS) tumors, as evidenced by promising preclinical studies and a favorable safety profile in humans. Its proposed benefits include potentiating the efficacy of existing chemotherapeutic agents and selectively targeting stem cell-like populations within tumors. Despite these advantages, curcumin's limited tissue distribution remains a major obstacle to its clinical application. Innovations in drug delivery systems or the development of curcumin analogs may enhance its bioavailability and therapeutic impact. Considering the poor prognosis associated with brain tumors and the stagnation in recent advancements in standard treatment modalities, rigorous clinical trials are imperative to validate curcumin's efficacy and explore its potential to improve patient outcomes [15].

**Fig. (2).** Structure of curcumin.

# MUSHROOMS

## Mushrooms as Antioxidant and Potential Anticancer Agents in Brain Cancer

Mushrooms represent the most advanced class of fungi within the plant kingdom and are renowned for their extensive therapeutic properties. These include antihyperlipidemic, anticancer, immunoregulatory, anti-inflammatory, and cardioprotective effects, largely attributed to their rich content of secondary metabolites [25]. Nutritionally, mushrooms are a dense source of essential compounds, including unsaturated fatty acids, ergosterol, and bioactive molecules, such as β-glucans, triterpenoids, and antioxidants. Their potent antioxidant activity observed in both wild and edible varieties is primarily due to the presence of vitamins, flavonoids, tocopherols, phenols, and carotenoids. Notably, Northern Serbian mushroom varieties, such as Agaricus bisporus (AbHW) and Ganoderma resinaceum (GrHW), have demonstrated antitumor, antiproliferative, and lipid peroxidation-inhibiting properties in their hot water extracts [25]. Numerous mushroom species are traditionally recognized for their medicinal value, offering immunomodulatory, antidiabetic, cardiovascular, hepatoprotective, and antitumor benefits [25]. The antitumor activity of Ganoderma lucidum is particularly linked to its immune-enhancing and antioxidant properties, as well as its role in chemoprevention [26].

The lipid profile of mushrooms varies by species. Lentinula edodes (shiitake) mushrooms exhibit a relatively high content of Polyunsaturated Fatty Acids (PUFA) and sugars compared to Saturated Fatty Acids (SFA). In contrast, oyster and king oyster mushrooms contain significant levels of Monounsaturated Fatty Acids (MUFA), along with PUFA and SFA [27]. The antioxidant, chelating, and free radical scavenging properties of Taiwanofungus salmoneus, a higher basidiomycete, have been utilized in the fermentation of cooked cereal grains, such as wheat and oats. In RAW 264.7 murine macrophages, the fermented extract significantly inhibited the production of interleukin-1β, interleukin-6, and tumor necrosis factor-α. Similarly, Inonotus obliquus, a medicinal fungus traditionally used in cancer therapy, has shown strong antioxidant activity [25]. Its methanol extract demonstrated potent free radical scavenging effects against DPPH and ABTS radicals and was found to contain three types of inonoblins and three variants of phelligridins (BIOENERERED 11247) [25]. A recent study suggests that daily consumption of approximately 18 grams of mushrooms may enhance cellular antioxidant capacity and reduce the risk of malignancy [28]. Certain species, including Lactarius deliciosus and Coprinus comatus, have shown notable anti-glioma activity by inducing cell cycle arrest in the G1 and G2/M phases and inhibiting metalloproteinase activity.

Oxidative stress is a key contributor to neurodegenerative diseases, such as Alzheimer's, leading to cellular and DNA damage. Mushrooms, rich in antioxidants, help mitigate oxidative stress and protect neuronal integrity. Varieties such as Coriolus versicolor and Hericium erinaceus are particularly effective in reducing neuronal stress, promoting the expression of nerve growth factors, and inhibiting the formation of neurotoxic proteins. Bioactive compounds like hericenones and erinacines found in Hericium species offer neuroprotective effects and may play a role in preventing neurodegeneration. In addition to their antioxidant properties, mushrooms are rich in peptides, dietary fibers, oligosaccharides, and other bioactive substances. These components contribute to a wide range of health benefits, including anti-inflammatory, anticancer, and anti-aging effects. Their complex molecular profile supports liver, brain, and cardiovascular health, underscoring their value as functional foods for holistic well-being.

## Berberine

Berberine (BBR), a naturally occurring isoquinoline alkaloid, is widely used in Traditional Chinese Medicine and can be extracted from various medicinal plants, including Phellodendron chinense Schneid, Phellodendron amurense, Coptis japonica, Hydrastis canadensis, and Berberis aristata [38, 39]. Berberine exhibits a broad spectrum of pharmacological activities, including lipid-lowering, hypoglycemic, anti-inflammatory, and antimicrobial effects [29, 30]. Notably, it has demonstrated potent anticancer properties, capable of suppressing the proliferation of cancer cells in gastric, bladder, colon, glioma, breast, melanoma, pancreatic, endometrial, and lung cancers [31 - 33].

Mechanistically, berberine exerts its anticancer effects by modulating key signaling pathways, such as MAPK, Erk1/2, JNK, AKT, and Wnt/β-catenin [34, 35]. In various human cancer cell lines, berberine directly binds to DNA, inducing double-strand breaks that disrupt gene transcription and lead to cell cycle arrest [40]. Neuroblastoma (NB), the most common extracranial solid tumor in children, originates from neural crest cells during embryonic development. Beta-Adrenergic Receptors (β-ARs), which belong to the G-Protein-Coupled Receptor (GPCR) family, consist of three subtypes: β1-, β2-, and β3-AR. The therapeutic potential of β-blockers in NB has been explored through both *in vitro* and *in vivo* experimental models [36].

Among these, propranolol—a nonselective β-blocker—has shown significant anticancer activity by activating the p53 and p73 signaling pathways, thereby inducing apoptosis and inhibiting tumor growth, survival, and proliferation in neuroblastoma cell lines derived from fifteen human samples. Other β-blockers with notable anticancer effects in NB include nebivolol (a selective β1-blocker) and carvedilol (a mixed α/β-blocker) [37]. These findings underscore the therapeutic promise of berberine and β-blockers in cancer treatment, particularly in targeting molecular pathways and inducing cell cycle arrest and apoptosis in various malignancies [38, 39].

Berberine (BBR) has demonstrated potent anticancer activity through multiple mechanisms, including the induction of apoptosis and autophagy. Studies have shown that BBR disrupts mitochondrial membrane potential, leading to impaired mitochondrial respiration and dysfunction across various cancer types [40]. In glioma cell lines U343 and MIA PaCa-2, BBR treatment enhances BECN1 activity and upregulates LC3 expression, thereby promoting autophagy [41]. These findings suggest that BBR targets diverse cellular pathways to inhibit cancer cell growth and progression (Fig. **3**), supporting its potential as a nutraceutical agent that may be used alone or in combination with conventional cancer therapies. Choi et al. reported that BBR induces apoptosis in SK-N-SH neuroblastoma cells (expressing p53) by increasing pro-apoptotic Bax protein, cleaved caspase-3, and cleaved PARP, while reducing anti-apoptotic Bcl-2 expression. However, its apoptotic effects were less pronounced in SK-N-MC cells lacking p53, indicating a p53-dependent mechanism [42].

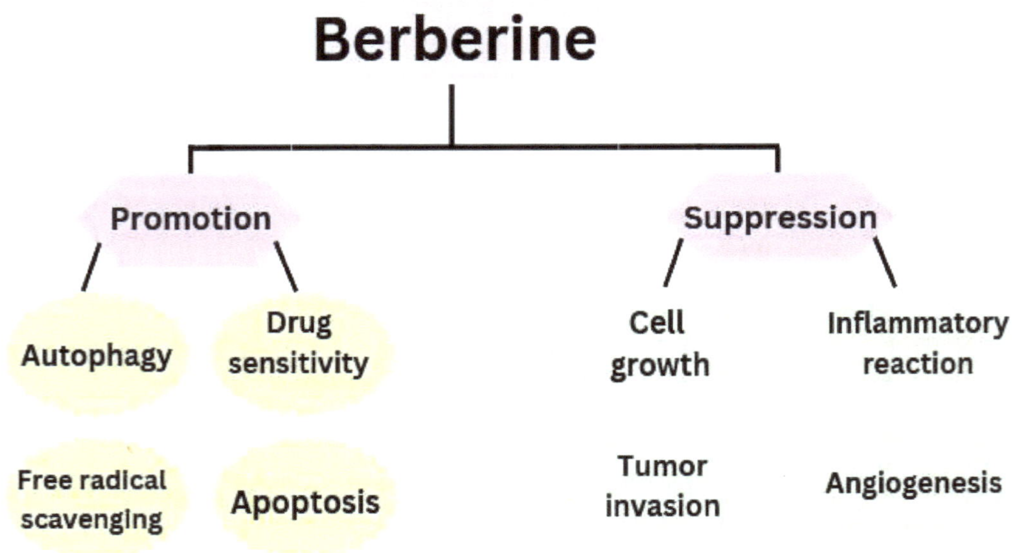


**Fig. (3).** Diagrammatic depiction of BBR-regulated pathways relevant to cancer.

BBR also suppresses Epithelial-Mesenchymal Transition (EMT) in Neuro2A cells by inhibiting PI3K/Akt and Ras-Raf-ERK signaling pathways, activating p38

MAPK, and downregulating TGF-β through modulation of TGF-β receptors II and III. It promotes EMT reversal by converting mesenchymal markers (vimentin and fibronectin) into the epithelial marker E-cadherin [43]. In Neuro2A and SH-SY5Y neuroblastoma cells, the combination of BBR with Solid Lipid Curcumin Particles (SLCP) enhances cytotoxicity without inducing toxicity [44]. Furthermore, the co-administration of curcumin, BBR, and arsenic trioxide ($As_2O_3$) significantly increases cytotoxic effects against neuroblastoma tumor cells. BBR also promotes neuronal differentiation in Neuro2A cells by upregulating key neuronal markers, including Microtubule-Associated Protein 2 (MAP2), β-tubulin III, and Neural Cell Adhesion Molecule (NCAM). Simultaneously, it downregulates stemness markers, such as CD133, β-catenin, Notch2, MYCN, Sox2, and nestin, thereby reducing tumor stemness and promoting differentiation [43].

Further investigations into BBR's impact on cancer cell survival revealed increased Cytochrome C expression in BBR-treated cells, as shown by immunocytochemistry, indicating mitochondrial involvement. Flow cytometry using Annexin V/PI staining confirmed that BBR strongly induces both early and late-stage apoptosis in U87 and U251 glioblastoma cells. Western blot analysis demonstrated decreased levels of Bcl-2 and increased expression of Bax, Cytochrome C, and cleaved caspase-3, suggesting that BBR reduces cell viability primarily through Cytochrome C-mediated apoptosis [45]. BBR also induces autophagy in Glioblastoma (GBM) cells *via* the AMPK/mTOR/ULK1 signaling pathway. It dephosphorylates mTOR and increases AMPK levels in a dose-dependent manner, thereby enhancing autophagy. Additionally, BBR elevates phosphorylated Beclin-1 (Ser93), a key autophagy marker. Inhibition of AMPK or knockdown of Beclin-1 significantly reduced BBR-induced autophagy, confirming its dependence on this pathway. BBR also increased phosphorylated ULK1 levels, reinforcing its role in autophagy induction through the AMPK/mTOR/ULK1 cascade [45].

The effect of autophagy suppression on BBR-induced apoptosis in glioma cells was further evaluated. *In vitro*, the autophagy inhibitor Bafilomycin A1 (Baf) enhanced BBR-induced apoptosis. *In vivo*, BBR significantly reduced tumor growth in a U87 mouse model, and its combination with Chloroquine (CQ), another autophagy inhibitor, further improved therapeutic efficacy compared to BBR alone. Western blot analysis confirmed increased p-AMPK and decreased p-mTOR levels following BBR treatment. Immunohistochemistry revealed elevated LC3B expression and reduced proliferation in the combination treatment group. Importantly, no hepatotoxicity was observed in treated mice [45]. Beyond its anticancer properties, BBR exhibits neuroprotective effects, making it a promising candidate for preventing neurodegenerative diseases. Notably, BBR has been shown to inhibit Monoamine Oxidases (MAOs), enzymes implicated in the pathogenesis of Alzheimer's disease and highly expressed in neuroblastoma cells [46, 47].

## Ganostile

Mushrooms such as Ophiocordyceps sinensis, Ganoderma lucidum, and Hericium erinaceus are among the most extensively studied in mycotherapy for their medicinal properties. Ganoderma lucidum, known as Ling Zhi in China and Reishi in Japan, typically grows on deciduous oak and chestnut trees. It belongs to the Basidiomycota class of perennial saprophytic fungi [48]. G. lucidum contains a diverse array of bioactive compounds, including triterpenoids, polysaccharides, nucleosides, sterols, and alkaloids. Among these, polysaccharides and triterpenoids are considered the principal constituents responsible for its anticancer effects [49, 50]. G. Lucidum Polysaccharides (GLPs) enhance immune function by activating natural killer cells, macrophages, dendritic cells, and T and B lymphocytes. This immunostimulatory activity promotes phagocytosis and cytokine production, contributing to its antitumor efficacy [51]. Additionally, GLPs exhibit antiangiogenic and direct cytotoxic effects, thereby suppressing tumor growth [52]. High-molecular-weight polysaccharides, particularly β-D-glucans, have demonstrated significant antitumor activity in both *in vitro* and *in vivo* models. These effects are attributed to their immunomodulatory, antiproliferative, pro-apoptotic, antimetastatic, and antiangiogenic properties [53]. G. lucidum is also a key component of the dietary supplement Ganostile, which includes extracts from Astragalus membranaceus, Eleutherococcus senticosus, and Echinacea purpurea. E. senticosus, sometimes referred to as E. purpurea, belongs to the Araliaceae family [54]. It contains a variety of bioactive compounds, including phenols, lignans, flavonoids, and polysaccharides, which contribute to its therapeutic potential in managing cerebrovascular diseases, ischemic heart disease, immune regulation, neurasthenia, and hypertension [55].

Immunolabeling of Proliferating Cell Nuclear Antigen (PCNA) revealed strong nuclear staining in control cells, indicative of active cell cycling, mitotic activity, and an organized actin cytoskeleton. Following treatment with Ganostile, PCNA localization remained nuclear but was accompanied by nuclear membrane rupture and cytoplasmic degradation [56-60]. Treatment with cisplatin alone or in combination with Ganostile increased PCNA levels, suggesting elevated proliferation. In contrast, the combination of Ganostile with Pt(IV)Ac-POA significantly reduced PCNA expression, indicating suppressed proliferation. Pt(IV)Ac-POA treatment induced nuclear and cytoskeletal disorganization, including disruption of actin filaments and the presence of pycnotic nuclei [61-64]. When combined with Ganostile, the actin cytoskeleton was further disintegrated, and cells exhibited marked shrinkage [55]. Fig. (**4**) illustrates the cellular changes observed under different treatment conditions.

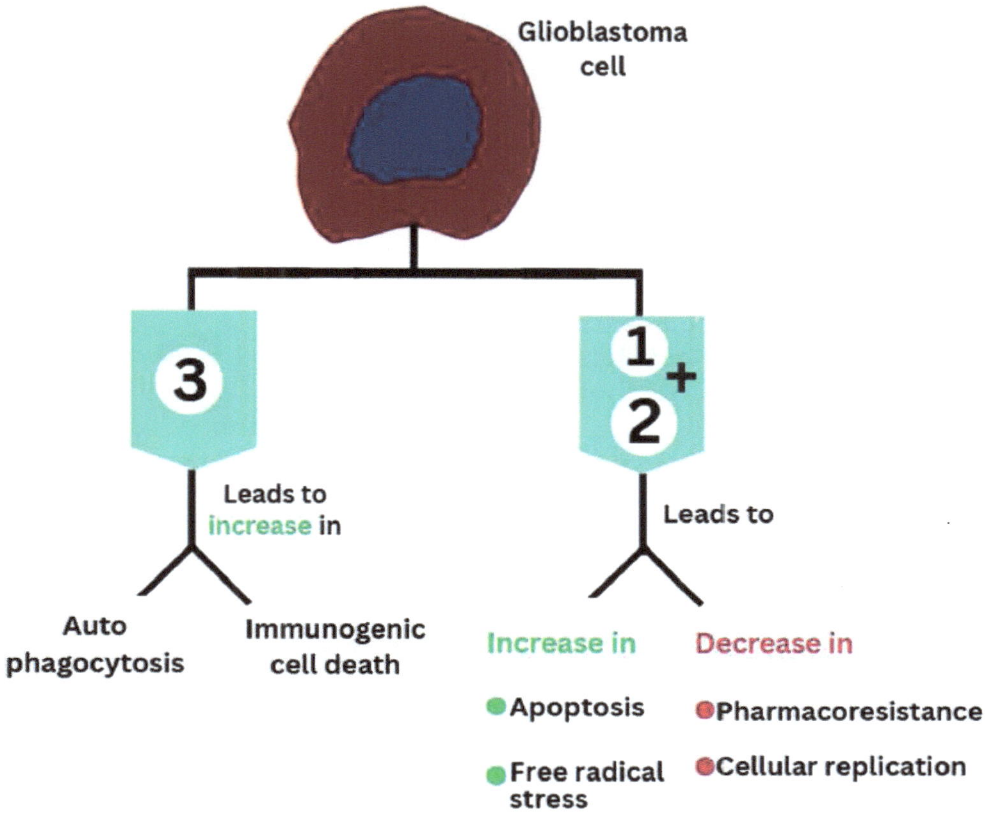


**Fig. (4).** Demonstration of the key findings (1- Pt(IV)Ac-POA ;2- Ganostile ; 3- Cisplatin).

Ganostile and NOS2-targeted treatments reduced NOS2 expression, suggesting decreased nitrogen radical stress and cytoskeletal disruption. Conversely, cisplatin combined with Pt(IV)Ac-POA elevated NOS2 levels, indicating increased oxidative stress [65-69]. Superoxide Dismutase 2 (SOD2) expression was upregulated across all treatments, with the highest levels observed in the Ganostile combination group, reflecting an enhanced oxidative stress response. Cyclooxygenase 2 (COX2) levels decreased following Ganostile and platinum treatments, while COX4 expression significantly increased with Ganostile co-treatment, suggesting altered mitochondrial function and distribution [55, 70]. Control cells exhibited low LC3b expression. Ganostile treatment increased LC3b levels and its colocalization with lysosomes, indicating enhanced autophagy. The combination of cisplatin and Pt(IV)Ac-POA further increased LC3b-lysosome overlap, with the highest levels observed in the Ganostile co-treatment group, suggesting robust autophagosome-lysosome fusion.

Diffuse p62 distribution was noted in control cells, which was significantly reduced by Ganostile. Cisplatin and Pt(IV)Ac-POA treatments increased p62 expression, whereas Ganostile co-treatment reversed this effect, maintaining p62 levels comparable to those of the controls. PINK1 levels remained low in control and Ganostile-treated cells but increased slightly with cisplatin and Ganostile. The combination of Pt(IV)Ac-POA and Ganostile significantly elevated PINK1 expression, indicating enhanced mitophagy. Parkin levels were low in control and Ganostile-treated cells, with no significant change following cisplatin treatment. However, Pt(IV)Ac-POA combined with Ganostile markedly increased Parkin expression, suggesting strong mitophagy activation [55]. This study investigated the combined effects of platinum-based chemotherapy and Ganostile, a myco-phytotherapeutic supplement, on glioblastoma cells. Key findings include: Synergistic Antitumor Activity – Combined treatments resulted in higher cell mortality and reduced proliferation compared to individual treatments; Apoptosis Induction – Increased cleaved caspase-3 and decreased PCNA labeling indicated enhanced apoptotic activity; Morphological Alterations – Significant changes in cellular morphology and organelle structure were observed; Autophagy and Mitophagy Activation – Elevated expression of LC3b, p62, PINK1, and Parkin confirmed activation of autophagic and mitophagic pathways; Oxidative Stress Modulation – Increased oxidative stress markers (NOS2, SOD2) and altered antioxidant responses were noted; Necrotic Cell Death – Transmission electron microscopy revealed features of ferroptosis and necroptosis, particularly in response to combination therapies [55].

Although Ganostile alone did not exhibit sufficient potency to be considered a standalone therapeutic agent, its use as an adjuvant significantly enhanced the efficacy of cisplatin and Pt(IV)Ac-POA, demonstrating a positive synergistic effect [54].

U251 MG glioblastoma cells and primary human fibroblasts were cultured under standard conditions. U251 MG cells were maintained in EMEM supplemented with essential nutrients, while fibroblasts were cultured in D-MEM with Fetal Bovine Serum (FBS) and additional supplements. Ganostile, a phyto-therapeutic supplement containing various plant extracts and excipients, was used at a concentration of 5 mg/mL in combination treatments. To determine the optimal dose of Ganostile, an MTS assay was performed on U251 cells and human fibroblasts. Cells were treated with Ganostile concentrations ranging from 0 to 100 mg/mL for 48 hours. Following treatment, MTS solution was added, and cells were incubated for three hours. Absorbance was measured at 490 nm to assess cell viability, which was calculated as a percentage relative to the control cells. All experiments were conducted in triplicate to ensure statistical reliability [54]. The findings suggest that Ganostile enhances the efficacy of platinum-based chemotherapy in the treatment of glioblastoma. While not potent enough as a monotherapy, its synergistic effects with cisplatin and Pt(IV)Ac-POA offer a promising therapeutic strategy. Further research is warranted to explore the *in vivo* applicability of these results and to investigate the individual contributions of Ganostile's components in both *in vitro* and *in vivo* tumor models [54].

## miRNA

miR-9 DNA methylation at CpG islands influences chromatin remodeling and gene expression, notably affecting non-coding RNAs, such as miR-9. In Glioblastoma (GBM), miR-9 functions as a tumor suppressor by downregulating STMN1, a gene involved in microtubule dynamics during cell cycle progression. Reduced miR-9 expression correlates with tumor growth in GBM's heterogeneous cell populations [71]. miR-9 also modulates chromatin structure by repressing the EZH2 axis, facilitating transcription factor binding such as REST (RE1 Silencing Transcription Factor), FOXP1 (Forkhead Box Protein P1), CREB (cAMP Response Element-Binding Protein), and NF-1 (Neurofibromin 1), thereby influencing proliferation, migration, and tumorigenicity. Additionally, miR-9 is involved in neurogenesis and has been linked to Alzheimer's and Parkinson's diseases, affecting cell survival and differentiation. Restoring miR-9 expression in glioma cells reduces tumorigenicity and supports regenerative medicine approaches, including somatic cell reprogramming into functional neurons.

The miRNA cluster—miR-34abc (comprising miR-34a, miR-34b, and miR-34c)—is a critical tumor suppressor frequently inactivated in GBM due to CpG island hypermethylation, resulting in uncontrolled cell proliferation. In Alzheimer's Disease (AD), miR-34a targets cyclin D1 to prevent aberrant neuronal cell cycle re-entry, a process disrupted by amyloid β1–42 peptide. In Parkinson's Disease (PD), downregulation of miR-34bc is associated with mitochondrial dysfunction and neuronal death. Collectively, miR-34abc represents a promising therapeutic target for both neurodegenerative diseases and brain tumors [71]. The miR-29 family (miR-29a, miR-29b, and miR-29c) regulates DNA methylation by targeting *de novo* DNA methyltransferases DNMT3A and DNMT3B. Reduced miR-29 activity leads to the overexpression of these enzymes, resulting in aberrant DNA methylation and an increased risk of glioblastoma. miR-29 also inhibits glioblastoma cell proliferation by targeting the PDPN gene. In neurodegenerative conditions, such as Spinocerebellar Ataxia 17, miR-29 downregulation disrupts normal brain function and contributes to disease progression [72]. Despite its tumor-suppressive role, the relationship between miR-29, neurodegeneration, and cancer remains complex and warrants further investigation [73].

miR-124—predominantly expressed in the brain—is essential for neurodevelopment, synaptic morphology, and neuroprotection. It maintains neuronal identity by repressing the expression of non-neuronal genes. In brain cancers, such as glioblastoma and medulloblastoma, reintroduction of miR-124 reduces tumor cell proliferation and invasion [74]. Conversely, diminished miR-124 expression in neurodegenerative diseases contributes to neuronal death and disease progression. Due to its dual role as a neuroprotective and tumor-suppressive agent, regulating miR-124 levels may offer a viable therapeutic strategy for treating brain-related disorders [75].

## Green Tea

Ongoing research continues to explore the curative potential of green tea and its bioactive constituents in cancer treatment. Many cancer patients currently use this "miraculous herb" as a complementary therapy to enhance the efficacy of conventional treatments. Green tea, derived from the leaves of Camellia sinensis, is recognized for its antitumor, cardioprotective, and neuroprotective properties. One clinically approved proteasome inhibitor, Bortezomib (BZM), is used primarily in the treatment of multiple myeloma and mantle cell lymphoma [76, 77]. BZM functions as a selective antagonist of the 26S proteasome, a multi-subunit protein complex responsible for degrading ubiquitinated proteins targeted for removal [78]. Proteasome inhibition by BZM leads to the accumulation of misfolded and obsolete proteins, triggering the Unfolded Protein Response (UPR) and inducing Endoplasmic Reticulum (ER) stress [79, 80]. This stress response operates as a cellular yin-yang system: under mild stress, it promotes survival and homeostasis, but under severe stress—such as that induced by BZM—it initiates apoptosis [81, 82]. A key modulator of ER stress is Glucose-Regulated Protein 78 kDa (GRP78), which possesses anti-apoptotic properties, including inhibition of caspase-7 activation [83]. Elevated GRP78 levels have been associated with chemoresistance in cancer cells, making it a strategic target for sensitizing tumors to chemotherapy [84]. Notably, (-)-epigallocatechin gallate (EGCG), a major polyphenol in green tea, has been shown to bind to and inhibit GRP78, thereby enhancing tumor cell sensitivity to chemotherapeutic agents, such as temozolomide and etoposide [85, 86]. Green tea contains a diverse array of polyphenols, including EGCG, (-)-epicatechin gallate (ECG), (-)-epigallocatechin (EGC), and (-)-epicatechin (EC) [74]. EGCG, in particular, exhibits structural flexibility that enables it to interact with proteins, lipids, and nucleic acids, functioning as a chemical facilitator [75]. A comprehensive summary of nutraceutical-based anticancer drugs is presented in Table **1**.

Despite its therapeutic promise, EGCG has been shown to interfere with the cytotoxic effects of BZM. In both *in vitro* and *in vivo* studies, EGCG or green tea extract significantly suppressed BZM-induced proteasome inhibition, ER stress, and tumor cell death. Importantly, this antagonistic effect was observed at EGCG concentrations achievable in humans [74]. In Multiple Myeloma (MM) cell lines, BZM effectively induced cell death within 48 hours, whereas EGCG alone had minimal impact on cell viability. However, when combined with BZM, EGCG completely abrogated BZM-induced apoptosis, preserving 100% cell viability. EGCG also inhibited BZM-induced cell cycle arrest and apoptotic morphology, as confirmed by microscopy [74]. Similar protective effects were observed in glioblastoma cell lines U251 and LN229. MTT assays revealed that EGCG increased the IC50 of BZM, indicating reduced cytotoxicity. Colony formation assays further demonstrated that EGCG allowed long-term survival and proliferation of glioblastoma cells exposed to BZM. These findings suggest that EGCG may attenuate the therapeutic efficacy of BZM in both multiple myeloma and glioblastoma [74] (Fig. **5**).

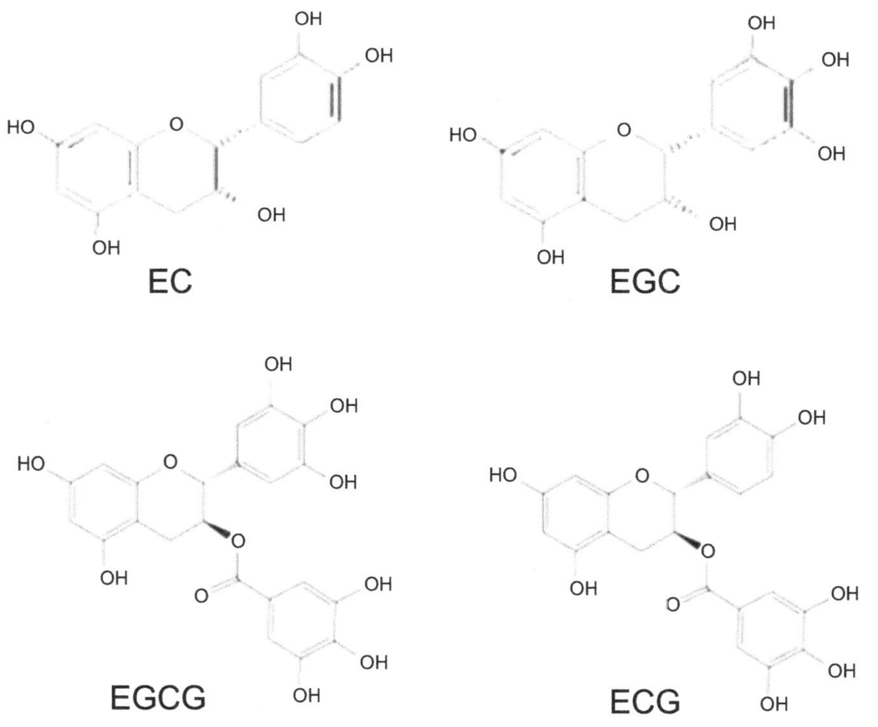


**Fig. (5).** The structures of EC, EGC, EGCG, and ECG.

MTT assays were employed to evaluate the effects of various green tea-derived mixtures on cell proliferation and survival, specifically to determine whether whole Green Tea Extract (GTE) and individual polyphenols other than EGCG could counteract the cytotoxic effects of Bortezomib (BZM). The results indicated that all tested polyphenols exhibited some degree of cytoprotective activity against BZM-induced cytotoxicity. However, the protective effects of (-)-epigallocatechin (EGC), (-)-epicatechin (EC), and (-)-epicatechin gallate (ECG) were significantly less potent than those of (-)-epigallocatechin gallate (EGCG) [74]. Green tea, its concentrated extract (GTE), and specific constituents such as EGCG are widely recognized for their health-promoting properties. These extracts are not only consumed by healthy individuals for general wellness but are also increasingly used by cancer patients as part of Complementary And Alternative Medicine (CAM) regimens. Many patients adopt these practices without consulting their physicians, aiming to enhance conventional treatments or mitigate side effects. As a result, numerous clinical trials are underway to systematically investigate the therapeutic potential of green tea and its constituents, with the goal of optimizing their medicinal applications.

However, findings from recent studies raise concerns about the interaction between green tea polyphenols and BZM. The anticancer efficacy of BZM was significantly compromised in the presence of GTE and certain polyphenolic components. Specifically, when green tea polyphenols were present, BZM failed to inhibit proteasome activity, thereby preventing the induction of ER stress and subsequent apoptosis in tumor cells [74]. Mechanistically, EGCG—characterized by its 1,2-diol functional group—binds directly to the boronic acid moiety of BZM, neutralizing the drug's cytotoxic effects across various tumor cell lines, including glioblastoma and multiple myeloma. Importantly, EGCG's antagonistic action is not attributed to its antioxidant properties but rather to a specific chemical interaction with BZM. Unlike other antioxidants, EGCG can completely abolish BZM's therapeutic efficacy, even at low concentrations, making it a highly potent inhibitor of BZM activity. These findings underscore the need for caution when combining green tea or its polyphenols with proteasome inhibitors in cancer therapy. While green tea offers numerous health benefits, its interaction with chemotherapeutic agents like BZM may compromise treatment outcomes. Further research is essential to delineate safe usage guidelines and to inform clinical practice regarding the integration of nutraceuticals with conventional cancer therapies.

## Ashwagandha

Withaferin A (WA), a natural compound extracted from Withania somnifera, commonly known as ashwagandha, has demonstrated significant anticancer potential. Among the most aggressive and physiologically active brain tumors originating from the central nervous system is Glioblastoma Multiforme (GBM), classified as WHO grade IV astrocytoma. Extracts from Withania somnifera leaves, referred to as i-Extract, are available in various alcoholic forms and contain bioactive constituents, such as Withaferin A, Withanone, and Withanolide A. Traditionally used in Indian medicine, Withania somnifera has long been recognized for its broad therapeutic applications and safety profile. Scientific studies have demonstrated that WA possesses a range of pharmacological properties, including antidiabetic effects [87], hepatoprotective activity against acetaminophen-induced liver injury [88], and anticancer effects across multiple malignancies, such as breast, prostate, non-small cell lung, pancreatic, and mammary gland cancers [89 - 96]. WA has been reported to inhibit tumor cell proliferation, induce cell cycle arrest, and promote apoptosis [91, 92]. However, the precise molecular mechanisms underlying its anticancer effects remain under investigation.

In glioblastoma cell lines U87 and U251, WA significantly inhibited cell proliferation and induced cell cycle arrest at the G2/M phase. *In vivo* studies further confirmed WA's ability to suppress GBM tumor growth. Mechanistically, WA activates the ATF4–ATF3–CHOP axis, contributing to both apoptosis and cell cycle arrest. Comparative analysis revealed that WA more effectively inhibited the proliferation of U87 and U251 cells than normal astrocytes (HA1800), with distinct IC50 values indicating selective cytotoxicity.

WA treatment led to elevated markers of apoptosis, including increased levels of cleaved PARP1 and caspases, in a time- and dose-dependent manner. The use of Emricasan, a pan-caspase inhibitor, significantly reduced WA-induced cytotoxicity, suggesting that WA promotes apoptosis *via* the intrinsic pathway.

Further analysis demonstrated that WA increased the proportion of U87 and U251 cells in the G2/M phase and reduced the expression of cyclins A and B, as well as phosphorylated CDK1, indicating suppression of CDK1 activity. The G2/M arrest was primarily mediated through p21 overexpression, independent of p53. Knockdown of p21 partially reversed cell cycle arrest and restored cell viability, implicating the p21–Bad pathway in WA-induced apoptosis. WA's anticancer effects in GBM are mediated through multiple mechanisms: Intrinsic Apoptotic Pathway: Activation of pro-apoptotic proteins Bad and Bim. Cell Cycle Arrest: p21-dependent, p53-independent G2/M phase arrest. ER Stress Response: Activation of the ATF4–ATF3–CHOP axis. In additional studies using rat and human glioma cell models, the active constituents of ashwagandha leaf extract Withaferin A, Withanone, and Withanolide A were evaluated. Low doses of i-Extract combined with WA induced cellular senescence, while higher concentrations triggered apoptosis. These findings support the therapeutic potential of i-Extract and its components in glioma treatment [97]. Moreover, treatment with a combination of extract and Withanone was reported to activate the tumor suppressor p53 pathway, further enhancing its anticancer efficacy [98, 99]. These findings underscore Withaferin A's potential as a potent anticancer agent, particularly in glioblastoma therapy. Its ability to induce apoptosis, arrest the cell cycle, and activate stress response pathways suggests that WA could be effectively integrated into combination treatment strategies. Further research is warranted to explore its clinical applicability and synergistic potential with existing therapies.

## Silibinin

A naturally occurring polyphenolic flavonoid called silibinin (also known as silybin) was extracted from milk thistle (Silybum marianum) seed extracts, whose structure is shown in Fig. (**6**). Experimental evidence demonstrates silibinin's potent ability to inhibit the aggressive and migrating traits of tumour cells, exhibiting both *in vitro* and *in vivo* positive effects against cancer [100, 101].

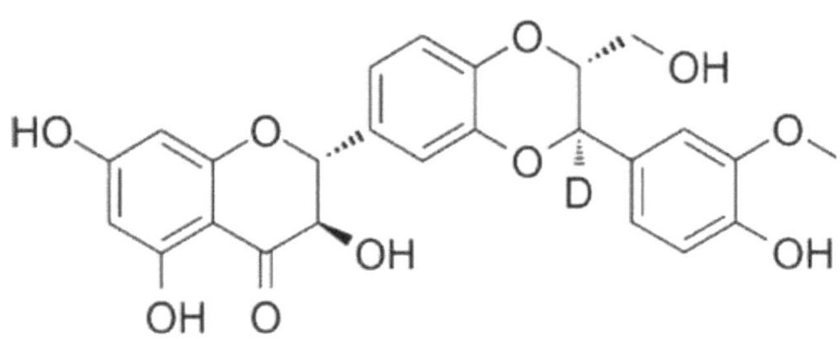


**Fig. (6).** Structure of Silibinin.

In 2014, a 62-year-old woman with lung adenocarcinoma and brain metastases experienced worsening neurological symptoms despite initial treatment with chemotherapy and whole brain radiotherapy.

Following the addition of the experimental nutraceutical silibinin (Legasil) to her regimen, she exhibited significant clinical and radiological improvement, with reduced brain lesion volumes and decreased edema. Similarly, a 67-year-old man with lung adenocarcinoma and a cystic brain metastasis experienced clinical improvement and reduced brain lesion size after adding Legasil to his treatment, despite stable pulmonary disease.

Both patients demonstrated enhanced performance status and continued with second-line treatments.

Patients with Brain Metastases (BM) from Non-Small Cell Lung Cancer (NSCLC) typically face a poor prognosis, with survival often limited to approximately one month without treatment. Standard therapeutic approaches for patients with limited metastases include stereotactic radiosurgery and Whole-Brain Radiation Therapy (WBRT) [102]. Recent clinical trials have demonstrated that combining cisplatin-pemetrexed chemotherapy with WBRT can extend median survival to approximately 12.6 months. Silibinin, a plant-derived flavonolignan, has emerged as a promising candidate for BM treatment due to its ability to inhibit Signal Transducer and Activator of Transcription 3 (STAT3), a key regulator of cancer progression [102]. Preclinical studies suggest that silibinin may alleviate BM-related symptoms and enhance brain-specific antitumor activity. However, clinical validation remains pending. Major challenges include silibinin's poor water solubility and limited bioavailability, although high-dose formulations and intravenous delivery have shown potential. Further research is needed to establish its therapeutic role and explore its integration with existing treatment modalities.

Despite growing interest in nutraceuticals for brain tumor management, several barriers hinder their clinical translation:

Glioblastomas and other brain tumors exhibit extensive genetic, epigenetic, and phenotypic diversity. This heterogeneity leads to variable treatment responses, rendering single-agent therapies—including nutraceuticals—less effective across patient populations. Differences in genetic mutations, molecular pathways, and tumor microenvironments contribute to treatment resistance and limit uniform efficacy. Additionally, Cancer Stem-Like Cells (CSCs) within gliomas drive recurrence and progression. To overcome these challenges,approaches, such as identification of patient-specific biomarkers are crucial for tailoring nutraceutical-based therapies. Integration of nutraceuticals with targeted therapies, immunotherapy, and conventional treatments may enhance therapeutic outcomes. Furthermore, use of genomic, transcriptomic, and proteomic profiling can help optimize nutraceutical combinations for specific tumor subtypes.

Tumors may develop resistance to nutraceuticals by activating alternative survival pathways or altering the uptake and metabolism of these compounds. Chronic exposure can lead to adaptive resistance, undermining therapeutic efficacy. Rotational or cyclical administration of nutraceuticals can help prevent adaptation. Combination with chemotherapy or small-molecule inhibitors can enhance efficacy. Co-administration of efflux pump inhibitors can help improve intracellular drug retention.

Many nutraceuticals, including curcumin and resveratrol, suffer from poor solubility, rapid metabolism, and low systemic availability. These limitations reduce absorption in the gastrointestinal tract, resulting in extensive first-pass metabolism and decreased therapeutic concentrations. Moreover, many compounds struggle to penetrate the Blood-Brain Barrier (BBB), restricting their direct impact on brain tumors.

Use of advanced drug delivery systems, such as nanoparticles, liposomes, and micelles, to improve solubility and systemic circulation. Chemical modifications or conjugation with carrier molecules to enhance BBB permeability and bioavailability. To fully realize the therapeutic potential of nutraceuticals in brain tumor treatment, future research should prioritize: Synergistic Approaches – Investigating combinations of nutraceuticals with chemotherapy, radiotherapy, and immunotherapy, Advanced Delivery Systems – Employing nanoparticles, hydrogels, and lipid-based carriers to enhance bioavailability and BBB penetration, Biomarker-Driven Therapies – Identifying molecular markers to predict patient responses and personalize nutraceutical regimens, and Clinical Validation – Transitioning from *in vitro* and animal models to well-designed clinical trials to establish safety, efficacy, and standardized usage guidelines. By addressing these challenges, nutraceuticals can evolve from complementary agents to integral components of brain cancer treatment strategies, offering improved therapeutic outcomes and potentially extending survival rates for patients with aggressive brain tumors.

**Table 1. A list of nutraceuticals with anticancer potential in brain cancer.**

| Name of Nutraceuticals | Cell Signalling Proteins | Cell line | References |
|---|---|---|---|
| 1) Curcumin | ERK, JNK, p53, DAPK1 | U87 (Human Glioma Cell Line) | [21] Qian *et al.*, 2015. |
| 2) Curcumin | VEGFA | Pituitary Adenocarcinoma Cell Line | [7] Klinger *et al.*, 2016 |
| 3) Mushroom powder | Metallo Proteinases | Glioma Cell Line | [27] Ba DM *et al.*, 2021 |
| 4) Berberine | TGF-β, p38 MAPK | SH-SY5YNB | [43] Maiti *et al.*, 2019 |
| 5) Gasotile | Caspase 3, COX2 | Glioblastoma Cell Line | [54] Gaiaschi *et al.*, 2024 |
| 6) Withaferin A | P53 | U87 glioblastoma Cell Line | [94] Hahm *et al.*, 2013 |

## CONCLUSION

Nutraceuticals exhibit significant potential as adjunctive agents in the treatment of brain cancer due to their diverse biological activities, including antioxidant, anti-inflammatory, and pro-apoptotic effects, as well as their ability to modulate key cell signaling pathways. Their integration with conventional therapies, such as chemotherapy, radiotherapy, and immunotherapy, may offer synergistic benefits, enhancing therapeutic efficacy while minimizing adverse effects. By targeting molecular mechanisms involved in tumor progression and resistance, nutraceuticals could contribute to more personalized and effective treatment strategies. However, their clinical application remains constrained by challenges, such as poor bioavailability, limited blood-brain barrier penetration, and insufficient clinical validation. To fully harness their therapeutic potential, future research must prioritize well-designed clinical trials, mechanistic investigations, biomarker-driven personalization, and advanced formulation technologies. With rigorous scientific evaluation and strategic integration into oncology practice, nutraceuticals may transition from supportive supplements to integral components of brain cancer therapy, ultimately improving patient survival and quality of life.

## REFERENCES

[1] Sung H, Ferlay J, Siegel RL, *et al.* Global cancer statistics 2020: GLOBOCAN estimates of incidence and mortality worldwide for 36 cancers in 185 countries. CA Cancer J Clin 2021; 71(3): 209-49. [http://dx.doi.org/10.3322/caac.21660] [PMID: 33538338]

[2] Aminianfar A, Vahid F, Shayanfar M, *et al.* The association between the dietary inflammatory index and glioma: A case-control study. Clin Nutr 2020; 39(2): 433-9. [http://dx.doi.org/10.1016/j.clnu.2019.02.013] [PMID: 31153672]

[3] Yoshida T, Maoka T, Das SK, *et al.* Halocynthiaxanthin and peridinin sensitize colon cancer cell lines to tumour necrosis factor-related apoptosis-inducing ligand. Molecular cancer research. MCR 2007; 5(6): 615-25. [http://dx.doi.org/10.1158/1541-7786.MCR-06-0045] [PMID: 17579120]

[4] Suh Y, Afaq F, Johnson JJ, Mukhtar H. A plant flavonoid fisetin induces apoptosis in colon cancer cells by inhibition of COX2 and Wnt/EGFR/NF- B-signaling pathways. Carcinogenesis 2008; 30(2): 300-7. [http://dx.doi.org/10.1093/carcin/bgn269] [PMID: 19037088]

[5] Choi EJ, Park JS, Kim YJ, *et al.* Apoptosis-inducing effect of diketopiperazine disulfides produced by Aspergillus sp. KMD 901 isolated from marine sediment on HCT116 colon cancer cell lines. J Appl Microbiol 2011; 110(1): 304-13. [http://dx.doi.org/10.1111/j.1365-2672.2010.04885.x] [PMID: 21122037]

[6] Miller EC, Giovannucci E, Erdman JW Jr, Bahnson R, Schwartz SJ, Clinton SK. Tomato products, lycopene, and prostate cancer risk. Urol Clin North Am 2002; 29(1): 83-93. [http://dx.doi.org/10.1016/S0094-0143(02)00020-4] [PMID: 12109359]

[7] Al-Wadei HAN, Ullah MF, Al-Wadei M. GABA (γ-aminobutyric acid), a non-protein amino acid counters the β-adrenergic cascade-activated oncogenic signaling in pancreatic cancer: A review of experimental evidence. Mol Nutr Food Res 2011; 55(12): 1745-58. [http://dx.doi.org/10.1002/mnfr.201100229] [PMID: 21805621]

[8] Cucarull B, Tutusaus A, Hernáez-Alsina T, *et al.* Antioxidants threaten multikinase inhibitor efficacy against liver cancer by blocking mitochondrial reactive oxygen species. Antioxidants 2021; 10(9): 1336. [http://dx.doi.org/10.3390/antiox10091336] [PMID: 34572967]

[9] Panic N, Nedovic D, Pastorino R, Boccia S, Leoncini E. Carotenoid intake from natural sources and colorectal cancer: a systematic review and meta-analysis of epidemiological studies. Eur J Cancer Prev 2017; 26(1): 27-37. [http://dx.doi.org/10.1097/CEJ.0000000000000251] [PMID: 26960163]

[10] Selvakumar P, Badgeley A, Murphy P, *et al.* Flavonoids and other polyphenols act as epigenetic modifiers in breast cancer. Nutrients 2020; 12(3): 761.
[http://dx.doi.org/10.3390/nu12030761] [PMID: 32183060]

[11] Walker Blake C, Mittal Sandeep. Therapeutic potential of curcumin for the treatment of brain tumours. Oxid Med Cell Longev 2016; 9324085.
[http://dx.doi.org/10.3390/ijms21249435]

[12] Chio CD, Luca FD, Starvaggi J, Previti S, Zappalà M, Ettari R. The role of nutraceuticals in chemoprevention and their therapeutic effects when used in combination with synthetic drugs. Curr Med Chem 2025; 32(34): 7596-623.
[http://dx.doi.org/10.2174/0109298673348105250102040623] [PMID: 39945263]

[13] Salami M, Salami R, Aarabi MH, *et al.* Targeting glioma cells with nutraceuticals: Therapeutic effects based on molecular mechanisms, new evidence and perspectives. Mini Rev Med Chem 2023; 23(11): 1167-92.
[http://dx.doi.org/10.2174/1389557522666220531151137] [PMID: 35642112]

[14] Kocaadam B, Şanlier N. Curcumin, an active component of turmeric ( *Curcuma longa* ), and its effects on health. Crit Rev Food Sci Nutr 2017; 57(13): 2889-95.
[http://dx.doi.org/10.1080/10408398.2015.1077195] [PMID: 26528921]

[15] Klinger Neil V, Mittal Sandeep. Therapeutic potential of curcumin for the treatment of brain tumours 2016.
[http://dx.doi.org/10.1155/2016/9324085]

[16] Anand P, Sundaram C, Jhurani S, Kunnumakkara AB, Aggarwal BB. Curcumin and cancer: An "old-age" disease with an "age-old" solution. Cancer Lett 2008; 267(1): 133-64.
[http://dx.doi.org/10.1016/j.canlet.2008.03.025] [PMID: 18462866]

[17] Seyithanoğlu MH, Abdallah A, Kitiş S, *et al.* Investigation of cytotoxic, genotoxic, and apoptotic effects of curcumin on glioma cells. Cell Mol Biol 2019; 65(3): 101-8.
[http://dx.doi.org/10.14715/cmb/2019.65.3.15] [PMID: 30942162]

[18] Gersey ZC, Rodriguez GA, Barbarite E, *et al.* Curcumin decreases malignant characteristics of glioblastoma stem cells *via* induction of reactive oxygen species. BMC Cancer 2017; 17(1): 99.
[http://dx.doi.org/10.1186/s12885-017-3058-2] [PMID: 28160777]

[19] Wang X, Deng J, Yuan J, *et al.* Curcumin exerts its tumor suppressive function *via* inhibition of NEDD4 oncoprotein in glioma cancer cells. Int J Oncol 2017; 51(2): 467-77.
[http://dx.doi.org/10.3892/ijo.2017.4037] [PMID: 28627598]

[20] Fratantonio D, Molonia MS, Bashllari R, *et al.* Curcumin potentiates the antitumor activity of Paclitaxel in rat glioma C6 cells. Phytomedicine 2019; 55: 23-30.
[http://dx.doi.org/10.1016/j.phymed.2018.08.009] [PMID: 30668434]

[21] Bangaru MLY, Chen S, Woodliff J, Kansra S. Curcumin (diferuloylmethane) induces apoptosis and blocks migration of human medulloblastoma cells. Anticancer Res 2010; 30(2): 499-504.
[PMID: 20332461]

[22] Zhao J, Wu J, Zhang L, Chen Y, Yu S, Zhao Y. Curcumin pretreatment and post-treatment both improve the antioxidative ability of neurons with oxygen-glucose deprivation. Neural Regen Res 2015; 10(3): 481-9.
[http://dx.doi.org/10.4103/1673-5374.153700] [PMID: 25878600]

[23] Qian Y, Ma J, Guo X, *et al.* Curcumin enhances the radiosensitivity of U87 cells by inducing DUSP-2 up-regulation. Cell Physiol Biochem 2015; 35(4): 1381-93.
[http://dx.doi.org/10.1159/000373959] [PMID: 25792385]

[24] Wu B, Yao H, Wang S, Xu R. DAPK1 modulates a curcumin-induced G2/M arrest and apoptosis by regulating STAT3, NF-κB, and caspase-3 activation. Biochem Biophys Res Commun 2013; 434(1): 75-80.

[http://dx.doi.org/10.1016/j.bbrc.2013.03.063] [PMID: 23545262]

[25] Anusiya G, Gowthama Prabu U, Yamini NV, *et al.* A review of the therapeutic and biological effects of edible and wild mushrooms. Bioengineered 2021; 12(2): 11239-68. [http://dx.doi.org/10.1080/21655979.2021.2001183] [PMID: 34738876]

[26] Xu JW, Zhong JJ. Genetic engineering of *Ganoderma lucidum* for the efficient production of ganoderic acids. Bioengineered 2015; 6(6): 357-60. [http://dx.doi.org/10.1080/21655979.2015.1119341] [PMID: 26588475]

[27] Reis FS, Barros L, Martins A, Ferreira ICFR. Chemical composition and nutritional value of the most widely appreciated cultivated mushrooms: An inter-species comparative study. Food Chem Toxicol 2012; 50(2): 191-7. [http://dx.doi.org/10.1016/j.fct.2011.10.056] [PMID: 22056333]

[28] Ba DM, Ssentongo P, Beelman RB, Muscat J, Gao X, Richie JP Jr. Higher Mushroom consumption is associated with lower risk of cancer: a systematic review and meta-analysis of observational studies. Adv Nutr 2021; 12(5): 1691-704. [http://dx.doi.org/10.1093/advances/nmab015] [PMID: 33724299]

[29] Ruan H, Zhan YY, Hou J, *et al.* Berberine binds RXRα to suppress β-catenin signaling in colon cancer cells. Oncogene 2017; 36(50): 6906-18. [http://dx.doi.org/10.1038/onc.2017.296] [PMID: 28846104]

[30] Zou K, Li Z, Zhang Y, *et al.* Advances in the study of berberine and its derivatives: a focus on anti-inflammatory and anti-tumor effects in the digestive system. Acta Pharmacol Sin 2017; 38(2): 157-67. [http://dx.doi.org/10.1038/aps.2016.125] [PMID: 27917872]

[31] Ma W, Zhu M, Zhang D, *et al.* Berberine inhibits the proliferation and migration of breast cancer ZR-75-30 cells by targeting Ephrin-B2. Phytomedicine 2017; 25: 45-51. [http://dx.doi.org/10.1016/j.phymed.2016.12.013] [PMID: 28190470]

[32] Sun Y, Yu J, Liu X, *et al.* Oncosis-like cell death is induced by berberine through ERK1/2-mediated impairment of mitochondrial aerobic respiration in gliomas. Biomed Pharmacother 2018; 102: 699-710. [http://dx.doi.org/10.1016/j.biopha.2018.03.132] [PMID: 29604589]

[33] Asemi Z, Behnam M, Pourattar MA, Mirzaei H, Razavi ZS, Tamtaji OR. Therapeutic potential of berberine in the treatment of glioma: insights into its regulatory mechanisms. Cell Mol Neurobiol 2020. [http://dx.doi.org/10.1007/s10571-020-00903-5] [PMID: 32557203]

[34] Jabbarzadeh Kaboli P, Rahmat A, Ismail P, Ling KH. Targets and mechanisms of berberine, a natural drug with potential to treat cancer with special focus on breast cancer. Eur J Pharmacol 2014; 740: 584-95. [http://dx.doi.org/10.1016/j.ejphar.2014.06.025] [PMID: 24973693]

[35] Yue X, Zhao Y, Xu Y, Zheng M, Feng Z, Hu W. Mutant p53 in cancer: accumulation, gain-of-function, and therapy. J Mol Biol 2017; 429(11): 1595-606. [http://dx.doi.org/10.1016/j.jmb.2017.03.030] [PMID: 28390900]

[36] Wolter J K, Wolter N E, Blanch A, *et al.* Anti-tumour activity of the beta-adrenergic receptor antagonist propranolol in neuroblastoma Oncotarget 2014; 5(1): 161-72.

[37] Calvani M, Subbiani A, Bruno G, Favre C. Beta-blockers and berberine: A possible dual approach to contrast neuroblastoma growth and progression. Oxid Med Cell Longev 2020; 2020: 1-11. [http://dx.doi.org/10.1155/2020/7534693] [PMID: 32855766]

[38] Chen X-W, Di Y M, Zhang J, Zhou Z-W, Li C G, Zhou S F. Interaction of herbal compounds with biological targets: a case study with berberine The Scientific World Journal 2012; 31. [http://dx.doi.org/10.1100/2012/708292] [http://dx.doi.org/10.1100/2012/708292]

[39] Tillhon M, Guamán Ortiz L M, Lombardi P, Scovassi A I. Berberine: new perspectives for old

remedies Biochem Pharmacol 2012; 84(10): 1260-7.
[http://dx.doi.org/10.1016/j.bcp.2012.07.018]

[40] Wang Y, Kheir MM, Chai Y, *et al.* Comprehensive study in the inhibitory effect of berberine on gene transcription, including TATA box. PLoS One 2011; 6(8): e23495.
[http://dx.doi.org/10.1371/journal.pone.0023495] [PMID: 21887260]

[41] Agnarelli A, Natali M, Garcia-Gil M, *et al.* Cell-specific pattern of berberine pleiotropic effects on different human cell lines. Sci Rep 2018; 8(1): 10599.
[http://dx.doi.org/10.1038/s41598-018-28952-3] [PMID: 30006630]

[42] Choi MS, Yuk DY, Oh JH, *et al.* Berberine inhibits human neuroblastoma cell growth through induction of p53-dependent apoptosis. Anticancer Res 2008; 28(6A): 3777-84.
[PMID: 19189664]

[43] Naveen CR, Gaikwad S, Agrawal-Rajput R. Berberine induces neuronal differentiation through inhibition of cancer stemness and epithelial-mesenchymal transition in neuroblastoma cells. Phytomedicine 2016; 23(7): 736-44.
[http://dx.doi.org/10.1016/j.phymed.2016.03.013] [PMID: 27235712]

[44] Maiti P, Plemmons A, Dunbar GL. Combination treatment of berberine and solid lipid curcumin particles increased cell death and inhibited PI3K/Akt/mTOR pathway of human cultured glioblastoma cells more effectively than did individual treatments. PLoS One 2019; 14(12): e0225660.
[http://dx.doi.org/10.1371/journal.pone.0225660] [PMID: 31841506]

[45] Wang J, Qi Q, Feng Z, *et al.* Berberine induces autophagy in glioblastoma by targeting the AMPK/mTOR/ULK1-pathway. Oncotarget 2016; 7(41): 66944-58.
[http://dx.doi.org/10.18632/oncotarget.11396] [PMID: 27557493]

[46] Ji H-F, Shen L. Molecular basis of inhibitory activities of berberine against pathogenic enzymes in Alzheimer's disease Sci World J 2012; 4.
[http://dx.doi.org/10.1100/2012/823201]

[47] Ugun-Klusek A, Theodosi TS, Fitzgerald JC, *et al.* Monoamine oxidase-A promotes protective autophagy in human SH-SY5Y neuroblastoma cells through Bcl-2 phosphorylation. Redox Biol 2019; 20: 167-81.
[http://dx.doi.org/10.1016/j.redox.2018.10.003] [PMID: 30336354]

[48] Qu L, Li S, Zhuo Y, Chen J, Qin X, Guo G. Anticancer effect of triterpenes from *Ganoderma lucidum* in human prostate cancer cells. Oncol Lett 2017; 14(6): 7467-72.
[PMID: 29344190]

[49] Ferreira ICFR, Heleno SA, Reis FS, *et al.* Chemical features of Ganoderma polysaccharides with antioxidant, antitumor and antimicrobial activities. Phytochemistry 2015; 114: 38-55.
[http://dx.doi.org/10.1016/j.phytochem.2014.10.011] [PMID: 25457487]

[50] Du GH, Wang HX, Yan Z, Liu LY, Chen RY. [Anti-tumor target prediction and activity verification of Ganoderma lucidum triterpenoids]. Zhongguo Zhongyao Zazhi 2017; 42(3): 517-22.
[PMID: 28952258]

[51] Sohretoglu D, Huang S. Ganoderma lucidum polysaccharides as an anti-cancer agent. Anticancer Agents Med Chem 2018; 18(5): 667-74.
[http://dx.doi.org/10.2174/1871520617666171113121246] [PMID: 29141563]

[52] Xu Z, Chen X, Zhong Z, Chen L, Wang Y. Ganoderma lucidum polysaccharides: immunomodulation and potential anti-tumor activities. Am J Chin Med 2011; 39(1): 15-27.
[http://dx.doi.org/10.1142/S0192415X11008610] [PMID: 21213395]

[53] Wang J, Yuan Y, Yue T. Immunostimulatory activities of β-d-glucan from Ganoderma Lucidum. Carbohydr Polym 2014; 102: 47-54.
[http://dx.doi.org/10.1016/j.carbpol.2013.10.087] [PMID: 24507254]

[54] Wang R, Shi L, Liu S, *et al.* Mass spectrometry-based urinary metabolomics for the investigation on

the mechanism of action of Eleutherococcus senticosus (Rupr. & Maxim.) Maxim. leaves against ischemic stroke in rats. J Ethnopharmacol 2019; 241: 111969.
[http://dx.doi.org/10.1016/j.jep.2019.111969] [PMID: 31125596]

[55] Gaiaschi Ludovica. A phyto-mycotherapeutic supplement, namely *Ganostile*, as effective adjuvant in brain cancer management: an*in vitro*study using U251 human glioblastoma cell line Int J Mol Sci 2024; 25(11): 6204.
[http://dx.doi.org/10.3390/ijms25116204]

[56] Cooper TA, Wan L, Dreyfuss G. RNA and disease. Hhmi 2009; 136(4): 777-93.
[PMID: 19239895]

[57] Saito Y, Saito H. MicroRNAs in cancers and neurodegenerative disorders. Front Genet 2012; 3: 194.
[http://dx.doi.org/10.3389/fgene.2012.00194] [PMID: 23056009]

[58] Connolly BS, Lang AE. Pharmacological treatment of Parkinson disease: a review. JAMA 2014; 311(16): 1670-83.
[http://dx.doi.org/10.1001/jama.2014.3654] [PMID: 24756517]

[59] Dunkel P, Chai CLL, Sperlágh B, Huleatt PB, Mátyus P. Clinical utility of neuroprotective agents in neurodegenerative diseases: current status of drug development for Alzheimer's, Parkinson's and Huntington's diseases, and amyotrophic lateral sclerosis. Expert Opin Investig Drugs 2012; 21(9): 1267-308.
[http://dx.doi.org/10.1517/13543784.2012.703178] [PMID: 22741814]

[60] Dye RV, Miller KJ, Singer EJ, Levine AJ. Hormone replacement therapy and risk for neurodegenerative diseases. Int J Alzheimers Dis 2012; 2012: 1-18.
[http://dx.doi.org/10.1155/2012/258454] [PMID: 22548198]

[61] Moraes WAS. 2015.Current pharmacological and non-pharmacological therapies for neurodegenerative diseases.

[62] Moreno JA, Halliday M, Molloy C, *et al.* Oral treatment targeting the unfolded protein response prevents neurodegeneration and clinical disease in prion-infected mice. Sci Transl Med 2013; 5(206): 206ra138.
[http://dx.doi.org/10.1126/scitranslmed.3006767] [PMID: 24107777]

[63] Weissmiller AM, Wu C. Current advances in using neurotrophic factors to treat neurodegenerative disorders. Transl Neurodegener 2012; 1(1): 14.
[http://dx.doi.org/10.1186/2047-9158-1-14] [PMID: 23210531]

[64] Wu YP, Chen WS, Teng C, Zhang N. Stem cells for the treatment of neurodegenerative diseases. Molecules 2010; 15(10): 6743-58.
[http://dx.doi.org/10.3390/molecules15106743]

[65] Young AB. Four decades of neurodegenerative disease research: how far we have come! J Neurosci 2009; 29(41): 12722-8.
[http://dx.doi.org/10.1523/JNEUROSCI.3767-09.2009] [PMID: 19828782]

[66] Singh SK, Hawkins C, Clarke ID, *et al.* Identification of human brain tumour initiating cells. Nature 2004; 432(7015): 396-401.
[http://dx.doi.org/10.1038/nature03128] [PMID: 15549107]

[67] Godlewski J, Ferrer-Luna R, Rooj AK, *et al.* MicroRNA signatures and molecular subtypes of glioblastoma: the role of extracellular transfer. Stem Cell Reports 2017; 8(6): 1497-505.
[http://dx.doi.org/10.1016/j.stemcr.2017.04.024] [PMID: 28528698]

[68] Li X, Yang L, Chen LL. The biogenesis, functions, and challenges of circular RNAs. Mol Cell 2018; 71(3): 428-42.
[http://dx.doi.org/10.1016/j.molcel.2018.06.034] [PMID: 30057200]

[69] Patel AP, Tirosh I, Trombetta JJ, *et al.* Single-cell RNA-seq highlights intratumoral heterogeneity in primary glioblastoma. Science 2014; 344(6190): 1396-401.

[http://dx.doi.org/10.1126/science.1254257] [PMID: 24925914]

[70] Plun-Favreau H, Lewis PA, Hardy J, Martins LM, Wood NW. Cancer and neurodegeneration: between the devil and the deep blue sea. PLoS Genet 2010; 6(12): e1001257. [http://dx.doi.org/10.1371/journal.pgen.1001257] [PMID: 21203498]

[71] Ghosh S, Kumar V, Mukherjee H, Lahiri D, Roy P. Nutraceutical regulation of miRNAs involved in neurodegenerative diseases and brain cancers. Heliyon 2021; 7(6): e07262. [http://dx.doi.org/10.1016/j.heliyon.2021.e07262]

[72] Sparreboom A, Cox MC, Acharya MR, Figg WD. Herbal remedies in the United States: potential adverse interactions with anticancer agents. J Clin Oncol 2004; 22(12): 2489-503. [http://dx.doi.org/10.1200/JCO.2004.08.182] [PMID: 15197212]

[73] Richardson MA, Sanders T, Palmer JL, Greisinger A, Singletary SE. Complementary/alternative medicine use in a comprehensive cancer center and the implications for oncology. J Clin Oncol 2000; 18(13): 2505-14. [http://dx.doi.org/10.1200/JCO.2000.18.13.2505] [PMID: 10893280]

[74] Golden EB, Lam PY, Kardosh A *et al.* Green tea polyphenols block the anticancer effects of bortezomib and other boronic acid-based proteasome inhibitors. Blood 2009; 113(23): 5927-37. [http://dx.doi.org/10.1182/blood-2008-07-171389]

[75] Kuzuhara T, Suganuma M, Fujiki H. Green tea catechin as a chemical chaperone in cancer prevention. Cancer Lett 2008; 261(1): 12-20. [http://dx.doi.org/10.1016/j.canlet.2007.10.037] [PMID: 18068893]

[76] Richardson PG, Sonneveld P, Schuster MW, *et al.* Bortezomib or high-dose dexamethasone for relapsed multiple myeloma. N Engl J Med 2005; 352(24): 2487-98. [http://dx.doi.org/10.1056/NEJMoa043445] [PMID: 15958804]

[77] Ludwig H, Khayat D, Giaccone G, Facon T. Proteasome inhibition and its clinical prospects in the treatment of hematologic and solid malignancies. Cancer 2005; 104(9): 1794-807. [http://dx.doi.org/10.1002/cncr.21414] [PMID: 16178003]

[78] Groll M, Berkers CR, Ploegh HL, Ovaa H. Crystal structure of the boronic acid-based proteasome inhibitor bortezomib in complex with the yeast 20S proteasome. Structure 2006; 14(3): 451-6. [http://dx.doi.org/10.1016/j.str.2005.11.019] [PMID: 16531229]

[79] Fahy BN, Schlieman MG, Mortenson MM, Virudachalam S, Bold RJ. Targeting BCL-2 overexpression in various human malignancies through NF-κB inhibition by the proteasome inhibitor bortezomib. Cancer Chemother Pharmacol 2005; 56(1): 46-54. [http://dx.doi.org/10.1007/s00280-004-0944-5] [PMID: 15791457]

[80] Meusser B, Hirsch C, Jarosch E, Sommer T. ERAD: the long road to destruction. Nat Cell Biol 2005; 7(8): 766-72. [http://dx.doi.org/10.1038/ncb0805-766] [PMID: 16056268]

[81] Fribley A, Wang CY. Proteasome inhibitor induces apoptosis through induction of endoplasmic reticulum stress. Cancer Biol Ther 2006; 5(7): 745-8. [http://dx.doi.org/10.4161/cbt.5.7.2971] [PMID: 16861900]

[82] Fribley A, Zeng Q, Wang CY. Proteasome inhibitor PS-341 induces apoptosis through induction of endoplasmic reticulum stress-reactive oxygen species in head and neck squamous cell carcinoma cells. Mol Cell Biol 2004; 24(22): 9695-704. [http://dx.doi.org/10.1128/MCB.24.22.9695-9704.2004] [PMID: 15509775]

[83] Reddy RK, Mao C, Baumeister P, Austin RC, Kaufman RJ, Lee AS. Endoplasmic reticulum chaperone protein GRP78 protects cells from apoptosis induced by topoisomerase inhibitors: role of ATP binding site in suppression of caspase-7 activation. J Biol Chem 2003; 278(23): 20915-24. [http://dx.doi.org/10.1074/jbc.M212328200] [PMID: 12665508]

[84] Lee AS. GRP78 induction in cancer: therapeutic and prognostic implications. Cancer Res 2007; 67(8):

3496-9.
[http://dx.doi.org/10.1158/0008-5472.CAN-07-0325] [PMID: 17440054]

[85] Ermakova SP, Kang BS, Choi BY, *et al.* (-)-Epigallocatechin gallate overcomes resistance to etoposide-induced cell death by targeting the molecular chaperone glucose-regulated protein 78. Cancer Res 2006; 66(18): 9260-9.
[http://dx.doi.org/10.1158/0008-5472.CAN-06-1586] [PMID: 16982771]

[86] Pyrko P, Schönthal AH, Hofman FM, Chen TC, Lee AS. The unfolded protein response regulator GRP78/BiP as a novel target for increasing chemosensitivity in malignant gliomas. Cancer Res 2007; 67(20): 9809-16.
[http://dx.doi.org/10.1158/0008-5472.CAN-07-0625] [PMID: 17942911]

[87] Lee J, Liu J, Feng X, *et al.* Withaferin A is a leptin sensitizer with strong antidiabetic properties in mice. Nat Med 2016; 22(9): 1023-32.
[http://dx.doi.org/10.1038/nm.4145] [PMID: 27479085]

[88] Jadeja RN, Urrunaga NH, Dash S, Khurana S, Saxena NK. Withaferin-A reduces acetaminophen-induced liver injury in mice. Biochem Pharmacol 2015; 97(1): 122-32.
[http://dx.doi.org/10.1016/j.bcp.2015.07.024] [PMID: 26212553]

[89] Thaiparambil JT, Bender L, Ganesh T, *et al.* Withaferin A inhibits breast cancer invasion and metastasis at sub-cytotoxic doses by inducing vimentin disassembly and serine 56 phosphorylation. Int J Cancer 2011; 129(11): 2744-55.
[http://dx.doi.org/10.1002/ijc.25938] [PMID: 21538350]

[90] Woo SM, Min K, Kim S, *et al.* Axl is a novel target of withaferin A in the induction of apoptosis and the suppression of invasion. Biochem Biophys Res Commun 2014; 451(3): 455-60.
[http://dx.doi.org/10.1016/j.bbrc.2014.08.018] [PMID: 25117439]

[91] Kim G, Kim TH, Hwang EH, *et al.* Withaferin A inhibits the proliferation of gastric cancer cells by inducing G2/M cell cycle arrest and apoptosis Oncol Lett 2017; 14(1): 416-22.

[92] Lv TZ, Wang GS. Antiproliferation potential of withaferin A on human osteosarcoma cells *via* the inhibition of G2/M checkpoint proteins. Exp Ther Med 2015; 10(1): 323-9.
[http://dx.doi.org/10.3892/etm.2015.2480] [PMID: 26170956]

[93] McKenna MK, Gachuki BW, Alhakeem SS, *et al.* Anti-cancer activity of withaferin A in B-cell lymphoma. Cancer Biol Ther 2015; 16(7): 1088-98.
[http://dx.doi.org/10.1080/15384047.2015.1046651] [PMID: 26020511]

[94] Liu X, Chen L, Liang T, Tian XD, Liu Y, Zhang T. Withaferin A induces mitochondrial-dependent apoptosis in non-small cell lung cancer cells *via* generation of reactive oxygen species. J BUON 2017; 22(1): 244-50.
[PMID: 28365961]

[95] Hahm ER, Singh SV. Withaferin A-induced apoptosis in human breast cancer cells is associated with suppression of inhibitor of apoptosis family protein expression. Cancer Lett 2013; 334(1): 101-8.
[http://dx.doi.org/10.1016/j.canlet.2012.08.026] [PMID: 22935676]

[96] Koduru S, Kumar R, Srinivasan S, Evers MB, Damodaran C. Notch-1 inhibition by Withaferin-A: a therapeutic target against colon carcinogenesis. Mol Cancer Ther 2010; 9(1): 202-10.
[http://dx.doi.org/10.1158/1535-7163.MCT-09-0771] [PMID: 20053782]

[97] Shah N, Kataria H, Kaul SC, Ishii T, Kaur G, Wadhwa R. Effect of the alcoholic extract of Ashwagandha leaves and its components on proliferation, migration, and differentiation of glioblastoma cells: Combinational approach for enhanced differentiation. Cancer Sci 2009; 100(9): 1740-7.
[http://dx.doi.org/10.1111/j.1349-7006.2009.01236.x] [PMID: 19575749]

[98] Widodo N, Kaur K, Shrestha BG, *et al.* Selective killing of cancer cells by leaf extract of Ashwagandha: identification of a tumor-inhibitory factor and the first molecular insights to its effect.

Clin Cancer Res 2007; 13(7): 2298-306.
[http://dx.doi.org/10.1158/1078-0432.CCR-06-0948] [PMID: 17404115]

[99] Widodo N, Takagi Y, Shrestha BG, Ishii T, Kaul SC, Wadhwa R. Selective killing of cancer cells by leaf extract of Ashwagandha: Components, activity and pathway analyses. Cancer Lett 2008; 262(1): 37-47.
[http://dx.doi.org/10.1016/j.canlet.2007.11.037] [PMID: 18191020]

[100] Deep G, Agarwal R. Antimetastatic efficacy of silibinin: molecular mechanisms and therapeutic potential against cancer. Cancer Metastasis Rev 2010; 29(3): 447-63.
[http://dx.doi.org/10.1007/s10555-010-9237-0] [PMID: 20714788]

[101] Siegel AB, Stebbing J. Milk thistle: early seeds of potential. Lancet Oncol 2013; 14(10): 929-30.
[http://dx.doi.org/10.1016/S1470-2045(13)70414-5] [PMID: 23993379]

[102] Bosch-Barrera J, Sais E, Cañete N, *et al.* Response of brain metastasis from lung cancer patients to an oral nutraceutical product containing silibinin. Oncotarget 2016; 7(22): 32006-14.
[http://dx.doi.org/10.18632/oncotarget.7900] [PMID: 26959886]

# SUBJECT INDEX

www.ingramcontent.com/pod-product-compliance
Lightning Source LLC
LaVergne TN
LVHW070950210926
848482LV00002B/158
* 9 7 9 8 8 9 8 8 1 1 7 6 1 *